What Risk?

edited by Roger Bate

OXFORD AUCKLAND BOSTON JOHANNESBURG MELBOURNE NEW DELHI

Butterworth-Heinemann
Linacre House, Jordan Hill, Oxford OX2 8DP
225 Wildwood Avenue, Woburn, MA 01801-2041
A division of Reed Educational and Professional Publishing Ltd

A member of the Reed Elsevier plc group

First published 1997
Revised paperback 1999

British Library Cataloguing in Publication Data
A catalogue record for this book is available from the British Library

Library of Congress Cataloguing in Publication Data
A catalogue record for this book is available on request

ISBN 0 7506 4228 9

Printed and bound in Great Britain by
Biddles Ltd, Guildford and King's Lynn

Contents

Foreword

David Davis MP*

Democracies are notoriously risk-averse forms of government. Parliaments always prefer to attack failure than to celebrate success. And a vigorous and competitive free press catalyses this process of criticism: good news sells no newspapers. That is part of the reason that historians looking back at the eighties and nineties might reasonably describe them as the decades of the health scare. Today, people are remaining healthier and living longer than ever before. Yet every so often there is an outbreak of panic about some health threat or other, which more often than not turns out to be either massively overstated or ill founded.

This excellent collection of studies and papers by world-class authorities in their fields, brought together by Roger Bate, does an outstanding job of shaking up many of those shibboleths. For example, if we were to believe the news media, we might be forgiven for believing that we were facing a cancer epidemic. Yet if we exclude the cancers caused by smoking, and allow for the fact that people are living longer, cancer levels have dropped by some 15% since 1950. Since the 1950s government agencies have acted to regulate use and emissions of vast numbers of synthetic chemicals. The result of this is that we ingest each day only about one tenth of a milligram of the 200 chemicals that governments rate as the most dangerous.

Yet in every cup of coffee that we drink there are at least 19 carcinogens. Indeed, the act of cooking our food – one of the most healthy innovations in human history – adds some 2000 mg of carcinogens to our diet every day. There are large numbers of apparent natural hazards, from natural pesticides in plants, through dioxins released by forest fires, to radon emissions in stone buildings, that cannot be eliminated or even reasonably avoided. The truth is that the process of life itself is a risky business, but one that our bodies are well equipped to deal with.

The perspective that is missing from much modern discussion is that all health risks are relative. Passive smoking is very much in the news currently. However, the risks are similar to those incurred in our normal diet: the probable risk from passive smoking is less than that which arises from eating 50 g of mushrooms every week. But assessment of risk is one of the weaker areas of human judgement. Fear of any particular threat is probably more proportional to the size of the headline than the size of the risk. Certainly people seem to overestimate personal risk for the more dramatic and more horrible causes of death. Ghastly pictures have more impact on our perception of risk than ghastly statistics. And the more dramatic the apparent threat, the quicker we seem to accept poorly tested hypotheses.

---

* Former UK Science Minister

This weakness in serious assessment of risk, in conjunction with democracies' habit of attacking failure, very often leads to behaviour by senior decision makers that is more blame avoidance than risk minimisation. This is one of the main reasons we see public officials so often operating on what is known as the precautionary principle. Indeed, in 25 years' time when we can take a calmer and more empirical view of the BSE crisis, this may well be the most serious criticism that might be levelled at the British government's decision making.

The response to this may be 'so what, at least we are reducing some risks'. In truth, this precautionary behaviour can cost lives. The American Environmental Protection Agency estimates that its regulations, aimed at maintaining very low levels of synthetic chemicals in the environment, cost the American economy $140 billion. That means that the USA spends more than 100 times as much to prevent one hypothetical and highly uncertain death from pollution, than it spends to save a life through medical intervention. If this is true of America, which is relatively lightly regulated, how much more true will it be of Western European countries.

Similarly, the effect of tight regulation on the use of pesticides is undoubtedly to increase the cost of production of fruit and vegetables. But the best anti-cancer strategy that exists is a diet with a high proportion of fruit and vegetables. When we introduce regulations that make them more expensive, the poorest, and hence most vulnerable people, can afford less. To put this real risk in context, the quarter of society that eats the least fruit and vegetables has double the cancer rate of the quarter that eats the most.

This book demonstrates that the environmental absolutism that leads to these extreme regulatory regimes is based on obsolete science. In essence, much of the policy creation in the last few decades has been based on a number of fallacies. Firstly, if a substance is synthetic, it is considered to be more likely to be toxic than if it is 'natural'.

This overlooks the fact that, for example, most plants defend themselves against pests and predators by generating a range of natural pesticides. These are just as likely to be carcinogenic as any synthetic compound – hence the carcinogens in ordinary coffee. So the 'natural versus synthetic' assumption is just wrong. Secondly, it has been assumed that if a substance is carcinogenic when eaten in large quantities by rats, it will be carcinogenic when humans are exposed to any quantity of it. This is akin to knowing that a temperature of 200°C can cause severe burns, and concluding that a there is cause to worry about temperatures of 20°C. It does not recognize that there is a threshold below which the defence mechanisms and metabolic processes of normal human beings can cope perfectly well.

When the basis for much regulation was laid some thirty years ago, the science was insufficiently clear to show whether there were safe thresholds for most modern carcinogens. This book argues, with some force, that modern science demonstrates the existence of thresholds. Indeed, it demonstrates that for some carcinogens there is a level of exposure that leads to minimum cancer risk which is greater than zero – what they term a 'hormetic effect'. The most surprising examples of this are for radiation. A study of radon levels in homes showed that cancer levels were lower in people exposed to a small quantity of radon than in those exposed to none. Similar results were found in recently declassified Soviet data on 10 000 people exposed to radiation from nuclear weapons material.

The book concludes that governments should aim resources at reducing the major risks, because many precautionary measures only target the minor risks. New technologies, although often introducing new risks, often serve to replace more harmful older technology. This data, from such eminent and authoritative sources, casts severe doubt on the wisdom of the absolutist approach to cancer risks in the environment today. It ought also to give pause to many thinkers in other areas of scientifically based policy. For example, is the risk handling strategy adopted by the food and drug approval agencies of the West anything like optimal for the welfare of our people? I suspect not. Are the real impacts on the global human population of many of our well-intentioned environmental measures really thought through? I doubt it.

This book sometimes makes hard reading precisely because it challenges conventional wisdom, and forces us to reassess much that we take for granted. That reassessment is a vital precursor to a truly rational environmental policy. As a sobering reminder of how wrong our prejudices and presumptions can be, this book should be read by every politician. As a cautionary reminder of how systematic we ought to be in the assessment of risk before making policy judgements, it should be read by every government scientist and technical policy adviser. And as a shocking reminder of how harmful being enthusiastically wrong can be, it should be read by every eco-warrior and professional pundit.

Preface

Science and the media

This book is about the likely health effects resulting from the emission of small quantities of potentially harmful toxins into the environment and how these substances should be controlled. Many such hazardous substances have been identified and many more probably exist. The particular substances under discussion were chosen because of their relatively high public profile – each has been, at one time or another, the subject of considerable media attention that intensified public fear.

Most of the hazards were newly identified at the time of this media attention, often because they were the subject of epidemiological investigations, or because new technology enabled measurement of toxins at lower concentrations than had earlier been possible. Some were always present but we were not aware of them or their effects, some are by-products of new tech-nology. In most cases, the science of the alleged causal link between toxin and effect is still being formulated, so that an absolute answer to the question of whether a particular substance is a health hazard at the level at which it is commonly present in the environment – or even at much higher levels – is often not available. Frequently, this inability to satisfy curiosity gives rise to alarm and then to suspicion, so that informed debate is overtaken by fears of conspiracy.

Recent campaigns highlighting the apparent lack of public understanding of science have focused on educational programmes in schools and universities. However, most adults become informed about science and technology through the media. These campaigns have largely ignored this fact and there has been little critical analysis of the way that science is portrayed by journalists or of the relationship between the two influential social institutions of science and the media.

For most people, science is understood through the filter of journalistic language and imagery. This filter has changed considerably since science journalism really took off in the 1950s. In the early days, journalists were, in many respects, simply retailers of science, presenting neatly packaged inform-ation to readers. However, as the science writers themselves became more sophisticated, and the mood of the times changed, they became more critical of the science they were asked to interpret. By the 1960s journalists were discussing the mixed blessings of science, and by the mid-1970s the environ-mental and consumer movements had begun to speculate on the potential risks to human health from the products of technology. This speculation was further fuelled by several technological disasters: the Bhopal and Seveso chemical spills, the explosion of the space shuttle, 'Challenger' and, of course, the nuclear meltdown at Chernobyl. Since then there has been a tendency to welcome new discoveries, such as surgical breakthroughs, but to be highly critical of even the smallest environmental and public health threat.

Many scientists have been concerned that information from pressure groups is treated uncritically by the media. The 'Brent Spar' issue, where Greenpeace duped television producers with doctored video coverage, has awakened the electronic media to the problems of being spoon-fed news by pressure groups. Concerned for their image as much as anything else, the media in the late 1990s is attempting to find balance in scientific reporting.

Scientific objectivity arises from empirical testing of theories, which are revised in the light of new evidence and/or better theories. However, empirical testing is not the way that the media create objectivity. Journalists accept that it is not their role to achieve objectivity, but they are expected to approach the ideal of neutrality and unbiased reporting by balancing diverse points of view, by presenting all sides fairly, and by maintaining a clear distinction between news reporting and editorial opinion. However, this ideal is rarely lived up to. The media often fail to present an objective view; they can present a very selective sample of theories and empirical tests and often fail to interpret these correctly, misunderstanding the science and the statistics. And they do so because of prejudice, time pressure and ignorance. It often frustrates scientists that, for the media, balance becomes synonymous with objectivity. However, some scientists use this to their advantage, claiming that the majority view (which they propound) must be correct.

Balanced reporting by journalists is necessary but not sufficient if the public is to be accurately informed of the science relating to any particular scare. Of course, the ability of a journalist to balance a debate requires that he or she be aware of all relevant opinions. However, certain opinions may be contrary to a prevailing orthodoxy, political position or ideological hegemony, in which case there may be reluctance to raise dissenting opinions for fear of retribution against those holding this opinion. The worst examples this century were the debates on eugenics and Lysenkoism. Opinions which are not voiced in public will not be taken into account by the media and therefore only the perception of balance may be given, while the reality is considerably different.

The European Science and Environment Forum (ESEF) seeks to aid the debate by providing the media with rigorous analyses of current debates, bringing to the attention of the media and the public critical works in the scientific literature that may have been overlooked. Through books, briefings, speaking tours and conferences, ESEF members seek to preserve the integrity of science and to promote wider scientific literacy. This book is part of that effort.

Scientists often blame the media for exaggerating stories of alarm, but of course it is not just the media that like exciting 'positive' results. In a recent paper in the science journal Oikos (Csada et al. 1997), three Canadian biologists explained how research which is important, but not exciting or innovative, seldom makes it into the more prestigious scientific journals. Those journals rarely carry papers where the findings are largely 'negative'. For example,a researcher might analyse the data relating to the link between pesticide residues in apples and bladder cancer, and conclude that his results indicated no correlation. One would think that his findings would be useful for those working in similar fields. But such results are not exciting, and thus the chance of the paper being published in a top journal like Nature is remote – at least, that is the conclusion of Csada et al., who analysed 1812 scientific papers published between 1989 and 1995, picked at random from 40 biology journals. Only 9 per cent of the papers contained 'non-significant' results; the figure was even lower for the most prestigious journals.

Given the pressure on university researchers to publish – and in good journals – the bias against publishing 'negative' results has some worrying implications. First, it is likely that hypotheses tested will be conservative, because positive results will seem more likely in such cases. More outlandish hypotheses – ones that might broaden the scientific picture – will not be entertained. Second, researchers are likely to select carefully the data in search of a significant correlation. If the chance of being published is increased by showing a positive result, researchers will be tempted to trawl through the data until they find one – ignoring any negative correlations they encounter on the way. Careers may depend on such things.

It is because even the much-vaunted peer review process is far from pure (see Feinstein's paper) that debate in wider circles is so important. Imagine five identical epidemiological research projects analysing the links between pesticide residues and bladder cancer. Four find no correlation; one finds a correlation. If, because of publication bias, the latter project is published and the former research is ignored, the non-specialist scientist will become slightly worried about pesticide residues. A journalist with information only about the published study could unwittingly turn minor concerns into a grave, powerful discovery.

But, even if the above correlation did exist, the statistician's caveat is worth remembering: association is not causation. Many fat people drink diet cola but this does not imply that diet cola causes obesity. Most people understand this but many other claimed correlations, especially those for which people do not have personal experience, can lead people to jump to the wrong conclusions. Of course, people do not often deliberately choose to make mental errors or to remain ignorant of highly relevant facts. Too often, though, we seize the first plausible explanation offered. Once we have formed a belief, we are inclined to dismiss contrary evidence. We like to tell ourselves that we are superior to the people who burned witches centuries ago but we are still prone to the same basic mental errors: seeing patterns where there are none, assuming cause where there is only coincidence, and creating widespread alarm from scanty evidence.

There is no simple solution either to the degree of misinformation in the public domain or to the process that leads to panic. However, providing more reliable information is likely to help. Indeed, many of the papers in this book present information concerning particular hazards that should allay some of those fears. Meanwhile, the papers by Adams and Sandman provide insights into the nature of risk and of public perception that scientists would do well to consider; as the BSE fiasco has shown, scientists often fail to understand why people become scared.

High impact, low probability events – those we fear most, or are made to fear most?

Several of the authors in this book show that many common fears of environmental toxins are generally not well founded, and that the rarity of the harms from these toxins makes a harmful event simultaneously newsworthy and irrelevant for public policy. As I have argued elsewhere (Bate 1996), this is often due to vested interests who are more concerned about profit, kudos or publicity than truth. I discuss briefly one such example.

Toxic shock syndrome (TSS)

Examples of tragic, very rare events are seldom worse than the inexplicable loss of life of someone young. Toxic shock syndrome (TSS) can cause such a tragedy. In early 1997, a young girl, admitted in an emergency to a district hospital, suffered respiratory arrest, her lungs collapsed, septicaemia set in and she died following multiple organ failure. According to a UK national newspaper report of this death, the doctor told the girl's distraught mother that she had died of TSS and that this was probably caused by her use of tampons. Her mother cannot now bear to watch advertisements for sanitary products.

TSS is an extremely rare condition caused by a toxin produced by the bacterium Staphylococcus aureus. It afflicts men, women and children following surgery, burns, stings, severe injuries, and menstruation (both with and without tampon use). TSS was first described in the US in 1978, after a number of child cases were recorded. Early reports went largely unnoticed by the media, but by early 1980 multiple cases in the US led to significant media attention. At this time, the majority of the cases were reported in women, and a link was suggested between TSS, menstruation and the use of tampons. The issue was considered serious enough for the *New England Journal of Medicine* to publish an article about TSS which contained much of the same information that had appeared in news reports, something that normally never happens. The *NEJM* tries to publish only new information; the only other time in recent history that the *NEJM* has broken this convention on public health research was during the scare over AIDS.

In the late 1970s, a few years before this spate of TSS cases, a new type of tampon had come onto the American market. Called 'Rely', it worked in a different way from normal tampons in that it consisted of a perforated paper container (a bit like a tea-bag) which contained tiny pieces of synthetic sponge. It was more absorbent than any tampon ever used before and this led to changes in menstrual hygiene habits among users, with tampons left in place for much longer than normal. TSS seemed to be linked to these high absorbency tampons. As a result, high absorbency polyacrylate fibres were removed from all tampons in 1984, including 'Rely', and subsequently the number of cases began to fall. The US Centers for Disease Control (CDC) reported an immediate decline in cases, but not all studies demonstrated this. 'Rely' was never sold in Europe.

For the past 15 years in the UK there have been, on average, 20 confirmed cases of TSS, resulting in three deaths every year. Typically, one of these deaths is menstrually related. Most doctors will never see a case in their lifetimes: it is extraordinarily rare. There are 10 million menstruating women in UK, using a billion tampons every year and there is one related fatality. Nevertheless, relations of the victims are aggrieved and shocked to learn of this tiny risk. We are scared, as the communications expert Peter Sandman explains in his paper, because of the rarity and severity of the event. A TSS death is far more newsworthy than a death from a road accident.

TSS is relevant to this book because it provides a useful topic to study for the effects of the links between the media and scientific information and as a topic for the analysis of pressure-group activity.

After tampon manufacturers reacted to the increased number of reported menstrual TSS cases by taking suspect products such as 'Rely' off the market the issue died down for several years. It was sparked off again in the UK in 1988/9 by the discussion of synthetic contaminants, including dioxin and pesticide residues, in tampons. This time, the parents of unfortunate TSS cases had allies in their quest for blame. Environmental groups such as Friends of the Earth and the Women's Environmental Network were eager to show the potential harm of chemicals in hygiene products, and demanded only natural products. The US Environmental Protection Agency announced in 1988 that dioxin was a probable human carcinogen, giving these groups ample opportunity to push for products such as nappies and tampons to be free from dioxin. The media, ever alert for a public health scare, were provided with ample information from these groups. No mention was ever made of the fact that nearly half the TSS cases were in men.

The British television programme *World In Action* contributed to public alarm by pointing out that dioxin – 'the world's most toxic compound' – had been found in tampons. National newspapers followed with headlines such as 'Chemical Health Threat to Women', 'Tampons Contain Deadly Poison', 'Dioxin fear for women', and 'Poison towels shock'. The Women's Environmental Network launched a paper entitled 'The Sanitary Protection Scandal'. They called for a ban on chemicals, especially dioxin, in tampons. Hundreds of articles were written over the next six months, heightening public concern, encouraging companies to change the formulation of their products and making medical authorities pay attention to the issue.

Women undoubtedly changed their hygiene habits after the scare. There was a significant fall in tampon use, just as there had been in the early 1980s. Usage subsequently fell further, especially after the introduction of thin, hi-tech towels in 1993, and continues to fall today. According to most research, this is primarily due to the fear of TSS from tampons.

But was the hysterical prose that led to this change of habit really necessary, or did it just alarm people without reason? Furthermore, is there any evidence that dioxin in tampons was responsible for TSS or even constitutes a measurable risk?

As Dr Müller explains in his paper, dioxins are highly toxic compounds, but pose no threat at background environmental (ambient) concentrations. But was this knowledge widely available in 1988? Dr John Greig, senior research scientist at the UK Medical Research Council at the time of the scare said: 'We all have dioxins in us. If a baby is breastfeeding it is taking in more dioxins through its mother's milk than it does through its nappies'. Around the same time the CDC stated that, 'there is no indication that the use of rayon fibres [and their greater potential dioxin concentration] in tampons increase the risk of TSS, compared to cotton fibres'. The US EPA considered that at least 97 per cent of dioxin ingestion by human beings is from diet.

Recent independent tests of tampons from leading manufacturers show that it is not possible to detect dioxin at even one part per trillion. There is no meaningful dioxin risk from tampons or nappies.

Nevertheless, green groups and politicians wanting media exposure, manufacturers who produce cotton-only tampons, and the media themselves continue to discuss the dioxin tampon issue as though it were a real threat. But given that TSS is potentially fatal, surely this is not a bad reaction? On the contrary, such a reaction can be very bad for society, including the risk of TSS caused by other factors: if synthetic compounds in tampons are not responsible for menstrual TSS but attract all the attention, then other causes may not be investigated.

As the authors in this book consistently argue, many scares are just that and nothing more. It is a book about suggested causes of cancer which draws the unfashionable conclusion that it may be wasteful or even pointless to look at environmental causes. Environmental pollutants may lead to other problems, such as impacts on reproductive processes, but not to cancer. There appears to be a threshold below which environmental toxins are not harmful. Whilst some occupational and accidental exposures may be above such levels, ambient concentrations are far below.

We hope that greater media awareness of the opinions of the authors in the book will lead to debates being more balanced.

Please note that there is some overlap and repetition of similar points between papers, allowing the papers to be read individually. This overlap often reinforces authors' opinions, but also highlights differences.

References

Bate, R. (1996) *Energy and Environment* , **7**(4),323–331

Csada, R., James, P. and Espie, R., (1997) *Oikos*, **76**(3),591

Biography

Roger Bate is Director of the Environment Unit at the Institute of Economic Affairs and a Director of ESEF. He is the author of several academic papers on science policy and economic issues, and has published numerous articles in papers such as the *Wall Street Journal* and the *Sunday Times,* and has appeared frequently on television and radio. He is a fellow of the Royal Society of Arts.

Executive summary

This book examines the empirical evidence concerning the impact of low-doses of certain toxins on human health. Key findings of the papers are:

Direct extrapolation from high doses to lower doses ignores the possibility that there exist doses at which no harmful effects occur (Wilson, Fournier). Such 'threshold' doses seem to exist for almost all toxins, including benzene (Weetman), environmental tobacco smoke (Nilsson), ionizing radiation (Jaworowski), asbestos (Fournier), and dioxin (Müller). 'The dose makes the poison.'

Epidemiological studies are particularly prone to statistical biases (Wilson, Feinstein, Fournier, Weetman, Nilsson, Jaworowski, Pershagen).

US regulations limiting environmental exposure to benzene (for example by limiting its use in gasoline) were based on studies that significantly over-estimated benzene's risk to human health (Weetman).

US EPA estimates of the risk to human health posed by environmental tobacco smoke (ETS) are so laden with biases that they should not be used as the basis for public policy. The least biased studies of ETS suggest that it is unlikely to be a major cause of increased risk of lung cancer in non-smoking women. Moreover, focusing on ETS as a possible cause of lung cancer may detract attention from more important causes (Nilsson).

The carcinogenic effects of synthetic pesticides are minuscule compared to the benefits, in terms of protection from cancer, resulting from eating fruits and vegetables. Regulating the use of pesticides increases the cost of food and thereby reduces the consumption of fruits and vegetables. Such regulations therefore tend to harm the poor and are not necessarily in the public interest (Ames and Gold).

These findings are likely to challenge the reader's understanding of risk regulation. This is because for a long time it has been politically correct to assume that substances that are toxic at high doses are also toxic at low doses and very low doses and even at barely perceptible doses. It turns out that the ancient wisdom, 'the dose makes the poison' is a more accurate depiction of the real world. Of course, the past half-century has seen the identification of numerous chemicals that are highly toxic. But humans have adapted to a world filled with toxins and most of us are able successfully to defend ourselves against the levels of toxins we are likely to encounter in the environment. Few of the new toxins are present in large enough doses in the environment to overload our defences. Of course, exceptions exist; but these are rare and their use is not likely to be common: since they would present an even greater hazard to the persons in occupation involving more immediate contact with them, firms avoid or restrict their use lest industrial injury suits should follow. In addition, people with immune deficiencies have lower thresholds and should perhaps be protected. But it is impossible for the whole of society to be 'protected' for the sake of a small minority of 'at risk' people. Moreover, many of the environmental hazards to which we are all exposed are naturally occurring, such as radon gas produced by radium in the earth, pollen released from plants, and dioxins released during forest fires. It would clearly be impossible to eliminate all environmental hazards (although this remains the ostensible objective of certain environmental pressure groups). A better policy would be to protect those people who are particularly susceptible by enabling them to live in environments where the hazards are less common and ensure that, as far as possible, the rest of us are not exposed to levels that are higher than, say, 10 per cent of the median threshold level (Fournier).

Substances known or believed to be carcinogenic or otherwise hazardous at high levels of exposure are not necessarily hazardous at low levels (Wilson, Feinstein, Fournier, Weetman, Nilsson, Jaworowski, Ames, Pershagen, Müller). Regulations that ignore the likely existence of a threshold effect impose costs on society without commensurate benefits. Several authors discuss the reasons for the lack of scientific rigour that plagues epidemiology and policy.

Epidemiologists face peer pressure to avoid publishing results that are 'negative', i.e. indicate that there is no association between a suspected causal agent and an ailment (Pershagen, Feinstein).

Epidemiologists have a strong incentive to show positive results: their funding is contingent on conforming with received wisdom (Feinstein, Kealey).

Despite the ready availability of scientific evidence, the system by which this evidence is made available to politicians is flawed – with biases entering the information flow at many stages – so the advice given to politicians may not be based on the best available evidence. Policies such as the eradication of lead from gasoline and the reduction in emissions of sulphur dioxide in Britain/Europe were not based on sound science (Everest).

Blame avoidance seems to have been the main aim of the politicians responsible for handling public health panics such as Legionnaire's disease, Swine 'flu, Chernobyl, BSE, and AIDS (Craven and Stewart). Actions taken to avoid blame in response to a panic are unlikely to improve matters.

Sandman argues that the media may increase public unease about a particular issue even if their intention is to present unbiased reporting. The reason for this is that individuals respond to negative information (and journalists feed on this fear by ensuring that headlines are scary), so even if the balance of the information presented in a news story would seem 'objectively' to be reassuring, many people will remain unconvinced and for some the fear may worsen.

Finally, Adams argues that some risks (those that are directly perceptible or perceptible through the lens of science) may successfully be managed through regulations. However, many environmental hazards are 'virtual risks' – they are the subject of intense debate by the scientific community and they may or may not present a (small) risk to the public – so regulation is unlikely to prove justifiable.

Acknowledgements

In a book with so many authors there are always numerous people to thank. All the authors have produced very interesting papers that have made my job as editor relatively simple. I would like to thank Julian Morris, Andrew Grice and Barry Sweetman for their editorial suggestions. The Earhart Foundation was generous in providing financial support for my work, as was The Märit and Hans Rausing Charitable Foundation for providing financial support for the project.

For their inspiration or advice in general, and specifically with respect to this project, I would like to thank: John Adams, Bruce Ames, John Blundell, Frits Böttcher, Mary Douglas, John Emsley, David Fisk, Michael Fumento, Mike Gough, David Murray, Matt Ridley, Jim Secord, Fred Smith, Mike Thompson, Sophie Valtat and Aaron Wildavsky. I would especially like to thank Lorraine Moody, this book would not have been possible without her unstinting efforts.

The views in this book are, of course, those of the individual authors and not those of the European Science and Environment Forum. ESEF is delighted to present this work in conjunction with Butterworth-Heinemann, and I would like to thank Michael Forster, the commissioning editor at BH for all his efforts on our behalf.

Biographies

John Adams

Dr Adams is a reader in geography at University College, London. He was a member of the original board of directors of Friends of the Earth and has participated in numerous debates about environmental risks over the past 25 years – an experience which has provided him with a close-up view of the stereotypical responses to risk described in his paper. His book, Risk (UCL Press 1995) was described by *The Economist* as 'beguiling' and by *Nature* as 'extremely counterintuitive'.

Bruce N. Ames

Dr Ames is a Professor of Biochemistry and Molecular Biology and Director of the National Institute of Environmental Health Sciences Center, University of California, Berkeley, CA 94720. He is a member of the National Academy of Sciences and was on their Commission on Life Sciences. He was a member of the National Cancer Advisory Board of the National Cancer Institute (1976-82). His many awards include: the General Motors Cancer Research Foundation Prize (1983), the Tyler Prize for environmental achievement (1985), the Gold Medal Award of the American Institute of Chemists (1991), the Glenn Foundation Award of the Gerontological Society of America (1992), the Lovelace Institutes Award for Excellence in Environmental Health Research (1995), the Honda Foundation Prize for Ecotoxicology (1996) and the Japan Prize (1997). His 380 publications have resulted in his being the 23rd most-cited scientist (in all fields) (1973-84).

Barrie M. Craven

Dr Craven is an economist and a graduate of the University of Hull and the University of Newcastle upon Tyne. He has published in the field of monetary economics in the Journal of Monetary Economics and in the Manchester School. More recently he has researched public policy and health care issues. Current research is focused on the resource strategies associated with AIDS where results are published in several journals including the Journal of Public Policy and Financial Accountability and Management. He has taught at Curtin University, Western Australia and a Cal Poly in California.

Marie-Louise Efthymiou

Professor Marie-Louise Vernallet-Efthyniou received her Doctorate in Medicine in 1963, a certificate of special studies in cardiology in 1963, and a certificate in occupational medicine in 1964. She has been a Professor since 1983 and Chief of Service of the Centre Anti Poison de Paris and of Occupational Medicine since 1984. At the time of writing this, Professor Efthymiou has been responsible for 475 communications and has directed or presided over 138 theses. She is responsible for the Centres for Pharmacovigilance and Toxicovigilance and for Occupational Medicine.

David Everest

Dr David Everest graduated from University College London. After a period on the staff of the then Battersea Polytechnic – now University of Surrey – Everest joined the Scientific Civil Service in 1956 at the former National Chemical Laboratory. The work involved advising the UK Atomic Energy Authority on the procurement of uranium, thorium and beryllium, and the development of extraction methods of these metals from low grade ores. In 1965 he joined the National Physical Laboratory, becoming successively Head of the Division of Inorganic and Metallic Structure and the Division of Chemical Standards. He joined the former Department of Industry in 1977 on the research customer side and in 1979 became Chief Scientific Officer Environmental Pollution at the Department of the Environment. After retiring in 1986, Everest has been a Research Associate of the UK Centre for Economic and Environmental Development, a visiting Research Fellow at the University of East Anglia and Editor of the journal Energy & Environment.

Alvan R. Feinstein

Dr Feinstein is Sterling Professor of Medicine and Epidemiology at the School of Medicine, Yale University. He has published numerous articles in scientific journals and is one of the world's leading epidemiologists.

Etienne Fournier

Professor Fournier was the Head of Clinical Toxicology at the Fernand Widal Hospital in Paris from 1982 until his recent retirement from full-time service. He continues to be involved in research and supervision of studies. He has worked closely on the creation and development of poison antidote centres.
He is the Honorary Head of the Lariboisièr-Saint-Louis faculty and a member of the National Academy of Medicine, the International Union of Toxicology, the French Society of Toxicology and the Society of Legal Medicine.

He is both a doctor of medicine and of physical sciences; he has publications on poisons and their treatment dating back to 1958, and is a national expert on human toxicology.

Lois S. Gold

Dr Lois Swirsky Gold is Director of the Carcinogenic Potency Project at the National Institute of Environmental Health Sciences Center, University of California, Berkeley, and a Senior Scientist at the Berkeley National Laboratory, Berkeley. She has published 80 papers on analyses of animal cancer tests and implications for cancer prevention, interspecies extrapolation and regulatory policy. The Carcinogenic Potency Database (CPDB), published as a CRC handbook, analyses results of 5000 chronic, long-term cancer tests on 1300 chemicals. Dr Gold has served on the Panel of Expert Reviewers for the National Toxicology Program, the Boards of the Harvard Center for Risk Analysis and the Annapolis Center, and was a member of the Harvard Risk Management Group.

Zbigniew Jaworowski

Zbigniew Jaworowski, MD, PhD DSc is a multi-disciplinary scientist and professor emeritus from the Central Laboratory for Radiological Protection in Warsaw. He has recently been a guest scientist at the Institute for Energy in Kjeller, Norway. He has studied pollutions with radionuclides and heavy metals and served as a chairman of the United Nations Scientific Committee on the effects of atomic radiation (UNSCEAR).

Terence Kealey

Dr Terence Kealey is a clinical biochemist at Cambridge University. He has published numerous scientific papers and writes frequently for newspapers including the Daily Telegraph. He has also published the popular science book, The Economic Laws of Scientific Research (MacMillan 1996).

Hans E. Müller

Prof. Dr med. Dr rer.nat. Hans E. Müller trained as chemist and physician, taking doctorates in these areas at the University of Mainz, Germany. He specialized in immunology and medical microbiology and became professor at the University of Bonn in 1973. In 1975 he became head of the Public Health Laboratory in Braunschweig and turned to problems of hygiene and public health before he retired 1995. He has published over 350 scientific articles and five books.

Joan Munby

Dr Munby initially qualified in pharmacology, biochemistry and physiology from Aston University, Birmingham. Her PhD. was in endocrinology. At present Dr Munby is senior lecturer in pharmacology in the School of Health Sciences at the University of Sunderland. Amongst her present interests is environmental risks to health, a topic to which she has contributed several publications. Dr Munby served as assistant editor of Indoor Environment from 1990 to 1996.

Robert Nilsson

Professor Robert Nilsson is senior toxicologist at the National Swedish Chemicals Inspectorate, the central agency for chemicals control in Sweden. In addition, he holds a position as adjunct professor of molecular toxicology and risk assessment at the University of Stockholm. He has acted as technical advisor to the International Program on Chemical safety (IPCS), Geneva, for the production of Health and

Safety Guides, and is a member of the Executive Board for International Society of Regulatory Toxicology and Pharmacology as well as being a member of the Editorial Board for the journal published by this society. He has on a number of occasions served as an expert in various programmes conducted by the OECD Environment Directorate as well as by WHO, and has initiated and/or co-ordinated several research projects supported by the US Chemical Industry and CEC DGXII.

Together with his colleagues at the Karolinska Institute he developed a new specific and sensitive clinical method for the early diagnosis of toxic tubular kidney damage. His current activities involve epidemiological and mechanistic studies on the mode of carcinogenic action of arsenic.

Göran Pershagen

Dr Pershagen was born in 1951, gained his MD in 1980, and his PhD in 1982 from the Karolinska Institute, Stockholm. He has been Professor at the Department of Epidemiology, Institute of Environmental Medicine, Karolinska Institute since 1987. Since 1991, he has been the Head of Department of Environmental Medicine of the Stockholm County Council. Dr Pershagen has served repeatedly as a consultant to WHO, IARC and CEC, and is presently the Acting-President of the International Society for Environmental Epidemiology. He is the author of more than 100 international scientific publications in environmental epidemiology.

Peter M. Sandman

Dr Sandman has been a Rutgers University faculty member since 1977. He founded the Environment Research Program (ECRP) at Rutgers in 1986, and was its Director until 1992. During that time, ECRP published over 80 articles and books on various aspects of risk communication, including separate manuals for government, industry and the mass media. Now based at Massachusetts, Dr Sandman retains his academic affiliations as Professor of Environmental and Community Medicine at the Robert Wood Johnson Medical School. He holds a BA (Psychology) from Princeton University and an MA (Communication) and received his PhD in Communication from Stanford University in 1971.

Gordon Stewart

Gordon Thallon Stewart MD, FRCP, FRCPath, FFPHM, FRSS, DTM and H. Dr Stewart is the emeritus Professor of Public Health at the University of Glasgow and a Consultant in Epidemiology and Preventive Medicine. During WWII, he served as a Medical Officer in the Royal Navy at sea and in SE Asia. After the War, he held appointments in medicine, laboratory medicine and epidemiology in Aberdeen, Liverpool and London before moving to America in 1963 as a Senior Fellow of the National Science Foundation. During his nine years in America as Professor of Epidemiology, he organised an epidemiological and control programme dealing with the abuse of drugs, with the support of the National Institute of Mental Health. After returning to Glasgow, as Professor of Epidemiology, he became involved in self-help programmes, and with the effects of routine vaccination of children. He initiated, with the help of the Department of Health, the continuing long-term evaluation of the risk-benefit of immunization, and helped launch the Vaccine Damage Act, 1979.

Dr Stewart has served as a Visiting Professor, lecturer and examiner at various universities in North America, Europe, Africa and Asia. Since 1952, he has often worked for the World Health Organisation, UNICEF, New York City Health Department, as well as for UK health authorities and some industries. His expertise in infectious diseases contributed to the development and trials of new antimicrobial drugs, notably new penicillins and cephalosporins in the 1950s and 1960s. Since 1968, he has been increasingly active in assessing a wider range of hazards to the public health, notably drug abuse, other forms of health-damaging behaviour and, latterly, AIDS. Some of this work is continuing into his retirement.

Fritz Vahrenholt

Dr Vahrenholt is a Senator in the City Environmental Authority of Hamburg, Germany, a post he has held since June 1991. His academic background is in chemistry and he has been a member of the German Academic Federation since 1969. Between 1972 and 1974 he undertook postdoctoral research at the University of Münster and the Max Planck Institut in Mulheim. In 1974 he was made Head of the Chemical Industries Section in the Federal Office for the Environment in Berlin, and in 1976, became the Head of the Department of the Environment, Waste Economy and Emissions. In the Hessen State Ministry for State Development, Environment, Agriculture and Forestry in Wiesbaden. Prior to his appointment as Senator, he was a Privy Councillor in the City Environmental Authority in Hamburg and the Head of the Office of the Senate of the City of Hamburg.

Donald F. Weetman

Professor Weetman qualified in applied pharmacology from Chelsea College, London. After working in the pharmaceutical industry, he entered academic life at Barking Regional College, before moving to Sunderland in 1968. Professor Weetman was awarded a personal chair in pharmacology in 1990, and he retired a year later to become an independent consultant on health and education matters. He was formerly the chief technical adviser on drug testing to the Ministry of Science and Technology of the Republic of Korea under a United Nations Industrial Development Organization programme.

Professor Weetman has served on numerous committees and editorial boards of scientific journals. He was the founding editor-in-chief of *Indoor Environment*. He is author of over 300 original papers and technical reports. Professor Weetman and Dr Munby are currently writing a book on research methodology in the biomedical sciences, which is partly based on the research methods module of the MSc Health Sciences degree at Sunderland.

James D. Wilson

Dr Wilson is a Senior Fellow at Resources for the Future, and leader of the risk analysis program within the Center for Risk Management. An organic chemist by training, he holds degrees from Harvard and the University of Washington (PhD, 1966). Joining the Monsanto Company in St. Louis, he worked several years doing bench research, from which he moved first into research management and then into environmental policy. His tenure as research manager included a several-year stint working on a 'dioxin' problem, and included managing the interface between one Monsanto business unit and product regulatory agencies. In 1983 he began to focus on science policy issues concerning application of risk analysis to human health and environmental regulatory decisions. He was President of the Society for Risk Analysis in 1993 and was named a Fellow of the Society in that year.

I Methodology

1 Thresholds for carcinogens: a review of the relevant science and its implications for regulatory policy

James D. Wilson

Summary

Regulation of carcinogens in the United States isbased on a 'no-threshold' policy. This assumes that there exists no level of exposure for which the possibility of causing harm is truly zero. The alternative threshold policy would assume that there exists some level of exposure at which no harm will come to anyone in a population so exposed. The no-threshold policy may have made sense when adopted thirty or more years ago, because the scientific evidence then available was not able to distinguish between these two opposing hypotheses, and no-threshold provided a greater margin of safety (although at some cost). Since then, our understanding of biological processes related to the birth and growth of cancer has greatly expanded. We now understand that two different biological processes can increase cancer risk. Increasing the rate at which cells divide (the mitotic rate) is one of these; increasing the rate at which mutations occur, independently of mitotic rate, is another. Correspondingly, two types of carcinogen may be identified: mitogenic carcinogens act predominantly by increasing the rate at which cells divide (increasing the probability of spontaneous mutation), whilst mutagenic carcinogens act directly on DNA (causing errors in replication).

Mitotic rate is under close physiological control, operated through a complex system including a variety of intercellular messenger molecules, thereby ensuring a relatively stable rate of cell division. Functions controlled remain within certain limits in the face of external stressers (such as potential carcinogens) must, by definition, exhibit a threshold in their response to small changes in such external stress. So long as the external stress induced by a mitogenic carcinogen is within the limits of the physiological control system, the mitotic rate will not change. All mitogenic carcinogens exhibit a threshold, the level being determined by the individual's physiological response.

For classical mutagenesis too, the weight of evidence favours the conclusion that thresholds exist. Evidence for no-threshold has almost no weight, while there is moderate evidence – primarily that cancer rates are not elevated in areas of high background radiation flux – that the mutation rate is under active physiological control. It is likely that mutagenic carcinogens also exhibit a threshold.

In light of this new understanding of the process of cancer formation, policy should no longer be based on the assumption that there is 'no threshold'. In developing regulatory rules, policy-makers should first distinguish between 'mitogenic' and 'mutagenic' carcinogens respectively, those that act predominantly by increasing the rate at which cells in certain tissues divide, and those that act directly on DNA. Mitogenic carcinogens should receive the same treatment as 'non-carcinogens'. At the US Environmental Protection Agency (EPA), at least, policies are changing to reflect this understanding. Mutagenic carcinogens should be regulated using the 'de minimus' approach, which relies on the concept of a 'practical threshold', and provides very adequate protection of public health.

Introduction

This chapter examines the scientific bases for two opposing theories about the process by which cancers start, and the public policy implications of that choice. For policy purposes, it is important to understand this process and how it is affected by human activities, particularly by new chemical substances whose use has transformed modern civilization. People come into contact with and absorb these substances, frequently at levels that are very small compared with amounts that cause discernible harm. The tools available to science only very rarely permit a direct determination of the likelihood that injury will occur from such very small levels of exposure. Thus policy-makers must rely on scientists' judgement of that likelihood, a judgement that is heavily influenced by prevailing theories of cancer and carcinogenesis. Over the past twenty-five years the consensus on these has evolved, and with that evolution has come a change in recommendations for policy.

The two opposing theories concern the behaviour of cancer 'dose–response' relationships. These relationships are expressed in terms of mathematical functions that describe how cancer incidence ('response') changes with levels of exposure ('dose') to carcinogens. In technical terms, the proposition we wish to examine concerns the behaviour of these functions as exposure becomes 'arbitrarily small' – but not zero. ('Arbitrarily small' means that we are allowed to choose a level of exposure that is as small as we wish, approaching very near to zero but not reaching that absolute value.) Here, we are interested in exposure values that are less than what can be measured experimentally, but not in any *specific* exposure value. One of these, the threshold theory, holds that increased incidence (over background) equals zero at some non-zero exposure. The other, no-threshold, holds that increased incidence equals zero only at zero exposure. Until very recently, public policy has been based on the no-threshold theory.

Until the 1950s it was a well-established tenet of toxicology that for any poison there exists a threshold dose below which exposure poses no risk. Late in that decade, regulatory scientists proclaimed that for carcinogens, 'no safe dose exists' (Fleming 1959; Flamm and Scheuplein 1989). This proclamation occurred near the beginning of three decades' worth of laws and regulations intended to protect human health and our environment. Now we are engaged in a substantial reconsideration of the policy edifice built in those decades, asking whether those laws and regulations should endure. As part of this reconsideration, it is appropriate to examine the scientific basis for the laws and regulations, to ascertain which parts are well founded and which stand on sand.

The no-threshold policy has always been controversial. Most biologists believe that thresholds always exist. Mathematicians point out, correctly, that the curves biologists use to represent reality, such as relationships between exposure and response, are uniformly continuous, infinite, monotonically increasing functions that do not admit a discontinuity such as a threshold. Further, the uncertainties in ascertainment of exposure and response are large enough to prevent any direct experimental test: the 'noise' is large compared to the 'signal' one hopes to detect. At the time the no-threshold policy was established, scientific theory could provide little guidance to policy-makers. The threshold and no-threshold options were equally plausible. Thus, regulatory scientists sensibly urged adoption of the more protective policy, based on the no-threshold theory.

Over the last twenty years the situation has changed. We now understand much better both how cancers start and grow and how biological functions are regulated. There is now a consensus that thresholds exist for some kinds of carcinogenic processes. (US EPA 1996). We suggest here that the weight of evidence favours the view that thresholds exist for all such processes. Policies should be adjusted accordingly. Doing so would not change the degree of health protection.

This chapter is intended as a contribution to science policy. The argument flows from a review of the current situation – recapitulating the meaning of threshold and policies that depend on its interpretation – followed by a summary of new science relevant to these policies and then some conclusions about policies. The discussion is based on a theory of how science contributes to policy formation, detailed in an appendix to this chapter.

Situations such as the threshold debate provide a rigorous test of science. Whether or not a threshold exists cannot be answered by direct test. It is beyond reach of the kind of experiment or study that would provide scientists with the strongest support for their beliefs. We cannot *know* how the dose–response curve actually behaves as it approaches arbitrarily close to zero effect. ('To know' something should be understood to mean 'to have great confidence in' some conclusion.) Because the conclusion is intended to inform public policy in which individual scientists may well have a stake, it becomes very important that all the evidence be reviewed, discussed, and weighed as fairly as can be done. It is equally important that any conclusion be accompanied by a frank discussion of any uncertainties: if an uncertainty is large and the confidence not high, public policy may best be served by a choice that does not reflect the current scientific consensus.

The meaning of threshold and current application to policy

Like many useful concepts, threshold carries somewhat different meanings to different specialities within science and the public health profession. To pharmacologists (with whom it appears to have originated) the idea is straightforward: there are always doses (levels of treatment) that are without discernible effect on those being given drugs. In this context, the theory of a threshold for activity is not controversial. It is, in fact, universally accepted and applied. Centuries of observation support the idea.

It is useful to refer to the pharmacologists' concept as a 'practical threshold' ('effective threshold' and 'threshold of regulation' are used synonymously in this chapter). It implies that responses which cannot be observed have no practical significance, but may admit the possibility of some response at low exposure, or predict an undetectably small one, yet for practical purposes these may be neglected. Much of our current policy edifice for dealing with carcinogens actually stands on this concept. It is, for instance, consistent with the 'no significant risk' standards common in environmental statutes (Byrd and Lave 1987; Barnard 1990; Cross et al. 1991).

To people dealing with public health policy, threshold carries a more stringent implication: at some non-zero exposure the response not only passes below the limit of detection, and not only approaches zero as a limit, but equals zero. We leave aside the issues posed by essential nutrients such as selenium, and the complications introduced into analysis of dose–response curves by competing causes of mortality. It can be taken as given for policy purposes that for truly essential nutrients, a threshold has to be assumed. There has been considerable discussion, not yet resolved, of this issue in the instance of a drinking-water standard for arsenic. Biologists appear to believe that the thresholds they observe are absolute, but the question has not been one of much interest; and for them the 'practical threshold' concept suffices.

There have been circumstances where, for public health officials, the practical threshold has not sufficed. These officials are charged to provide the public with credible advice regarding safety. Their credibility depends on their confidence in their understanding of the hazards facing the public. They can confidently declare exposures safe when no adverse effect can be identified, if a threshold exists. However, if they believe no threshold exists, their confidence may be much less. The practical threshold states only that no harm can be detected; such a prediction may well not be acceptable, because the ability to detect such harm is not very sensitive. Detectability of harm is established by the year-to-year variability in cancer rates; for most this is at least 2–3 per cent of the annual rate. For the more common cancers, these annual rates vary in the range of tens to thousands to hundreds of thousands per year. Thus exposure to some specific chemical might be causing thousands of deaths without detection. For example, in adult men in the United States, the death rate from bladder cancer varies randomly between 55 and 65 per hundred thousand per year. With close to 100 million adult men in the US population, between 55 000 and 65 000 deaths occur every year from this cause. Year-to-year increases and decreases of two or even three thousand deaths (5 per cent of the mean rate) occur frequently. It is not possible to identify a change of this magnitude that may be due to some particular cause, in the face of such fluctuations, unless it were to continue for many years. However, for rare cancers such as that caused by exposure to vinyl chloride, the detection limit is only a few deaths per year.

Half a century ago, as scientists began to give serious consideration to protecting against radiation and chemicals that cause cancer, it was regarded as not acceptable to the public that such a situation might be regarded as 'safe'. At that time theory suggested that there might not be a threshold for exposure to substances that may cause this disease. So the profession recommended basing policy on the assumption that no threshold exists. From this came the idea that, 'There is no safe exposure to a carcinogen'. Contributing to this idea was a lack of confidence among professionals in what should be considered 'safe'. They worried about absolutes. Lowrance (1976) has since shown that 'safe' is not an absolute, but relates to a notion of acceptable risk. In the 1960s the public health profession was not certain that *any* degree of cancer risk would be acceptable. Today we have a better understanding of people's tolerance for different kinds of risks, in different situations. (For instance, we know that more risk is tolerated when those at risk feel able to exert some control over their own fates, see Chapter 15.)

The 'no safe exposure' dogma reflects the paradigm of carcinogenesis that was current through the 1960s. This theory implied that all cancer occurs as a consequence of the reaction of 'carcinogens' with DNA (Olin et al. 1995). DNA reacts with typical carcinogens then recognized much as related low-molecular-weight model compounds do. Almost all such carcinogens come from the chemical family of 'electrophiles'. DNA-like molecules react with these in a way that gives rise to a direct proportionality between dose and effect. According to this theory, a no-threshold, linear relation between very small chemical exposure and cancer risk is quite reasonable.

The idea that any exposure to a carcinogen produces some risk to those exposed became widely established in policies of regulatory agencies in the United States between the late 1950s and the mid-1970s. Its first and most enduring formal expression was in the famous 'Delaney Clause' enacted as part of the 1958 amendments to the Food, Drug, and Cosmetic Act (FDCA). This clause prohibits approval and thus use of intentional ('direct') food additives that 'induce cancer'. Although symbolically this language has great weight, in practice it has seen almost no use (Prival and Scheuplein 1990). The Food and Drug Administration already had the duty to keep all harmful substances from food. It can and has used its general authority from FDCA to ban certain carcinogens in the food supply.

The FDA is also charged to assure that the food supply is adequate to our needs. This responsibility sometimes conflicts with the requirement that food be safe; FDA commonly balances these contradictory objectives by recognizing Paracelsus' ancient dictum 'the dose makes the poison', i.e. that sufficiently small amounts of the most deadly poison will do no harm. That is, FDA trades off certainty of safety for certainty of sufficiency to reach the needed balance. Under the general provisions of FDCA, FDA also can recognize the limitations of inferences for human response from experiments in animals, weighing the utility of such data in the regulatory balance. But with Delaney, Congress instructed FDA to adopt a different policy, assuming rigorously that safety could not be achieved with any amount of carcinogenic substance in food and taking at face value any results from animal experiments.

Beyond extending Delaney to food colours and animal drugs during the 1960s, Congress has not written the no-threshold idea into any of the many other health and safety laws. Nevertheless, the idea has pervaded regulatory approaches to carcinogens, and extension to developmental toxicity has been proposed. Note that no regulations based on a 'no threshold' assumption for endpoints other than cancer have been implemented (Rai and Van Ryzin 1985). Notable examples of its impact include US EPA's adoption of the 'linearized multistage' procedure for evaluating cancer risks, and the 'risk-specific dose' hazard indices used for environmental regulations, the distinction that that agency makes between 'carcinogens' and 'non-carcinogens' in drinking water standards, and so on. It underlay the OSHA 'Cancer Policy' proposed in 1979 (which has been neither withdrawn nor implemented) and continues to influence both OSHA's and FDA's regulatory approaches to substances called 'carcinogens'. It also is seen frequently in state laws and regulations, particularly California's 'Proposition 65'.

Over the years, FDA has whittled down the applicability of Delaney, especially for 'indirect' food additives. For these and for 'contaminants' FDA's charge to maintain an abundant and healthful food supply has led it to adopt a *de minimus* policy (Prival and Scheuplein 1990). Most foods are now known to contain insignificantly small amounts of naturally occurring carcinogens (NRC 1996); banning all carcinogen-containing foods would mean we would have nothing to eat! Thus FDA has moved to the 'practical threshold' to regulate carcinogens. US EPA once proposed to do likewise for pesticide residues (US EPA 1988), but that policy was struck down by the courts, and the Agency and interested parties are trying to work out an alternative.

The scientific case for a no-threshold assumption

Two kinds of evidence support the no-threshold policy. First, many sets of observations fit a non-threshold mathematical model. Second, if the underlying physical processes, either stochastic or otherwise, are describable by statistical methods, mathematics requires that any non-zero exposure implies a non-zero added risk. These arguments have some validity, and need evaluating in a science-based conclusion. Note that it has also been proposed that because background processes lead to cancer, exposures to carcinogens should be considered to supplement these processes, giving non-threshold, linear exposure-response functions (Crump et al. 1976). A similar argument has been made, perhaps more persuasively, by Crawford and Wilson (1995) for common adverse effects other than cancer. 'Additivity to background', as this is called, does imply a linear dependence of response on exposure, but it does not imply anything about the intercept of these straight lines. If no response occurs at some exposure, i.e. a threshold exists for the effect, the intercept will be non-zero.

Evidence from curve fitting

The first justification for assuming no threshold for carcinogenicity exists came from an observation on the effect of radiation treatment on single cells. The dose–response enabled a straight line to pass through the origin. Subsequently, the theory of mutation-caused cancer supplied a theoretical basis for analysing dose–response in this way. Still later, a second argument, based on treatment adding a small increment to an ongoing process, was raised (Crump et al. 1976) to justify the assumption. Yet these arguments at best permit the interpretation that no threshold exists; they permit equally well the contrary argument.

Categorically, the fact that a curve with zero intercept 'fits' a particular data set does not imply that a curve with non-zero intercept will not also yield an acceptable fit, especially to the kinds of data sets available in this field. Uncertainties are very large, compared to the intercepts expected on biological grounds. The original curve fitting was done with pencil and graph paper. To anyone who has done such manipulation, it is obvious that such a procedure can equally well produce a line that supports a threshold interpretation. Such data that can be fitted to a line of the form, $y = mx$ will also fit $y = mx + b$, with $b \neq 0$. In fact, if b is permitted to vary the uncertainty in the data sets available is such that computer-generated fits to this data will usually return $b \neq 0$, just by chance, even if the true value of b were zero. These data are simply too uncertain to allow distinguishing between threshold and no-threshold hypotheses. At best, the data permit setting a bound on the magnitude of a threshold.

Observations of dose–response from radiation-induced cancer do fit reasonably well to non-threshold curves (NRC 1990). Yet recent re-evaluations of some of the best of these data sets, using methods that allow the intercept to vary, have found better-quality fits are obtained with non-zero intercepts.

Stochastic processes

Many of the biochemical processes involved in the chain of events between exposure to a carcinogen and the start of a cancer are of a sort usually describable with statistical mechanics. Using these techniques one can calculate quantities such as the fraction of molecules from a particular dose of reactive carcinogen that will reach cellular nuclei. These techniques were originally developed to try to understand non-equilibrium properties of gases; the typical problem involved estimating the probability of a fast-moving elastic ball bouncing around inside a box passing through a small hole into an adjacent box in some period of time. This probability was found to depend on the speed with which the ball is moving and the size of the hole: the faster the speed (the higher its energy) and the larger the hole, the sooner the ball will pass through into the next box. With an assembly of such balls and knowledge of the distributions of their energies, one can calculate such things as the numbers of balls found in the adjacent box during successive periods, and thus understand the rate at which equilibrium is attained. In effect, the size of the hole can be equated to a barrier which the ball must surmount in order to pass from one state to another.

Entry of toxicants into the system (by inhalation, ingestion, absorption, etc.) is describable by statistical mechanics; the rate depends on such measurable parameters as adsorption coefficient, partition coefficients between blood and air, etc. Once into the blood, molecules are transported everywhere, some to be altered by chemical reactions. At the molecular level, the reaction of a mutagen and DNA leading to a carcinogenic transformation is best thought of as one in which the attacking mutagen is hindered by a series of barriers that are not infinitely high, not infinitely efficient at disarming this invader.

Chemists and engineers commonly use statistical methods to analyse this kind of process. These very successful methods make a fundamental assumption that the probability of overcoming all the barriers is random, or 'stochastic' and thus may become very small but never identically zero.

There are several of these barriers. One is metabolic transformation to an inactive substance, which includes reaction with protective elements such as the mucous of the respiratory tract and chemicals such as glutathione. It was puzzling, at first, that the chemicals that react most avidly with DNA components have relatively little carcinogenic or mutagenic potency *in vivo*. This was rationalized by noting that these chemicals are so indiscriminately reactive that at small doses they react with something else before reaching the cell nucleus, and at high doses they kill.

A second line of defence is DNA repair. Since DNA is constantly being damaged by various natural processes (mostly through oxidation), we have evolved very efficient enzymes that monitor the status of our DNA and repair it as necessary. Recently several genetic defects in DNA repair enzymes have been observed to be associated with an increased risk of cancer. Work by Ames and his collaborators suggests an impressively high rate of repair going on constantly: they find excreted in urine oxidative degradation products of DNA bases that imply repair of *ca.* 10^4 defects per cell per day (Ames and Saul 1986).

A third 'barrier' arises from the relatively small number of effective target sites for a mutagen, compared to the total sites available. To give rise to a cancerous cell, mutagens must attack one of the few ($\sim 10^3/\sim 10^7$) genes that code for proteins involved in regulation of cell division and differentiation

(Bradshaw and Prentis 1987; Knudsen 1985) or in DNA repair, and do so in a cell that is capable of division. (Terminally differentiated cells, for example, do not reproduce, so any mutations that may occur in them are not passed along to other cells.)

Viewed in conventional terms, and demonstrably true with treatments that produce observable effects, each of these processes is stochastic. One can readily estimate the probability that a molecule of a mutagen will get past each of these barriers and cause an effective mutation in a critical gene. At exposures leading to observable effects, these estimates account for the effects observed.

The probability that any one molecule will effect a cancerous transition in a cell capable of starting a cancer is vanishingly small 10^{-18} or less. Yet even this unimaginably small number is overwhelmed by the numbers of molecules that take part in an exposure capable of causing cancer. Because we can carry out experiments in mice, we know these numbers more precisely for these animals. Typically, for a moderately active carcinogen in a mouse, cancers will be caused by doses of ~50 mg/kg body mass. This corresponds to the order of 10^{19} molecules per mouse, or about 10^{10} per cell. Doses of this size usually result in one or a few tumours, the results of attacks on DNA from the ~10^6 cells capable of developing into a cancer that exist at any one time. Thus the 'yield' of tumours per molecule of carcinogen is the order of 10^{-19}.

Because 'effect' doses appear to scale with body mass, we infer that similar numbers of active molecules per cell are necessary for a single dose to cause cancer in humans. When the numbers of molecules are this large, stochastic models describe the process very well. We can readily account for differences in potencies among mutagenic substances by differences in metabolism, etc., that make them more or less capable of overcoming these barriers.

However, for most 'environmental' issues, policy-makers are concerned about rather different conditions, with exposures measured not in tens of milligrams per kilogram but micrograms or nanograms or picograms; not billions of molecules per cell but thousands or even less than one. For strong mutagenic carcinogens, the likelihood that a dose constituting some number of molecules will cause a cancer is less than the fraction it represents of the number known to cause cancer. From this it follows that at exposures totalling, say, a millionth of the number required for observation of tumours in a single-dose experiment, the likelihood of a cancer becomes vanishingly small. So if 50 mg/kg of a carcinogen causes tumours in any mouse examined, total exposures less than 50 ng/kg will cause tumours in (many) fewer than one mouse in a million. This line of reasoning supplies very strong support for the 'practical threshold' policies described above. If the likelihood of an exposure causing a cancer can be said to be negligibly small, the exposure can be considered safe (Lowrance 1976). At such small exposures, no effects can or will be observed. (In fact, as is noted below, the likelihood of cancer will be much smaller than this upper bound, because these carcinogens act as mitogens as well as mutagens, and their 'potency' is much less than proportional to dose, with decreasing dose.)

This straightforward reasoning underlies the initial application of 'quantitative risk assessment' to a health-regulatory situation – FDA's implementation of the 'DES proviso' from the early 1970s (Olin et al. 1995). The risk from extremely small exposures to the undoubted carcinogen DES was held to be negligible, and thus acceptable.

Mathematicians, however, correctly note that 'small' – even 'negligible' – is not the same as 'zero'. If the statistical-mechanical representation of the process is correct at these small exposures, some non-zero risk will remain. The question now becomes, is this representation exact, so that projections into the 'tails' of the underlying distributions are appropriate, or is it only an excellent representation of measurable reality? We have noted that the uncertainties are too large to permit discrimination between threshold and non-threshold behaviour on the basis of quantitative measurements. To examine this issue of low-level behaviour, we must take leave of mathematics and reconsider biology.

The evidence suggests that low-level behaviour differs. No known biological process is infinite; all are truncated by limits imposed by the physical world. Although the heights of men can be well represented by a normal distribution, and thus we can predict the frequency of 19-footers, in fact men the height of giraffes could not live. Our arteries would not stand the pressure required to pump blood to a brain so far from the ground. Similarly, no adult humans are smaller than about thirty inches. The distribution of heights is truncated, not infinite. As is argued below, the science of biological control suggests that these apparently stochastic processes are not really random but truncate at small exposures.

The proposition is made that stochastic processes require that dose–response curves describing such processes have no threshold. But we find that the science evinced to support the proposition actually stands on a tautologous assumption that is arbitrary, and not consistent with our other observations of nature. In sum: the choice of mathematical models used to extrapolate from the observable region into the void is ultimately arbitrary. Without a strongly supportable underlying theory, we can have little confidence that the extrapolation represents reality.

There exists essentially no reliable science that supports the no-threshold theory and excludes the threshold theory.

Scientific evidence for the existence of a threshold

In the previous section we examined the arguments supporting the no-threshold proposition, concluding that the evidence is equivocal and that support for the proposition is based on arbitrary, insupportable assumptions. Here we turn our attention to conclusions that can be drawn from modern biological theory. We review first the major change in the theory of carcinogenesis that occurred about 1980, and its implications. Next we review findings from biological control theory that have emerged in this same period. Together these two theories strongly constrain the mathematics used to model dose–response for cancer and other kinds of injuries.

The current theory of carcinogenesis holds that two kinds of processes can be involved in the development of cancers.

In 1981, Moolgavkar and Knudson described a theory of cancer that is capable of rationalizing essentially all that is known about this disease. Greenfield et al. (1984) independently developed essentially the same theory at the same time although the mathematical formulations of the two are different and complementary. Their theory unified the two long-contending theories of carcinogenesis, somatic mutation and uncontrolled growth, into a coherent whole. It recognized the requirement that heritable change in a cell's genotype must occur for a carcinogenic transformation to occur, and also recognized the importance accelerated cell division (mitosis) had for increasing in the rate at which such mutations occur. Cancer is much more likely to develop in tissues undergoing rapid growth, both because division must occur to 'fix' DNA damage as a mutation, and because a shorter interval between divisions may increase the likelihood that DNA damage will go unrepaired. Thus we can understand why cancers that occur only in childhood (e.g. Retinoblastoma) are found in tissues that stop dividing early in life, why rates of sex-organ cancers exhibit an age dependence that follows growth and development of these organs (rates increase dramatically following puberty), and why 'promoters' that do not affect mutation rates can nevertheless greatly speed the development of cancers. The classical somatic-mutation theory cannot by itself rationalize any of these observations (Moolgavkar and Venzon 1979; Moolgavkar 1983; Wilson 1989; Preston–Martin et al. 1990).

In the last decade there has been an extensive exploration of the implications of this theory, mainly by groups associated with Suresh Moolgavkar at the Fred Hutchinson Cancer Center, Sam Cohen at the University of Nebraska, and Curtis Travis at Oak Ridge National Laboratory. They have used mathematical modelling to explore the relative impact of mutagenic and mitogenic (cell-division rate) effects on the development of cancers, using data from both human experience (e.g. from uranium miners) and animal experiment (e.g. saccharin, liver-tumour promotion) (Moolgavkar et al. 1990; 1993; Ellwein and Cohen 1988; Cohen and Ellwein 1990; Travis et al. 1991). They have consistently found that, unaided, exposure to purely mutagenic conditions does not accelerate tumour formation very much. Purely mitogenic stimuli can significantly accelerate tumour formation, acting most strongly on the mutant, partially transformed cells that exist in all animals by the time of birth (Cohen and Ellwein 1990). However, the two kinds of effects act multiplicatively: tumour response rates increase dramatically in exposure regimes where both mitogenesis and mutagenesis occur. These findings have a number of other interesting implications for public policy (notably concerning how chemicals should be tested). Important for this present discussion is that there are two different physiological processes, interference with which can increase cancer risk. Strictly speaking, at exposure levels capable of causing observable increases in cancer incidence, these two effects interact, with increasing mitotic rate leading to an increased mutation rate (because of less effective DNA repair). However, at 'environmental' exposures, the separateness of the two effects can be distinguished. Of the two, speeding up cell division is quantitatively much more important than increasing the mutation rate (Cohen et al. 1991).

Mitogenesis is under strict biological control. The DNA damage repair rate appears also to be under close control. This implies that both cancer-causing processes have thresholds.

In this section we argue that the threshold in exposure–response curves arises from the existence of biological controls on both cell division and DNA repair rates. The mechanism by which mitogenesis is controlled comes primarily from outside individual cells (and thus cannot be understood from studies confined to the cellular level). The control on DNA repair is poorly understood; the existence of such control is inferred from several observations. We start with a few observations on control systems, noting the characteristics that necessarily imply the existence of a threshold in the 'controlled parameter'. This is followed by descriptions of the control systems for mitogenesis and DNA repair, that lead to conclusions about their exhibition of thresholds.

Characteristics of control systems

Control systems strive to maintain some parameter or variable within 'control limits' that are set externally to the system. (The simplest control systems employ only one limit, often called a 'set point'.) Parameters commonly controlled are temperature, pressure, and chemical composition or concentration. Three features are characteristic of this kind of control system: some means by which an actuator can affect the parameter to be controlled; a means by which a controller can sense the state of the system (usually by measuring the controlled parameter) and compare it to the control range or set point; and a means of signalling the actuator to take action to bring an out-of-control situation back within the control range. The means of sensing the state of the system, which involves the transfer of information from the system to the controller, is called a 'feedback loop'.

To aid understanding of these characteristics, consider the commonplace system that we use to keep our living spaces warm in cold weather. We employ some sort of heater that is cycled on and off by a controller called a 'thermostat'. We identify the air temperature we wish to maintain – the set

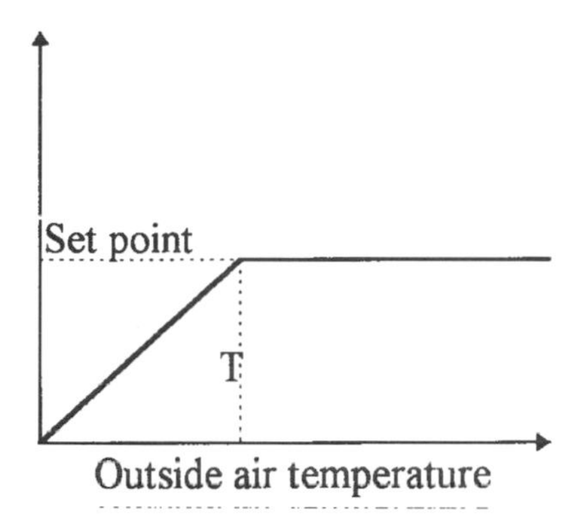

Figure 1.1

point – and adjust the thermostat so that when the air temperature drops below this point, the device signals the heater to turn on. When the air temperature is warmed above this point, the controller turns the heater off, returning the actuator to its normal state. In this example, the thermostat includes a means to sense when the air temperature needs correcting, by comparing it with the set point, and also a means to convey that information to the heater (in the form of an electrical signal); the agent of corrective action is, of course, the heater. In this case, the feedback to the controlling thermostat occurs in the form of an increase in the air temperature.

Note that the heater has some limit on its capacity to add heat to the living space: its 'full blast' rate. As long as the heat loss rate is less than the maximum output of the heater, this system will maintain a constant air temperature. But should the weather turn very cold and windy, and the rate of heat loss exceed the capacity of the heater, the air temperature will begin to drop below the set point. Were we to plot, say, the outside air temperature (holding wind speed constant) against the interior air temperature, the resulting curve would describe a threshold: a straight line of zero slope to that point where the heat loss rate equals the capacity of the heater, after which the interior temperature would decrease proportionately to any further decrease in the outside temperature (Figure 1.1).

Threshold behaviour is intrinsic to and necessary for a system that controls some parameter within limits and has a fixed capacity to accommodate to external conditions. Demonstrating that a system is under physiological control is sufficient to show that a threshold will exist in any relationship between measures of its state and measures of external stress.

Biological control systems Biological control systems work in almost exactly the same way as the mechanical system just described. For instance, the central controller of body temperature (thought to be in the hypothalamus)

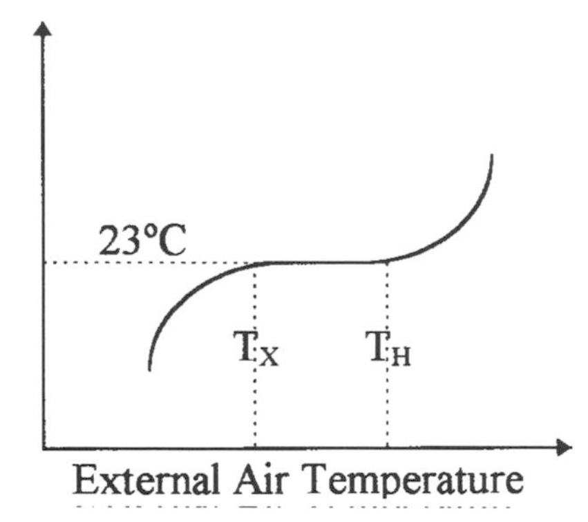

Figure 1. 2

responds to an increase in internal temperature by causing surface blood vessels to expand and by causing sweat to be released, both of which increase the rate at which heat is lost from the body. Instead of electrical signals, the body's central controller uses changes in the concentration of regulatory chemicals (many of them hormones) for example, to signal the need to expand or contract capillaries. (Excepting temperature, all biochemical signalling occurs by means of gradients in the concentrations of various chemicals.) We maintain a near-constant body temperature (within limits) with this system, but it can be overcome: there are limits on the heat load any individual can tolerate before body temperature rises with consequent adverse effects. This limit is the threshold.

Note that control of body temperature is quite precise. The hypothalamus must contain some sort of two-way controller. The rate at which we produce heat is variable, controlled by the amount of circulating thyroxin, the thyroid hormone. Our capacity to adjust to the external environment is obviously limited in both temperature directions: there exist two thresholds in the curve that relates internal body temperature to the external environment temperature (Figure 1.2).

There exist regulatory systems that do not act to maintain a controlled parameter within specified limits. For instance, a system might alter a parameter by some specified ratio or proportion. In industrial practice, such a device could be used for sampling, for instance, taking one can of beer out of every thousand that pass through a packing line. An example from electronics would be a device that takes an incoming signal and reduces its amplitude by some fraction before passing it on. Clearly such a system will have a capacity limitation, similar to that of a typical control system. An inflection point analogous to a threshold could be observed. However, the dose–response curve expected for a toxicant acting by perturbing such a system would be a 'hockey stick,' with a non-zero slope below the inflection point followed by a steeper straight line until saturation begins to set in (Figure 1.3).

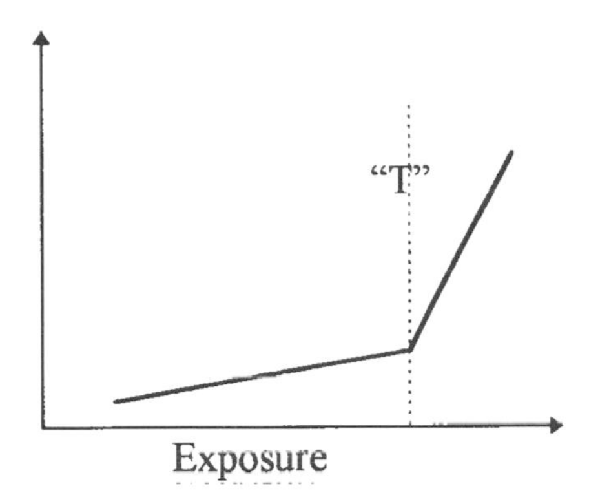

Figure 1. 3

Hormonal control Our bodies include control systems, that act on every physiological function. Most if not all involve hormones, the specialized chemicals synthesized in the 'endocrine' organs (pancreas, thyroid, hypothalamus, etc.) and released directly to the blood. None of these systems is understood completely, and evolution has ensured that complex interactions among them are the norm. Typically, control is not solely the responsibility of a single organ, but is shared. For instance, the pancreas increases the rate of insulin release in response to an increase in blood glucose, a decrease in blood insulin, and also in response to poorly understood signals coming from the intestines by way of the hypothalamus. (Insulin release increases in response to food, just before blood glucose begins to rise.) Control precision is increased by the liver; blood flowing through the liver exits containing less of each hormone than when it entered. Since signals consist of concentration gradients, this creates a constant 'sink,' comparable to the heat sink of a precise temperature control system, permitting fine control. As a consequence, hormones are continuously released to the blood stream.

Some mitogenic carcinogens act by perturbing hormonal control systems. For instance, they may act by increasing the rate of hormone destruction by stimulating enzyme activity. In rats dioxin and many other chlorinated hydrocarbons increase the activity of a mixed-function oxidase that destroys thyroid hormone, leading, at high enough exposures, to enlargement of the thyroid (goitre). In rodents, goitre predisposes to thyroid follicular cell cancer. (It is debatable whether it does so in humans.) Other chemicals act by interfering with hormone synthesis: in rodents thiourea and its analogues act by tying up one of the key enzymes in the synthesis of thyroid hormone, which also leads to goitre and thus thyroid cancer. 'Red Dye No. 3' causes the same response by interfering with a different enzyme (Olin et al. 1995). Oestrogen mimics increase endometrial and breast cancer risk by directly stimulating cell division (Preston-Martin et al. 1990). A progesterone antagonist might have the same effect, at least in the endometrium, since this hormone counteracts the growth-promoting effects of oestrogen in this tissue.

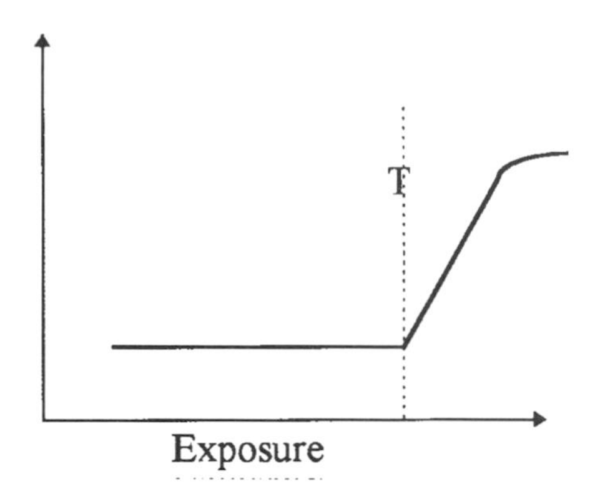

Figure 1. 4

Wilson (1995) derived from basic biochemistry and physical chemistry the nature of the dose–response curve for these effects (see Figure 1.4). This work showed that the equations describing the effects of chronic treatment with any substance that acts by perturbing a hormonal control system include a variable that does not depend on the toxicant dose, generally related to the variable on which the control system works. (For example, in the case of an insulin mimic, the effect of which would be reduced blood glucose, the rate at which glucose enters the blood stream is independent of treatment.) The magnitude of this independent term determines the magnitude of the threshold. This work demonstrates that the existence of a threshold is independent of the mathematics of receptor binding, and that it is a consequence of the existence of control itself.

Control of mitogenesis The system that regulates cell division and differentiation is now becoming reasonably well understood. The research leading to this understanding has been driven by the discovery that mutations in the genes that are involved in coding for the many proteins determining intercellular signalling – 'proto-oncogenes' – lead to cancer (Weinberg 1989). A theory of the control system has been described by Freeman and Wilson (1990). We know that the key features of control systems exist: a means to sense if control action is needed, a means to transmit this information to an agent of the needed action, and a capacity to act.

This control system is very complex, comprising two separate kinds of signalling pathways, one of which signals 'divide,' the other signalling 'stop'. In addition, each pathway includes redundancy and modulation possibilities. For instance, in cells programmed to divide throughout life, such as the stem cells of skin and intestinal lining and of the blood system, both pathways are at least doubly redundant, and three mutations are necessary to convert a normal cell into a first-stage cancer cell (Moolgavkar and Luebeck 1992). The primary signal transmitters come from a group of loosely similar large proteins called growth factors (e.g. epidermal growth factor and transforming growth factor-beta, or TGF-β), some dozens of which exist. Other substances, including oestrogen, progesterone, insulin, and thyroid hormones, cyclic adenosine monophosphate (cAMP), the Vitamin A derivative retinoic acid, and many others modulate the signals carried by these growth factors. Both the growth factors and modulators can be tissue-specific. Thus oestrogen and progesterone act primarily on reproductive-system tissues such as breast and endometrium. An increase in oestrogen concentration increases the likelihood that certain breast cells and the cells of the endometrium will divide; an increase in retinoic acid decreases that likelihood. However, control of mitogenesis appears to differ from the temperature control examples above in not having a central controller analogous to the thermostat/hypothalamus. This may be a system best characterized as having distributed control, in which many system components independently possess the ability to sense a deviation from the control limit and act to restore the system to normality. Individual cells capable of division possess the capacity to sense when division is required. At least some of the growth factors, including TGF-β, also form part of this surveillance system (Freeman and Wilson 1990).

Mitogenic carcinogens can increase the cancer rate either by directly increasing mitotic rate in an affected tissue, or indirectly by suppressing normal growth and thus providing an environment in which mutant cells can then grow. Included in the first category are oestrogen and its mimics; classical promoters such as phorbol ester that mimic natural intracellular messenger molecules, and substances that kill cells (e.g. chloroform). Tumours have long been known to be associated with scars: the cell division essential to healing provides an increased opportunity for the mutation needed to start a cancer. Indirect mitogens include dioxin, which stops the growth of skin and related tissues, and phenobarbital, N-2-fluorenylacetamide (2-AAF) and many other chemicals that stop division of rodent liver cells; continuous exposure to these kinds of carcinogens permits growth of mutant cells that do not respond to stop signals and thus are already one step along the road to cancer.

Therefore mitogenesis fulfils all the conditions of a control system: there exists a controller (in this case, distributed among all cells capable of division), a means of sensing the state of the system (i.e. whether to divide or not), a means of transmitting signals, and a means to effect desired changes. We know from a wide variety of observations that cell growth is strictly regulated. Thus it follows that all dose–response curves for mitogenesis exhibit a threshold.

Recall that the threshold question concerns the response of the control system to small externally derived stresses. For mitogenesis such stresses would consist of influx of chemicals that influence internal signal transmission, such as phorbol ester or dioxin, or of the proteins that are released when another cell dies, which increase the propensity of cells to divide (Freeman and Wilson 1990). We showed above that dose–response data are themselves too noisy to permit discrimination between threshold and no-threshold models. The biological theory just recounted conclusively demonstrates that in the case of carcinogens which act by increasing the rate at which certain cells divide, robust and redundant control mechanisms require a threshold. For such substances, scientific theory now recommends a policy that treats them in the same way as serious toxicants. As noted, EPA has begun to do this (US EPA 1996).

Control of DNA repair We postulate the existence of a system that controls DNA repair mainly from three kinds of indirect evidence. First, some of the components necessary for a control system are known. For instance, the enzymes that sense DNA defects and repair them provide a mechanism by which such control can be effected; defects in the proteins involved are observed to be associated with an increased risk of cancer. Second, interspecies comparisons show that the efficiency of DNA repair varies, with longer-lived species (e.g. sea turtles) having very efficient repair while short-lived species (e.g. mice) tolerating much more DNA damage. That we do not protect and repair DNA as effectively as turtles suggests there is an evolutionary advantage to maintaining some level of DNA alteration in the population. Given this, there must be some means of controlling the rate. Finally, cancer rates are not sensitive to the altitude at which people live, even though DNA damage rates increase with altitude, due to increased radiation. All of this strongly suggests the existence of some constant-rate control system for this parameter.

The existence of a DNA repair system capable of exercising control is necessary but not sufficient to demonstrate existence of a control system. That is, we know that DNA repair enzymes can both sense out-of-control situations and effect a return to the set point. There exists reserve capacity to repair damage that can be induced when the DNA-damage rate increases, also a necessary part of a control system. However, not enough is known to understand these processes in control-system terms: in particular, we do not seem to know the location of a controller for DNA repair, and we do not seem to understand what signals the need to mobilize the reserve DNA damage repair capacity.

Not knowing how DNA repair is controlled also means that it is not at all clear how this system may function to permit a constant rate of DNA damage to persist in cells that enter mitosis. There clearly exists 'fractional regulation': in humans, this includes the antioxidant systems, such as uric acid, vitamin C, etc., that protect against DNA oxidation (Ames and Saul 1986). Enzyme systems generally perform this kind of function, and such regulation might satisfy the needs of long-lived animals to minimize DNA damage (during reproductive years).

However, fractional regulation of this kind is not consistent with the very reliable observation that cancer rates are not a function of the altitude at which people live. In particular, leukemia, generally regarded as the cancer most susceptible to initiation by radiation, appears to be randomly distributed across the United States. People living at high altitude, e.g. in Denver, are subject to a significantly higher flux of ionizing radiation from cosmic rays and other extraterrestrial sources than are those of us who live near sea level.

If repair were not absolutely controlled to a constant background rate, Denverites should exhibit a higher mutation frequency than, say, Philadelphians, and a higher rate of leukemia. Yet death rates for leukemia are the same, on average, in the Rocky Mountain states as elsewhere, even though these states have overall the lowest death rates for all cancers.

The weight of evidence now available favours the proposition that mutation rates are under physiological control, via control on DNA repair. We observed that evidence for the no-threshold proposition for mutagenic carcinogens was vanishingly weak: that derived from study of exposure–response curves is limited by ability to distinguish between threshold and no-threshold behaviour at low exposure levels, while arguments based on 'additivity of exposure' and statistical mechanics are tautologous. The evidence favouring the threshold proposition is noisy and not terribly strong, deriving primarily from observations consistent with DNA repair being under physiological control. We might express our confidence in the correctness of the threshold theory in terms of odds in its favour; if so, these odds would probably be no better than 3:1.

Nevertheless, available science provides more support for threshold-based policies for dealing with mutagenic carcinogens than for policies based on the no-threshold assumption.

Implications for policy

In this section, we examine policy implications of the scientific theories described above. Note that this discussion only concerns small exposures to carcinogens, the arena in which scientific uncertainty is relatively high. Where exposure to carcinogenic chemicals is large enough to expect directly observable increases in cancer rates, there is no need to resort to extrapolation away from the observable range in order to understand the implications of policy choices. We have noted that few no-threshold policies are actually followed; the Delaney Clause for food additives and colours, and certain pesticide residues, and the dormant OSHA 'cancer policy' are the most prominent among these exceptions, at least in the US. Although in principle they are based on the no-threshold theory, in reality most regulatory policies rely on the practical threshold – the basis for *de minimus* risk, no significant risk, and negligible risk decision criteria. We noted that US EPA has begun to implement threshold-based policies for some mitogenic carcinogens (US EPA 1996).

At low exposures, mitogenic (non-mutagenic) carcinogens should be treated the same as other systemic toxicants.

US EPA's evolving policy is fully supported by current scientific understanding of how exposure to mitogenic substances increases cancer risk; it follows from cell replication being under robust biological control. This policy is now limited to a few well-studied cases; it should be generalized to apply routinely to all chemicals that do not exhibit mutagenic activity in a standard test battery.

Also tests for mutagenic activity are not absolute, in that these tests have a characteristic detection limit. Substances that do not exhibit activity in these tests could actually affect DNA, but have an undetectably small specific activity or potency. It is arguable that because we cannot be absolutely sure that substances are not mutagenic, policies should not assume that no activity exists. However, we noted above that mutagenic activity contributes much less to cancer potency than does the mitogenic activity. For those substances that have undetectably small mutagenic activity, neglecting this contribution will make no change in our confidence that sufficiently small exposures cause no harm.

At low exposures, risks from exposure to mutagenic carcinogens should be based on a practical threshold assumption.

Protectiveness of practical threshold based standards Most risks from exposure to carcinogenic substances are now evaluated using some 'practical threshold' approach. It is important to ask two questions of this kind of approach. One concerns the degree of protection actually afforded, and the other concerns what we actually receive from absolute, Delaney-style bans versus the practical threshold. We suggest that the benefit from a ban comes from reducing uncertainty that the control measures will actually prevent any injury. Yet there are costs associated with choosing absolute policies. Because approaches such as 'no significant risk' can reduce this uncertainty to extremely small values, policy-makers need to give very careful consideration to these costs, especially the indirect ones.

We concluded above that the weight of evidence favours a threshold for all cancer-causing processes, but that there is considerable uncertainty in this conclusion. We noted also that under the conditions of concern in environmental regulation, the very small numbers of active molecules available to attack susceptible cells becomes so small that we can be quite certain no cancer will result, even from a lifetime of exposure. Crudely, the one in a million standards conventionally set for exposures to mutagenic carcinogens represent >99 per cent confidence of no injury occurring in populations numbering tens of millions of people, even if there is no threshold. Our confidence in no injury is then further increased by the likelihood that a threshold exists. These plausible upper bound standards calculate a 95 per cent upper confidence limit on response, and so, because these are very small numbers, the nominal likelihood of injury in a single representative individual is $<2 \times 10^{-8}$. Taking 2:1 odds against no-threshold, the likelihood drops to $<10^{-8}$. At such odds, likelihood that no cancer will occur even in a population of 100 million people is more than 99 per cent.

A ban (where enforceable) presumably gives us 100 per cent confidence that no injury will occur from exposure to the banned substance. The practical-threshold approaches provide a greater than 99 per cent confidence that no injury will occur. Thus the difference between the two policies lies in this small (and not accurately measurable) difference in confidence that no one will be hurt.

Considerations on the value of bans: the case of Delaney The utility or value of the reduction in uncertainty obtained can be judged only by comparison with the benefits that may be foregone by a ban. This is clearly a societal value judgement, not a scientific one.

Taking pronouncements of Congress to represent the values of the American public, we can confidently conclude that for many substances and circumstances the losses that would accompany a ban are considered to exceed the value of the reduction in utility. For instance, wastes from industrial production that become air and water pollutants must meet the negligible risk standard; carcinogenic animal drugs must meet a similar standard. Industrial chemicals regulated under TSCA are regulated under a different kind of system in which 'unreasonable risk' is the operative term; this requires comparing the costs and benefits of a ban with other management alternatives, and may be more or less stringent than the 'negligible risk' standards. Since the mid-1960s Congress has considered the benefits of only one class of substances so low as to deserve a ban – PCBs – and has excepted saccharin from the Delaney ban.

From all accounts the understandable dread of cancer played a large role in passage of Representative Delaney's successive amendments to the Food, Drug, and Cosmetic Act. At the time these bits of legislation were passed, there existed a widespread sentiment that carcinogens just should not be part of the food supply (Fleming 1960). This sentiment presupposed that carcinogens were rare; now we know that they are common. Combined with public health professionals' belief that no level of exposure could be considered safe, Delaney appeared quite a rational policy. Dread of cancer remains.

Yet the food additive and food colour Delaney standards impose some rather high societal costs. These costs occur as a result of agencies flouting laws that make no sense (to those who staff these agencies). We noted that Delaney represented an unnecessary addition to the food additive laws that removed policy discretion from FDA and US EPA. This has caused considerable difficulty as more and more substances have been subjected to tests for carcinogenicity, and been found 'positive'. Many otherwise innocuous natural chemicals are found to 'induce cancer', in the strict interpretation of that phrase. Some of these are used as food additives. One is a principal component of the oil of oranges and lemons, *d*-limonene; when ingested by male rats in sufficient quantity, kidney cancers develop. *D*-limonene becomes a direct food additive by virtue of the process used in reconstituted orange juice, and it was recently approved by US EPA as a food-use pesticide. The scientific community is very confident that these male rat kidney tumours are not predictive of a response in humans, that their appearance means nothing for the safety of *d*-limonene (Olin et al. 1995). This confidence was cited by US EPA in its approval of the petition to sell the substance as a pesticide. Nevertheless, it seems that, prima facie, use of *d*-limonene as a direct food additive and a food-use pesticide both violate the Delaney Clause. Pesticidal use of the chemical is minor, but reconstituted orange juice is a very important product. It should be excusable to suspect that FDA has taken no action, having no wish to provoke a public outcry by banning orange juice.

The *d*-limonene situation provides the most direct confrontation between science and Delaney, but by no means the only one. Several minerals and complex sugars induce cancers in animals, and derivatives of all the essential amino acids and chemicals such as vitamin C would do so if tested according to the protocol that showed saccharin to be 'carcinogenic'. Strictly speaking, most of these have only been tested in experiments that produce changes in cancer precursors in the male rat bladder. These precursors are necessary to the subsequent development of tumours, and continued treatment with appropriate test agents has always led to the development of such tumours. Scientists expert in this field, however, are extremely confident that were these precursors to be subject to tests of the design that produces bladder tumours in male rats with saccharin, tumours would be induced.

Until US EPA was challenged over its authority to approve tolerances for carcinogenic pesticide residues, Delaney caused no more than discomfort among the scientists and policy advisers who had to work around it to satisfy FDA's other responsibilities. In 1986 a National Academy of Sciences committee had criticized US EPA for allowing certain old pesticides to remain on the market, after animal tests showed they may induce cancer, while failing to approve newer, apparently safer ones that were also 'possibly carcinogenic' (NRC 1986). US EPA responded by enunciating a *de minimus* risk policy, very similar to FDA's, that would allow it to achieve consistency in pesticide registrations (US EPA 1988). A California group successfully sued to overturn this policy, with the result that US EPA has moved to cancel many registrations, some of them for pesticides whose absence will harm American agriculture.

A clear consensus exists in the scientific community that presence of natural 'carcinogens' in food poses negligible or no risk to the public (NRC 1996). Nevertheless, strict interpretation of Delaney would seem to prohibit use of any 'natural carcinogen' as a 'direct' food additive. Since usual food processing techniques seldom cause these to be isolated and then returned to food, as is the case for *d*-limonene in orange juice, these natural carcinogens are considered food 'constituents' and not 'additives'. The Delaney provision then does not apply. FDA has not taken action to halt their use, appearing to deal with the dilemma by ignoring the existence of this information. Scientists would approve if FDA had concluded that strict application of the law would harm public health and do no good. However admirable this practice may be, it would seem to have doubtful legality. One might speculate that it has not been challenged because doing so would expose the contradictions in

Delaney and its clear obsolescence as policy.

Linear models for setting numerical exposure standards should be replaced with an equally protective, more transparent procedure.

Both Federal and state regulatory agencies commonly make use of a linear extrapolation procedure from results of animal tests to derive exposure standards for carcinogens. These make the assumption that a plausible upper bound on the real dose–response function can be approximated by drawing a straight line from test data to the origin. Justification for this approach has been the 'no safe exposure to a carcinogen' idea and the somatic mutation theory. As we have seen, this justification is now scientifically very weak, and the US Federal agencies at least have begun to use other approaches (US EPA 1996).

At the same time, scientists have shown (Gaylor 1989a; Krewski et al. 1993) that numerically identical exposure standards can be obtained from the same test data by a much simpler procedure, dividing the estimated 10 per cent exposure response in animals by a factor of 100 000. This parallels the exact procedure used for non-carcinogens, the only significant difference being that for non-carcinogens a divisor of 100 is normally employed (Dourson and Barnes 1988). Traditionally, the 'no observed adverse effect' exposure is employed as the basis for standard setting for non-carcinogens. The actual response to which this exposure corresponds will vary from test to test, but it is believed usually to be <5 per cent (Gaylor 1989b). Because of the variability in this response, a modified procedure called 'benchmark', first proposed by Crump (1984), is now being considered; at least one variant of it is exactly the same as this one.

This arbitrary divisor is a risk management tool, intended to assure that permitted exposures will be small enough to protect the most sensitive people. Technically, justification for this approach comes from the irreducible uncertainty of predicting how humans will respond when only information on animal responses is available. It can serve other ends as well, including policy-makers' judgements that the public wants or needs a greater degree of protection from a particularly dreaded disease such as cancer (cf. Dourson and Barnes 1988). Thus the use here of a very large safety factor can be justified as appropriate policy.

Setting standards by this method, instead of the way they are now set, would have two advantages. First, the process would be much more transparent. Current procedures involve use of a 'black box' – a computer program that very, very few people truly understand or can explain. If essentially the same result could be obtained by a simple, easily understood procedure, public understanding and regulatory agency credibility would be better served by the simpler one.

Second, it would return to policy-makers the means to trade off uncertainty against other factors that enter into public policy decisions. At present, the decision on how much uncertainty to tolerate lies almost exclusively in the hands of agency technical staff, often isolated from the political process. Science is almost always unable to specify exactly what levels of exposure will be truly safe, or what disease would occur from particular exposure scenarios. The uncertainties are much too large to allow highly confident statements about these matters. Yet, typically, policy-makers have no way of knowing how large these uncertainties may be. Were it to be recognized that the use of this enormous safety factor was convention and not science, the prerogative to modify it as necessary would pass to risk managers.

The regulatory agencies should give very serious consideration to abandoning their use of these 'linear extrapolation' methods in favour of this more open risk management practice. In doing so, however, they should be aware of the fact that risk management practices are the result of a social contract, not managerial whim, and that deliberation about the use and magnitude of the operative divisors should take place before any final use occurs. Deliberation including interested and affected parties, informed by input from experts, would greatly increase the likelihood of success.

Some science policy issues

The question of population thresholds

So far this discussion has centred on the response of an individual to exposure to a carcinogen. Not all individuals will respond in the same way, and that raises a policy issue for risk management to which we now turn. However, note that as populations are made up of individuals each of whom possesses a threshold of response to any carcinogenic agent, it follows that for any finite population a threshold will exist.

Consider first the case of mitogenic carcinogens. Adaptive capacity in the control mechanism for cell replication gives rise to a threshold in the response to any agent that acts only by affecting this replication rate. Obviously, individuals will differ in their adaptive capacity. For a population, the spectrum of differences will be describable in terms of a frequency distribution. There will always exist individuals who are, at any particular time, exceptionally susceptible to the effects of any stress. We need to know: how many of these individuals exist, and how great is their susceptibility? Is there a significant number of individuals in the general population who respond to infinitesimal increases?

Note that the distribution of susceptibilities in the population is likely to be approximately lognormal (number normally distributed as the logarithm of dose). This is commonly the case for other physiological functions, and we know nothing to suggest that mitogenic effects may differ in this regard. Given this the traditional dose–response function will be a probit or other logistic curve

(Sebaugh et al. 1991). This curve, describing the fraction of responders at any particular exposure, drops off very rapidly with decreasing exposure.

Further, the question of perfect versus truncated distributions arises here. As noted above, many biological phenomena can be described in practice by a normal or lognormal distribution, but the distributions are in fact truncated by physiological constraints. Because real distributions are almost always truncated, it is more plausible than not that the distribution of susceptibilities will also be truncated, and that an absolute threshold will exist in any finite population.

It is also worth noting that there exist processes which, in effect, remove the most sensitive individuals from consideration. Studies of mitogenic carcinogens have consistently shown that mitotic rates must be increased over long periods for any significant risk of cancer to occur (Moolgavkar et al. 1990; 1993;. Ellwein and Cohen 1988; Cohen and Ellwein 1990; Preston-Martin et al. 1990). Thus the extremely sensitive individuals must remain for a long time in a state in which their reserve control capacity is infinitesimal. Any individuals for whom this is true must be under severe stress. They must be chronically very ill or very severely malnourished. Practically speaking, such individuals are either at high risk of death from other causes or their risks from the normal environment are managed by means of hospitalization, or nursing-home care, or other institutional means. The risk management situations that require these very special individuals to be the focus of the kind of analysis normally intended to apply to the general public will be rare. This does not mean that the possibility can be ignored, but that usually it will not remain as a factor in the policy decision.

Mutagenic carcinogens pose a much more difficult set of policy issues that have not been given the attention they deserve. Two sources of heightened susceptibility exist.

First, it is plausible that the association observed between poor nutritional status (or, more properly, its surrogates, low income or socio-economic status) and higher rates of cancer death arise from the impact of this condition on the barriers to mutational change discussed above. There exists evidence that these people suffer higher rates of mutation than richer individuals (Jones 1995). This raises a very thorny issue for policy-makers facing site-specific decisions such as air permits for industrial facilities or clean-up levels for a toxic waste site near very poor neighbourhoods. Should the clean-up levels be stricter to offset these peoples' possibly increased susceptibility? This increase is not limited to cancer but relates to all causes of death. Clearly, increasing their income and thus improving nutrition represents the best policy means to improve their health, for stresses arising from site-related exposures are likely to be but a fraction of their total. It is not hard to imagine circumstances in which, given a particular level of expenditure, benefits to the community would be maximized by a minimal clean-up coupled with investments to improve their income. It is beyond the scope of this chapter to explore these issues, but further exploration would be useful.

A second group with increased susceptibility to cancer consists of several well-known sub-populations whose susceptibility is increased by virtue of genetically determined diminished DNA repair capacity. These include some small sub-populations very deficient in one or more kinds of DNA repair (e.g. people with xeroderma pigmentosum and 'fragile X syndrome'), and also the large fraction (~10 per cent) of the population that appears to have somewhat reduced repair effectiveness. These people appear to bear a very high proportion of the background cancer in the human population. (Knudson 1977 states that 70–80 per cent of cancer has some genetic component.)

Individuals who are highly repair-deficient cannot be considered part of the normal population because unless specific risk-management measures are taken they die very young. For instance, people with xeroderma pigmentosum inevitably develop malignant melanoma early in life unless they absolutely avoid sunlight.

At issue is the larger group, many of whom die of cancer in their 50s and 60s. It is not known if their DNA repair capacity is diminished, or is simply of less than normal effectiveness. If their repair capacity is normal, minimally elevated exposures to mutagens will not significantly increase their risk. But if this capacity is diminished, any extra exposure will increase risk. (More accurately, it will tend to increase risk of dying from cancer earlier rather than later.) In principle, this issue can be addressed through research, one that should be given a high priority as this sub-population is better identified and studied.

Non-cancer effects

This section has focused on cancer but many of the conclusions hold for other kinds of adverse effects as well. We showed that if toxic substances act by interfering with some physiological process that is closely regulated, such as cell replication, biological control implies a threshold determined by the capacity of the control system to compensate. This conclusion holds specifically for effects caused by substances that mimic or antagonize hormones and related chemical messengers. We note that oestrogens and the other sex hormones fall into this class; thresholds exist for the actions of 'environmental oestrogens'. Thresholds will also exist for developmental toxins that act by slowing cell division, as for example retinoic acid does.

Substances that act by straightforward cell killing also will exhibit threshold dose–response curves. These include strong acids and bases, and also organ-specific toxins such as carbon tetrachloride (liver) and cadmium salts (kidney). There will always exist some ability for tissues to resist these agents, some capacity to absorb damage without killing the cells. In addition, studies on dose–response of substances that accumulate in susceptible tissues, such as lead in nerve cells, may give the appearance of acting by no-threshold mechanisms. Lead, in particular, interferes with enzymes that play a role in development (among other things). Enzyme systems include spare capacity, and can adjust to stress. Low clearance rates may imply that low exposures cause adverse effects. Observing effects at low exposures does not simply imply that effects will occur at any exposure.

Yet not all toxic responses interface with enzymes. Some time-sensitive processes, such as mutations in critical control genes, are probably not controllable in the sense we have been describing. Mutation-caused developmental toxicity should not be assumed to exhibit threshold behaviour, even though the incidence of such effects is too small to be observed in most cases. Perhaps a quarter of cleft palate cases may be the consequence of mutations occurring in a fraction of the population that carries a defective gene for the growth factors, TGF-L. Inhibiting this growth factor results in cleft palate in mice; the fraction of births in the susceptible sub-population of mice that results in cleft palate is about that predicted on the basis of background mutation rates (J. D. Wilson, unpublished results). Nevertheless, this possibility should not be ignored in evaluating the health risks posed by mutagenic compounds.

Conclusions

Although no-threshold has been the accepted policy basis for regulating carcinogens for some forty years, it is based on scientific theories that have been supplanted. Newer theories and results support policies based on the threshold theory. The weight of scientific evidence supports the conclusion that there exists an absolute threshold of activity for carcinogens. The only credible evidence arguing that thresholds do not exist, at least for mutagenic carcinogens, relies on an assumption that all the processes that act as barriers to attack by mutagens act as probabilistic barriers, and that the probability distributions are infinite. It is more likely that they are truncated, and that the extra probability of a 'hit' resulting in a carcinogenic mutant cell does go to zero at some non-zero exposure. For non-mutagenic carcinogens, there is no credible argument supporting the no-threshold position that does not also support the threshold theory, and a strong basis for concluding that thresholds exist.

There exists a policy question concerning the variability of susceptibility to substances that increase mutation or cell replication rates; if some people have an infinitesimal reserve capacity, it could be concluded that, at the population level, the actual thresholds may occur at very small exposures. Such people must be remembered in doing risk analyses, but managing their risks separately from those of the general population may prove most satisfactory.

We observed that the no-threshold policy has been largely supplanted by a 'negligible risk' policy and see here that the ends implied by the combination of these two ideas can be achieved without making the no-threshold assumption. Thus the latter has no remaining policy utility. Since it is not supported by science, it should no longer form a basis for public policy.

The no-threshold policy seems to offer benefits to policy-makers: it makes certain kinds of regulatory decisions almost automatic, or at least much easier to decide than if the risk manager had to engage in the political act of balancing many interests against the uncertainty in analyses of risk. Accompanying this is a cost, borne, as is customary, not by the beneficiaries, but by the scientific community. If policy-makers claim that science – clearly obsolete science – drives them to make decisions desired on other grounds, science loses credibility. If for no other reason, we in the scientific community should urge that this no-threshold policy be abandoned.

Acknowledgments

I am very much indebted to Dan Byrd, Gail Charnley, Sam Cohen, R. A. 'Randy' Freeman, Mike Gough , Dick Hill, Alfred Knudson, Suresh Moolgavkar, Allan Poland, Bob Scheuplein, Beth Silverstein, and Curtis Travis for their contributions over many years to my education and thus to the ideas I have tried to describe here, and their patient review of earlier versions of this Chapter. The Monsanto Company, Washington Technical Information, Inc., and US EPA's Office of Policy and Program Evaluation provided support.

Appendix: How science contributes to policy formation

One observation made in the course of reviewing this threshold/no-threshold debate is the apparent difficulty people have in separating scientific inferences from policy conclusions. This author believes that the difficulty stems in part from scientists' neglect of uncertainty. Science only informs; it does not dictate to policy. Policy makers are perfectly free to ignore scientists' views of the world if they so choose. (Of course, they may look silly having done so, but that's their risk to choose.) Effective policy comes about as a consequence of deliberation among those interested and affected, deliberation informed by analysis of relevant scientific and other information. We distinguish 'scientific' information here because Western society has decided that it provides a reliable basis for making decisions (Ziman 1978). The three distinguishing features of science – making results public, competition among researchers, and testing theories against sensible reality – serve to produce knowledge that is considered more dependable than that produced any other way. We also note that decision-makers need knowledge that is beyond the power of science to inform, knowledge concerning values, sentiments, and political realities. These do not fall within the sensible physical world that is the domain of science.

Yet the utterances of scientists, however reliable, should not be mistaken for absolute truth. Scientific conclusions may be the closest approximation to reality available at any one time, but they need always to be regarded as uncertain. Early in this century the influence of logical positivism led to the conferral on science of a rarefied competence close to absolute truth, but we see things differently today: science is now recognized as very much a human pursuit, with all the limitations that implies. Notwithstanding their usual mode of discourse with non-scientists, scientists admit among themselves the uncertainties that always surround their conclusions. The data on which conclusions are based are always uncertain; the theories used to organize this data are always approximations. In fact, Shlyakhter (1994) considers that even the most careful of physicists tend to underestimate uncertainties.

It is very important for policy-makers to understand the limits of science in addressing any policy issue. It should be clearly understood that the utterances of scientists are really belief statements: they describe what scientists believe to be the best description of nature available at a particular moment.

There exists a very solid basis for having confidence in the practical utility of these beliefs. Over the last four centuries scientists have explored, tested, and hypothesized about all of the physical world that is accessible to study. The fundamental learnings from this study are embedded in and supported by an astonishingly intricate and strong web of observation and theory. At bottom, every strongly supported statement of science relies on thorough checks against physical reality: the more numerous these tests, the stronger the confidence that can be placed in any conclusion. Not all scientific statements are amenable to direct testing (although ingenious experimenters can often find ways to test theories that others think impossible). Our confidence in both those that are directly testable and those that are not depends strongly upon the weight of evidence supporting each proposition. A sufficient quantity of inferential or indirect evidence can add up to a strongly supportable conclusion.

Perhaps contrary to expectation, confidence in a conclusion is not necessarily the inverse of uncertainty surrounding that conclusion. Scientists may well be highly confident that something is correct within some range – a 'confidence interval'. For instance, suppose that some policy decision were to depend on the rate at which white men die from bladder cancer over the next five years. Suppose further that NCI has been measuring this rate for more than thirty years, and that it varies randomly around an average of 60 ± 5 deaths per year per 100 000 adult white males, where the '± 5' corresponds to two standard deviations of the year-to-year variability over the three decades. It is obviously impossible to know what the bladder cancer death rate will be a year or two hence, but we can make a prediction of what is likely to be from the NCI data, and by making one prior assumption. That assumption is that no influence will alter the underlying rate for the next several years, i.e. that the past predicts the future. This assumption is consistent with everything that is known about causes of bladder cancer. The policy-relevant scientific prediction would then be that the 'best estimate' is 60/100 000 deaths per year, with a >95 per cent confidence that the rate will be between 55 and 65 per 100 000. Because the year-to-year variability appears to be random, this prediction is uncertain; confidence that it will turn out to be exactly 60/100 000 will be only moderate, but confidence that it will fall within this range will be very high.

The policy options available may not be sensitive to this ~8 per cent uncertainty in the bladder cancer rate. If so, the impact on the decision to be made will be negligible. For instance, in the early 1960s there was very real uncertainty in our measurement of the distance between the surfaces of the Earth and Moon. Yet this was not at all important to the decision to proceed with the Apollo program, and would not have been important even if that uncertainty had been one hundred times larger. At the other extreme the uncertainties may be large enough that any of several policy options would be consistent with the scientific knowledge, for instance, conclusions from the social sciences are seldom precise enough to constrain decisions concerning such important social issues as the increasing rate of teenage pregnancies. Both the theories and observations that can be made include uncertainties so large that they provide little basis for informing choice among policy options.

The science which underlies this health risk assessment field ranges from very certain, from the point of view of policy, to quite uncertain. Predictions based on fundamental chemistry and biochemistry are very precise: for instance, all cells are destroyed by strong acids and strong bases. Conversely, predictions based on comparative toxicology may be very uncertain: knowledge of the dose at which any particular chemical causes, say, kidney failure in male rats, will not necessarily predict even whether humans will suffer the same effect, let alone at the same dose. (Nevertheless, there are sometimes means by which reasonably reliable predictions can be made, for specific decision circumstances.)

Scientists often speak of weight of the evidence regarding opposing scientific propositions. This phrase usually conjures up in people's minds the image of the scales of justice, an apt parallel. Howson and Urbach (1989) have convincingly demonstrated that the process of scientific induction can best be described as Bayesian, where evidence for and against any particular proposition translates into greater or smaller confidence in it. Scientists learn this mode of reasoning informally, by apprenticeship, and few understand or apply the underlying theory. Nonetheless, if a fair presentation of policy-relevant science is to be made, the justification for the way the evidence was weighed needs to be included.

If uncertainty is not acknowledged, as is often the case, we may be presented with 'duelling scientists'... When the uncertainty bounds do not constrain policy choices, advocates for one or another of the choices may, and often do, select the evidence that favours their preferred option and present it as though there were no uncertainty. Their opponents, not surprisingly, commonly employ the same tactic, selecting information and theory supporting their position. It is therefore exceedingly important for scientists to be candid and for policy-makers to understand the limitations on scientists' pronouncements. Realistically characterizing uncertainty will not banish the 'duelling scientists' phenomenon, but it will help those who wish to understand the underlying science and the degree to which it can inform a particular decision.

References and bibliography

Alvarez, J. L. and Seiler, F. (1996). *Health Physics* (in press).

Ames, B. N. and Saul, R. L. (1986). Oxidative DNA damage as related to cancer and ageing. In *Genetic Toxicology of Environmental Chemicals. A: Basic Principles and Mechanisms of Action*. Alan R. Liss, Inc.

Barnard, R. C. (1990). Some regulatory definitions of risk: interaction of scientific and legal principles. *Regulatory Toxicology and Pharmacology* **11**: 201-211.

Bradshaw, R. A. and Prentis, S. (1987). *Oncogenes and Growth Factors*. Amsterdam: Elsevier Science Publishers.

Byrd, D . M. and Lave, L. B. (1987). Significant risk is not the antonym of *de minimus* risk, in C. Whipple, ed., *De Minimus Risk*. New York: Plenum Publishing Company. Also D. M. Byrd and L. B. Lave,1987. Narrowing the range: a framework for risk regulators. *Issues in Science and Technology* (Summer), 92-100.

Cohen, S. M. and Ellwein, L. B. (1990). Proliferative and genotoxic cellular effects in 2-

acteylaminofluorene bladder and liver carcinogenesis: biological modelling of the ED01 study. *Toxicology and Applied Pharmacology* **104**: 79-93.
Cohen, S. M., Purtillo, D. T. and Ellwein, L. B. (1991). Pivotal role of increased cell proliferation in human carcinogenesis. *Modern Pathology* **4**: 371-382.
Crawford, M. and Wilson, R. (1995). Low-dose linearity: the rule or the exception? *Human and Ecological Risk Assessment* **2**: 305-330.
Cross, F. B., Byrd III, D. M. and Lave, L. B. (1991). Discernible risk – a proposed standard for significant risk in carcinogen regulation. *Administrative Law Review* **43**: 61-88.
Crump, K. S. (1984). A new method for determining allowable daily intakes. *Fundamental and Applied Toxicology* **4**: 854-871.
Crump, K. S., Hoel, D. G., Langley, C. H. and Peto, R. (1976). Fundamental carcinogenic processes and their implications for low dose risk assessment. *Cancer Research* **36**: 2973-2979.
Delpla, M. (1988). Comments on It is time to reopen the question of thresholds in radiation exposure responses by J. R. Totter. *Radiation Research* **114**: 578-579.
Dourson, M. and Barnes, D. (1988). Reference Dose (RfD):. description and use in health risk assessments. *Regulatory Toxicology and Pharmacology* **8**: 471-486.
Ellwein, L. B. and Cohen, S. M. (1988). A cellular dynamics model of experimental bladder cancer: analysis of the effect of sodium saccharin in the rat. *Risk Analysis* **8**: 215-221.
Environmental Protection Agency (1988). Regulation of pesticides in food: Addressing the Delaney paradox. Policy statement. *Fed. Reg.* **53**: 41103-41123.
Environmental Protection Agency (1996). Proposed guidelines for carcinogen risk assessment. *Fed Reg.* **61**: 17960-18011 (4/23/96).
Fleming, A. H. (1960). Testimony before the House Committee on Interstate and Foreign Commerce (86th Congress, 2nd Session) on H. R. 7624 and S. 2197, January 26, 1960.
Freeman, R. A. and Wilson, J. D. (1990). A two-channel hypothesis for regulation of cellular division and differentiation. *J. Theor. Biol.* **146**: 303-315.
Gaylor, D. W. (1989a). Preliminary estimates of the virtually safe dose for tumors obtained from the maximum tolerated dose. *Reg. Toxicol. Pharmacol.* **9**: 1-18.
Gaylor, D. W. (1989b). Quantitative risk analysis for quantal reproductive and developmental effects. *Environmental Health Perspectives* **79**: 229-241.
Greenfield, R. E., Ellwein, L. B. and Cohen S. M. (1984). A general probabilistic model of carcinogenesis: analysis of experimental bladder cancer. *Carcinogenesis* **5**: 437-445.
Howson, C. and Urbach, P. (1989). *Scientific Reasoning: The Bayesian Approach.* LaSalle, IL: Open Court Publishing.
Jones, I. (1995). Dr Irene Jones, Lawrence Livermore Laboratory, private communication.
Kitchin, K. T. and Brown, J. L. (1994). Dose-response relationship for rat liver DNA damage caused by 49 rodent carcinogens. *Toxicology* **88** 31-49.
Knudson, A. G., Jr. (1977). Genetic predisposition to cancer, in H. H. Hiatt, J. D. Watson, and J. A. Winstein, eds., *Origins of Human Cancer. Book A: Incidence of Cancer in Humans*. Cold Spring Harbor, NY: Cold Spring Harbor Press, pp. 45-52.
Knudson, A. G., Jr. (1985). Hereditary cancer, oncogenes, and anti-oncogenes. *Cancer Research* **45**: 1437-1443.
Knudson, A. G., Jr. (1995). *JNCI Monographs.*
Krewski, D., Gaylor D. W. and Lutz. W. K. (1995). Additivity to background and linear extrapolation, in S. Olin et al., eds., *Low-Dose Extrapolation of Cancer Risks: Issues*

and Perspectives. Washington, D C: ILSI Press., pp. 105-122.
Krewski, D., Gaylor, D. W., Soms, A. P. and Szyszkowicz, M. (1993). Correlation between carcinogenic potency and the maximum tolerated dose: implications for risk assessment, in National Research Council, *Issues in Risk Assessment*, Washington, D C: NAS Press, pp. 111-171.
Land, C. E. (1995). Studies of cancer and radiation dose among atomic bomb survivors. *J. American Medical Association* **274**: 402-407.
Lave, L. B. (1981). *The Strategy of Social Regulation: Decision Frameworks for Policy*. Washington, DC: The Brookings Institution.
Loeb, L. A. (1991). Mutator phenotype may be required for multistage carcinogenesis. *Cancer Research* **51**: 3075-3079.
Lowrance, W. W. (1976). *Of Acceptable Risk: Science and the Determination of Safety*. Los Altos, CA: William Kaufman, Inc.
Miller, B. A., Gloeckler Ries, L. A. Hankey, B. F. Kosary et al. (eds.) (1993). *SEER Cancer Statistics Review. (1973-1990)*. Bethesda, MD: National Cancer Institute of the National Institutes of Health; NIH Publication No. 93-2789.
Moolgavkar, S. H. (1983). Model for human carcinogenesis: action of environmental agents. *Environmental Health Perspectives* **50**: 285-291.
Moolgavkar, S. H. (1986). Carcinogenesis modelling: from molecular biology to epidemiology. *Annual Reviews of Public Health* **7**: 151-169.
Moolgavkar, S. H. (1988). Biologically motivated two-stage model for cancer risk assessment. *Toxicology Letters* **43**: 139-150.
Moolgavkar, S. H., Cross, F. T., Luebeck, G. and Dagle, G. E. (1990). A two-mutation model for radon-induced lung tumors in the rat. *Radiation Research* **121**: 28-37.
Moolgavkar, S. H. and Luebeck, E.G. (1992). Multistage carcinogenesis: population based model for colon cancer. *J. National Cancer Institute* **84**: 610-618.
Moolgavkar, S. H., Luebeck, E.G., Krewski, D. and Zielinski, J. M. (1993). Radon, cigarette smoke, and lung cancer: a reanalysis of the Colorado Plateau uranium miners data. *Epidemiology* **4**: 204-217.
Moolgavkar, S. H. and Knudson, Jr. A. G. (1981). Mutation and cancer: a model for human carcinogenesis. *J. National Cancer Institute* **66**: 1037-1052.
Moolgavkar, S. H., and Venzon, D. J. (1979). Two-event models for carcinogenesis: incidence curves for childhood and adult tumors. *Mathematical Biosciences* **47**: 55-77.
National Research Council. (1983). *Risk Assessment in the Federal Government: Managing the Process*. Washington, DC: NAS Press.
National Research Council. (1987). *Regulating Pesticides in Food: The Delaney Paradox*. Washington, DC: NAS Press.
National Research Council. (1990). *Health Effects of Exposure to Low Levels of Ionizing Radiation*. (BIER V). Washington, DC: NAS Press.
National Research Council. (1996). *Carcinogens and Anti-carcinogens in the Human Diet*. Washington, DC: NAS Press.
Olin, S., Farland, W., Park, C., Rhomberg, L. et al. (1995). *Low-Dose Extrapolation of Cancer Risks: Issues and Perspectives*. Washington, DC: ILSI Press.
Preston-Martin, S., Pike, M. C., Ross, R. K., Jones P. A. et al. (1990. Increased cell division as a cause of human cancer. *Cancer Research* **50**: 7415-7421.
Prival, M. J. and Scheuplein, R. J. (1990). Regulation of carcinogens and mutagens in food in the United States, in M. W. Pariza, H. -U. Aeschbacher, J. S. Felton, and S. Sato, eds, *Mutagens and Carcinogens in the Diet*. New York: Wiley-Liss Publishers, pp. 307-321.
Rai, K. and Van Ryzin, J. (1985). A dose–response model for teratological experiments

involving quantal responses. *Biometrics* **41**: 1-9.
Shimizu, Y., Kato, H. Schull, W. J. and Mabuchi, K. (1993). Dose-response analysis of atomic bomb survivors exposed to low-level radiation. *The Health Physics Newsletter* **1**: 1-5. Also J. L. Alvarez, Need for a new model for radiation risk assessment and management. *HPS Newletter* **XXIII**: 15-16.
Sebaugh, J. L., Wilson, J. D., Tucker, M. W. and Adams, W. J. (1991). A study of the shape of dose–response curves for acute lethality at low response: a 'megadaphnia' study. *Risk Analysis* **11**: 633-640.
Shaddock, J. G., Feuers, R. J. Chou, M. W. and Casciano, D. A. (1993). Evidence that DNA repair may not be modified by age or chronic caloric restriction. *Mutation Research* **301**: 261-266.
Shylakhter, A. I. (1994). Uncertainty estimation in scientific models: Lessons from trends in physical measurements, population studies, and energy projections, in B. Y. Ayyub and M. M. Gupta, eds, *Uncertainty Modelling and Analysis: Theory and Applications*, [publisher?] pp. 477-496. Also An improved framework for uncertainty analysis: accounting for unsuspected errors, *Risk Anal.* **14**: 441-477.
Thomas, R. G. (1995). Radium dial painters show a practical threshold. *HPS Newsletter* **XXIII** (6):9-10.
Travis, C. C., McClain, T. W. and Birkner, P. D. (1991). Diethylnitrosamine-induced hepatocarcinogenesis in rats: a theoretical study. *Toxicology and Applied Pharmacology* **109**: 289-304 (non-linear growth kinetics).
Weinberg, R. A. (1989). *Oncogenes and the Molecular Origins of Cancer*. Cold Spring Harbor, NY: Cold Springs Harbor Press.
Wilson, J. D. (1989). Assessment of the low-exposure risk from carcinogens: implications of the Knudson-Moolgavkar two-critical mutation theory, in C. C. Travis, ed., *Biologically-based Methods for Cancer Risk Assessment*. New York: Plenum.
Wilson, J. D. (1991). A usually-unrecognized source of bias in cancer risk estimations. *Risk Analysis* **11**: 11-12.
Wilson, J. D. (1994). Thresholds for adverse effects from toxicants affecting hormone economy arise at the organism level , in H. L. Spitzer, T. J. Slaga, W. F. Greenlee and M. McLain, eds, *Receptor Mediated Biological Processes: Implications for Evaluating Carcinogens*, New York, Wiley-Liss, pp. 223-236.
Ziman, J. (1978). *Reliable Knowledge: An Exploration Of The Grounds For Belief In Science*. Cambridge: Cambridge University Press.

2 Biases introduced by confounding and imperfect retrospective and prospective exposure assessments

Alvan R. Feinstein

Summary

The 'Gold Standard' approach to establishing a cause–effect relationship in clinical medicine is the (double blind) randomized trial. However, this approach is not considered ethically suitable for studying the impact of agents that are believed to be primarily noxious. Therefore, most epidemiological evidence comes from observations obtained from groups of individuals who were exposed during the natural circumstances of daily life. Moreover, many scientific principles that are not contingent on the use of a randomized trial are usually (although unnecessarily) ignored in many – perhaps most – epidemiological studies of risk factors for cancers, infarctions, atherosclerosis and other non-infectious diseases of societal importance. Particular problems include:

- First, the importance of establishing a cause–effect relationship is often rejected or disdained by prominent epidemiologists. In the relativistic world inhabited by these associationist epidemiologists, the sequence of action is irrelevant. The absurdity of this position is obvious: a person may suffer from a disease, be cured, then ingest some toxin and yet the toxin could still be blamed for the disease.
- Second, confounding factors receive insufficient attention. Such factors include the association (inverse correlation) between lung cancer and consumption of fruits and vegetables: if smokers eat fewer fruits and vegetables than non-smokers but consumption of these items is not taken into consideration, then the correlation between smoking and lung cancer may be upwardly biased. However, inclusion of many suspected confounding variables into a multivariate regression may lead to further biases if there are spurious correlations between 'confounders' and the dependent variable (e.g. incidence of lung cancer).
- Third, susceptibility bias has received almost no attention in epidemiological studies. Susceptibility bias occurs when factors that might predispose someone to be affected by the disorder under study are not taken into consideration.
- Fourth, detection bias may not be accounted for. Detection bias

occurs when different groups of researchers and pathologists (or the same researchers and pathologists at different times) invest differing degrees of effort into discovering particular ailments. For instance, technological developments that increase the probability of detecting a particular condition will tend to result in an increase in detection over time that is not correlated with an increase in incidence. Moreover, the use of death certificates (which are merely a passport to burial) as evidence in epidemiological studies is not justified.

- Finally, transfer bias, which occurs because of differences in the histories of individuals in case and control groups, is rarely appropriately accounted for in epidemiological studies.

The inadequacy of the science underlying many epidemiological studies would be worrying in any discipline but in a discipline that self-avowedly seeks to better the world it is of grave concern. The political influences on and goals of epidemiologists are obvious, inviting comparison with Lysenko's fantastic new scientific order. Routine methodological improvements will not come rapidly because the older generation of leaders in chronic disease epidemiology are themselves responsible for instituting, disseminating, building into the peer review process, coercing into grant approvals and using as stepping stones in their careers these very defective methods.

The epidemiological appraisal of risk factors is a cause–effect analysis in which the risk factor is defined as an exposure that is believed to cause or promote development of a particular disease. The exposure is usually an external entity – such as atmospheric pollution, diet or smoking – but can also be a person's internal biologic attribute, such as sex or family history.

The 'Gold-Standard' method of research for cause–effect analysis in human disease is the experiment called a *randomized trial*. During the past few decades, such trials have commonly been used to evaluate the cause–effect benefits of therapeutic agents, but randomized trials are seldom possible when the agent is believed to be a noxious risk factor. Many people would regard the trial as unethical if it is done because the agent is suspected of doing harm rather than good, and even if the ethical issue could be rationalized, the trial would seldom be feasible. Too few people, if prepared for a suitably informed consent, would volunteer to participate.

Consequently, data about risk factors seldom come from randomized trials; and epidemiologists have become accustomed to doing research with observational evidence, obtained and analysed in non-experimental designs that substitute for the desired trial that could not be done. The observational evidence is obtained for groups of people whose investigated events occurred in the natural circumstances of daily life.

The main point I want to make is that although a randomized trial contains many scientific principles beyond the randomized assignment itself, those principles are usually ignored or abandoned in many – perhaps most – epidemiological studies of risk factors for the cancers, infarctions, atherosclerosis and other non-infectious diseases that have major societal importance. Instead, epidemiologists tend to rely on mathematical models and assumptions that, while possessing all the majesty of elegant (and sometimes mysterious) statistics, lack the basic principles of scientific research (Feinstein 1988, 1989).

In a typical statistical report about an epidemiological risk factor, a particular disease is associated with a particular exposure, and the results are usually cited as risk ratios for the occurrence rates of disease in the exposed versus non-exposed groups. To avoid the threat of clarity, these ratios are also called by various other names, such as relative risks, and rate ratios. In the commonly used case-control studies, in which risk ratios cannot be determined, the comparative results are usually expressed as an odds ratio, which can mathematically approximate a risk ratio if the disease has a low rate of occurrence (such as below 0.01) *and* if the case–control results suitably represent the preceding cause–effect pathway.

Statistical approaches to confounding

The idea of a cause–effect pathway, however, is itself often rejected or disdained by prominent statistical epidemiologists (Cornfield, 1954; Greenland and Morganstern, 1988; Miettinen, 1988; Rothman, 1986) who often argue that all cause–effect relationships are associations, and that little or no attention need be given to the sequence and time direction in which events actually occurred. With this belief, everything can be analysed as a statistical association between the two main variables for outcome and exposure, and all other phenomena become regarded as covariates to be adjusted with diverse forms of matching, stratification, or, most commonly, multivariable analysis.

The main reason for the adjustments is to avoid or repair an entity called *confounding*, which is difficult to define but is something that compromises the statistical results, leading to distortions or misleading conclusions. For example, suppose statistics show that graduates of Ivy-League colleges have higher incomes and net worth in later life than graduates of other schools. The conclusion that Ivy-League education leads to financial benefits in later life would be confounded by the generally more favourable pre-college familial and socio-economic conditions of Ivy-League students. They may be starting out with more money, and may be given more along the way.

In the absence of attention to a scientific pathway, however, *confounding* is left without a specific source. It then becomes something like an early nineteenth century *miasma*, which was a non-specific vapour that did evil things, but is not otherwise identified (Feinstein 1985). Without a scientific strategy to guide the search, the customary statistical approach to confounding is to round up a set of available suspects: age, sex, occupation, height, weight, smoking, dietary habits, lifestyle , and almost anything else for which data have been collected, beyond the main information about the suspected exposure and disease. In a multivariate analysis of the relationship between the alleged exposure and the suspected disease, the additional data become covariates that presumably adjust the results for confounding.

Pathways of cause–effect events

For investigators who want to think about scientific biology rather than mathematical models, however, the phenomena of a cause–effect relationship do not occur merely as random events. Instead, the scientific reasoning for a cause–effect pathway begins with an entity that is constantly neglected or inadequately described in most statistical appraisals. This entity, to which I shall return again shortly, is the baseline state of the persons who become exposed or non-exposed. From their baseline state, they then undergo or decide to receive the exposure (or non-exposure), which can be performed with many accompanying phenomena that might be called co-exposures. After a period of time during which the exposure and its co-exposures may have continued or stopped, the persons are then found to have or not to have the disease that is the focus of the research. At this phase in the sequence of events, however, the people are still statistically unremarkable. They do not achieve statistical recognition and become the subjects of research until something happens to transform them from their anonymous state of natural existence to their mathematical immortality as units in a collected set of research data.

None of these movements from one location to another in the cause–effect pathway takes place randomly, and each movement can be accompanied by specific biases that produce confounding.

Susceptibility bias

In susceptibility bias, the persons who become exposed are prognostically more likely to develop the outcome event than those who are not exposed.The classical clinical example of susceptibility bias occurs in the therapy of cancer. Surgical treatment is generally reserved for operable patients, who usually have localized cancers *and* no major co-morbid diseases. The inoperable patients, who have metastatic cancer and/or major co-morbidity, are usually denied surgery and sent to receive radiotherapy or chemotherapy. These two groups of patients have strikingly different prognoses; and the operable group are inherently more likely to survive, regardless of treatment. Nevertheless, survival rates are constantly being compared for surgical versus non-surgical therapy, without regard to the confounding distortions of susceptibility bias.

In epidemiological research for risk factors, susceptibility bias is not easy to illustrate, mainly because the investigators have seldom looked for it and collected satisfactory data. For example, many doctors and patients know that the best general way to have a long life span is to 'choose' long-lived parents and other ancestors. Nevertheless, even though the data are easy to acquire, parental longevity is almost never recorded or analysed in epidemiological research. Suppose, however, that people with short-lived parents, suspecting that their own lives may also be short, decide to lead an enjoyable, hedonistic lifestyle. When they then die, as expected, earlier than other people, the reduced longevity may be attributed to their external rather than internal risk factors.

Another important susceptibility factor is the human psyche. Many doctors and patients can tell you that people who develop the heart attacks called myocardial infarctions are often tense and driven, with considerable repressed anger and hostility. If such patients also tend to smoke and to eat high fat diets, their heart attacks will be attributed to the external rather than the internal risk factors. Has this psychic feature been checked? No, not really. Instead of using well-constructed psychometric rating methods to identify anger, hostility, and tension, epidemiologists have obtained controversial results by using what may be the world's clumsiest and most oversimplified classification of the complexity of the human psyche. According to the customary epidemiological rating scale, which is also difficult to reproduce, every human psyche is either a Type A or a Type B. There are many kinds of races, ethnic groups, and even serum lipoproteins in human biology, but in epidemiological research for risk factors there have only ever been two types of human psyches. The quality of the associated science corresponds to what would happen if we divided all chemical exposures into being organic or inorganic.

Another neglected source of susceptibility bias is a factor that my colleagues and I have called protopathic (or early disease) bias (Horwitz and Feinstein, 1980). It arises when exposure (or non-exposure) is affected by a disease that has begun to produce clinical symptoms, but has not yet been recognized. For example, a middle-aged man who has developed the exertional chest pain of angina pectoris, but who has not told anyone about it, may deliberately avoid jogging or other vigorous forms of physical exertion. Later on, when he develops a myocardial infarction, it will be attributed to his slothful lifestyle. The early-disease protopathic problem is seldom recognized, because epidemiologists almost never ask about the baseline state features that can make people choose or avoid certain exposures.

These three types of susceptibility bias produce a confounding that cannot be recognized or adjusted with multivariate analysis, because the necessary data are not acquired and are therefore not available to be analysed.

Detection bias

The detection bias form of confounding arises when the outcome event is sought and identified with different vigour, methods, or criteria in the exposed and non-exposed groups. In clinical trials of therapy, particularly when a change of symptoms is the outcome event, detection bias is avoided not by randomization but by double-blind procedures in which neither the patient nor the doctor knows which treatment is being given. The problem arises in a different way in epidemiological studies of risk factors, where the outcome event is the development of a particular disease.

A fundamental difficulty here is that many chronic disease epidemiologists do not seem to know or understand chronic disease. They do not realize that many instances of cancer, coronary disease, and various forms of arteriosclerosis can exist in a clinically silent form, only discerned at a post mortem necropsy done after the patient has died of some other cause. A 90-year-old man, shot by a jealous lover, may show coronary arteries so occluded that one wonders how he was able to walk, let alone engage in the activity that led to his demise. The necropsy may also show that he had a cancer in the lung or in the colon that had previously been asymptomatic and unrecognized.

With so many silent cases of disease available for identification during life, all that is needed to produce detection bias is an exposure that is associated with increased diagnostic searches. Furthermore, all that is needed to raise the occurrence rates of any disease is an improved diagnostic method or wider dissemination and usage of an existing method (Feinstein and Esdaile, 1987).

For example, during the past decade, the increased occurrence rates of brain tumour, breast cancer, and pancreatic cancer have evoked many fears of cancer epidemics and have led to suspicions about risk factors such as electric power lines, microwave ovens, and cellular phones. Do these rising rates represent real increases in those diseases, or are they due to increased diagnosis during life, with the use of improved technologic methods, such as CAT scans of the head, mammography, and abdominal ultrasound? Knowledgeable clinicians would argue for the latter, but many chronic disease epidemiologists seem to be unaware of this possibility. One of my more heretical colleagues has said that many epidemiologists believe disease is something that occurs on a death certificate.

If you ask me to get an industry into trouble, I can promptly tell you how to do so: Set up a mammography screening programme. It will identify many breast cancers that were formerly undetected. The high rate of breast cancer in that industry, however, will not be compared against the rate found in some other industry that also has a mammography screening programme. Instead, with the customary epidemiological neglect of detection bias, the rate will be compared against the rate of breast cancer in the general public, for which mammography screening is not yet ubiquitous. The higher breast cancer rate in the industry will then be attributed not to mammography, but to employees' exposure to whatever product is made by their industry; widgets, chicken soup, dioxin, antibiotics, etc. In mere general studies, various cancers and other diseases become associated with working in or living near industrial sites, and no attention is given to the fact that people in these locations may also be more likely to receive the technologic testing that identifies the disease.

A profound neglect of detection bias is not only one of the major scientific deficiencies of current epidemiological research, but is quite scandalous. One study after another is made, often with contradictory results, about alcohol, food, smoking, oestrogens, or power lines as risk factors for breast cancer (and other cancers), and none of them give adequate attention to suitable methods that could remove the confounding of detection bias.

An even greater scandal, which I do not have time to go into, is the reliance on death certificate data for calculating incidence rates of disease (Gittlesohn and Royston, 1982). A death certificate is merely a civilized passport to burial. The only trustworthy information on a death certificate is the fact and date of death, and the age and sex of the deceased. For more than 50 years, every time someone has studied the causes of death listed on death certificates, the conclusion has been that the information is grossly inaccurate and unreliable, and that a wholly new system is needed for scientifically trustworthy data (Feinstein and Esdaile, 1987; Gittlesohn and Royston, 1982). Yet nothing has been done. In the midst of the magnificent scientific advances of molecular biology and medical technology in the twentieth century, the egregiously defective information on death certificates continues to be used for determining incidence rates of disease. The imageries are now augmented with computer-generated maps of the incidence of cancer in different regions.

Transfer bias

After the exposures and outcomes have occurred, the events must be counted in all pertinent subjects. In a randomized trial, or in an observational cohort study that follows people forward in time from their exposure (or non-exposure), the group of pertinent people is known and counted at the beginning of the study. If all of them are not suitably accounted for afterwards, the results can be biased when effects related to the exposure (or non-exposure) make people drop out or become lost to follow-up in disproportionate numbers in the two groups.

To avoid these transfer problems, a suitable scientific accounting is usually obtained with life-table analysis, or, in randomized trials, with so-called 'intentions to treat analyses'. In case–control studies, however, a cohort of exposed and non-exposed persons is not assembled. Not being themselves responsible for identifying a baseline state, the investigators often make no provision for what may have transpired, and which persons may have been lost, between the baseline state and the group assembled as cases and controls, at the end of the causal pathway, after everything has already happened.

Furthermore, in choosing the cases and controls, investigators can create biases above and beyond what already exists. The research may establish peculiar eligibility criteria that pertain for cases but not controls or vice versa, so that a form of exclusion bias is almost guaranteed (Horwitz and Feinstein, 1985).

Suppose a pharmaceutical company did a study, analysed the results, did not like them, and decided to discard members of the control group so that the subsequent results were much more favourable to the research hypothesis. The subsequent scandal would perhaps be front-page news. Yet perhaps the most glaring example of the production of transfer bias, however, was committed by one of the most prominent leaders of epidemiological research.

The strategy of changing the control group *after* the data was analysed was not done by a pharmaceutical company; it was done by one of the world's best known epidemiologists, in a famous (or infamous) case–control study that accused reserpine (a then popular blood pressure medication) of causing breast cancer (Armstrong et al., 1974). The authors of the paper did not hide what they had done: they said so in their published account of methods. Since the prominent epidemiologist has helped set the so-called standards for peer review, however, the paper was accepted and published in one of the world's leading medical journals. The accusation later turned out to be wrong, and reserpine was eventually exonerated, but as far as I know, the offending paper has never been retracted.

Exposure bias

I have saved exposure bias, as fourth source of confounding, for emphasis now, because it is specifically cited in the title assigned for this chapter. In most epidemiological studies, and in all case–control studies, the ascertainment of exposure is a retrospective process, conducted after the events have actually occurred. This ascertainment involves either a search through archival records, or a direct interview with the research subjects, or both. The process surely warrants special procedures to ensure objectivity, such as the double-blinding used to ascertain the outcome events in a clinical trial. For example, when my colleagues and I do case–control studies, we establish special methods of recruiting and interviewing subjects to keep interviewers from knowing who is a case and who is a control, and even to keep the interviewed subjects themselves from knowing whether they are cases or controls. We also set up a series of decoy hypotheses so that the interviewers will not know the particular exposure in which we are most interested (Feinstein, 1985).

In most epidemiological studies, however, no such efforts are made. In fact, one prominent leader in epidemiology has reportedly said that such efforts, which are now used routinely by myself and colleagues, are impossible. Consequently, in most studies, the interviewers usually know who is a case and who is a control, and they also know what hypothesis the investigator is trying to prove. This approach creates almost unlimited opportunities for interviewer bias by the researcher, and for recall bias by the subject. If you tell me what exposure you are looking for, and tell me the state of the person I am interviewing, I should be able to elicit a positive history in people whom you want to have it, and a negative history in people who should not. Yet almost no precautions are taken to avoid this type of bias in ascertaining exposure. Furthermore, since the bias is incorporated directly into the raw data, its existence is impossible to prove afterwards.

When the possibility of ascertainment is raised, many epidemiologists will retort, 'You haven't proved it'. This is a strange reply in scientific research, where the investigator's job is to take suitable precautions. For example, if a randomized trial is done without the precautions of pertinent double-blinding, the results will usually be rejected summarily, without anyone having to prove anything about bias. If a surgeon tries to enter an operating room with dirty hands, he or she will usually be promptly ejected, without anyone having to demonstrate any bacteria on the hands. In epidemiological chronic disease research, however, it is perfectly acceptable for the investigators to take no precautions against exposure bias, and then, if you complain, to demand you prove that bias has occurred.

I shall mention, only in passing, a separate problem in biased exposure. In many studies, the investigators do not identify, in advance, what they mean by *exposure*. Instead, they acquire the data, examine the results, and then demarcate exposure retrospectively in a way that is most favourable for whatever hypothesis they want to prove.

Reasons for problems

At this point, you are probably wondering whether things are really as bad as I have stated. I regret to say that yes, they are. If you then ask why things have got and have remained this way, I can offer several sets of explanations.

The first is that chronic disease epidemiology is still a relatively young and developing field, in which scientific methods are particularly difficult to use, because intact human beings cannot be studied with the same ease as inanimate substances, caged animals, or molecules. Because all domains of science develop at their own pace, there is absolutely no reason why this young field should be as methodologically well developed, sophisticated, and mature today as chemistry or physics. Even the marvels of molecular biology are only about 40 years old.

If scientific methodological errors still occur in chronic disease epidemiology, there is nothing shameful about them. As the field advances, methods will be improved. If the idea of shame is to be raised, however, we can surely be embarrassed by the folly of believing that chronic disease epidemiology is well advanced scientifically, and then giving serious credibility to results that would be instantly dismissed if presented in many other branches of science (Feinstein, 1988).

A second major reason for problems, is the difficulty of reproducing the research. If someone claims to have achieved cold fusion in a physics laboratory, physicists all over the world can go into their own laboratories and try to repeat the work. In epidemiology, however, you can never get exactly the same group of people to be examined in exactly the same place, in exactly the same way, and for exactly the same duration. For this reason, the problem of suitable methodology is particularly important. Because the *material* itself cannot be reproduced, the methods have to be particularly effective and persuasive. In this respect, however, many leaders of the field have been particularly delinquent. They have done relatively little to improve the shoddy methods I have described; they have sometimes been responsible for creating these methods; and they often reinforce the shoddiness in their work as peer reviewers for grant requests and journal manuscripts.

The third major problem, which leads to tolerance of methodological shoddiness, is a frequent focus on mathematical rather than scientific models for scientific research. Many chronic disease epidemiologists, for example, will argue that case–control studies and other epidemiological research structures are special kinds of scientific research that should be judged according to their own principles and standards, not with the same criteria used in other branches of science. Since an observational epidemiological study is not a randomized trial, and since chronic disease investigators do not regard these studies as a substitute for a randomized trial, the investigators will state that the work does not require and should not be judged by any of the other scientific standards that are applied in randomized trials. In the views of these investigators, case-control and other epidemiological studies can be approached with the rules of statistical inference, but no rules have been established for scientific inference.

A fourth major problem, which also encourages the tolerance of shoddy methods, is the role of social and political advocacy in public health. Clinical practitioners are often reproached for their God-like attitudes in believing they know what's good for individual patients. Public health epidemiologists, however, may sometimes show even greater self-deification in believing they know what's good for the world. Once they believe that something is evil – whether it be food, sloth, or fun – anything that can help to get rid of that evil is acceptable research, regardless of the scientific quality of methods. Although scientific investigation usually insists that the end is justified only if the means are justified in many forms of epidemiological research, the end justifies the means.

This advocacy has now become particularly troublesome, because it has begun to be used by governmental agencies, armed with regulatory authority, to pervert epidemiology into political science. In the former Soviet Union, biological science has not yet recovered from the intellectual degeneration it suffered when the scientific fantasies of Lysenko, which were compatible with Soviet political policy, became adopted as scientific policy. Instead of developing improved methods, epidemiological science in the United States, will probably go through a similar degeneration if it is transmogrified into a form analogous to Lysenkoism, used to advance the advocacy and political goals of governmental agencies. A recent oasis in the increasingly desolate scientific desert of certain public health agencies was the confession by one former agency director that he was wrong when he ordered an entire town in Missouri evacuated about a decade ago because of a risk factor that turned out not to be a threat. A few more confessions of that type might help science's reputation in the world of risk-factor opportunism and public health advocacy.

The last problem I shall cite is that if schools of medicine can (quite properly) be indicted for their failure to teach generalism and humanism, schools of public health – which have spawned so many chronic disease epidemiologists – can be indicted for their failure to teach disease or science or a scientific attitude of searching for errors. The students there seldom have any laboratory courses to give direct experience in doing experiments; there are almost no courses in pathology, physiology, or diagnosis to teach about disease; there is no exposure to autopsies that can teach about latent disease and diagnostic error; and, in fact, there is no established tradition of searching for error in teaching rounds or other special conferences. Several years ago, when Vandenbroucke and Pardoel (1989) published an intellectual autopsy to analyse the error in which amyl nitrite poppers were initially regarded as the cause for AIDS, I realized that it was the first such paper I had seen in the epidemiological literature.

Until improvements begin to appear routinely in the scientific quality of the work, we shall continue to be stuck with epidemiological claims about risk factors and to confront the problems of deciding when the claims warrant credibility. I do not think routine methodological improvements will come quickly, because I doubt that the older generation of leaders in chronic disease epidemiology will try to institute reforms in the defective methods that they have created, disseminated, built into the peer review process, coerced into grant approvals, and used as foundations and stepping stones for their careers and elevated status. There is always hope, of course, for the younger generation, and particularly for the generation that follows them. Furthermore, if things become too decadent in the US or in other English-speaking countries, there is always hope for other parts of the world.

Eventually truth and science will triumph. Until that splendid time arrives, however, we shall have to be careful to remember that statistical associations, even when embellished or adumbrated by multivariable analytic miracles, are not necessarily scientific evidence.

Acknowledgement

An earlier version of this chapter first appeared in *The Role of Epidemiology in Regulatory Risk Assessment* (John Graham, ed., 1995) Elsevier, pp 29–38.

References

Armstrong B, Stevens N, and Doll R. (1974). Retrospective study of the association between use of rauwolfia derivatives and breast cancer in English women. *Lancet*; **2**:672–675.

Cornfield J. (1954). Statistical relationships and proof in medicine. *Am. Stat.*, **8**:19–21.

Feinstein A. R. (1985). *Clinical Epidemiology. The Architecture of Clinical Research.* Philadelphia: W. B. Saunders.

Feinstein A. R. (1988). Scientific standards in epidemiologic studies of the menace of daily life. *Science*; **242**:1257–1263.

Feinstein A. R. (1989). Epidemiologic analyses of causation: the unlearned scientific lessons of randomized trials. *J. Clin. Epidemiol.*, **42**:481–489.

Feinstein AR and Esdaile JM. (1987). Incidence, prevalence, and evidence. Scientific problems in epidemiologic statistics for the occurrence of cancer. *Am. J. Med.*, **82**:113–123.

Gittlesohn A and Royston P. N. (1982). Annotated bibliography of cause-of-death validation studies. 1958–80. Washington DC: US Government Printing Office (Vital and Health Statistics. Series 2, No. 89) (DHHS Publication No. [PHS] 82–1363).

Greenland S. and Morganstern H. (1988). classification schemes for epidemiologic research designs. *J. Clin. Epidemiol.*, **42**:715–716.

Horwitz R. I. and Feinstein A. R. (1980). The problem of 'protopathic bias' in case-control studies. *Am. J. Med.*, **68**:255–258.

Horwitz R. I. and Feinstein A. R. (1985). Exclusion bias and the false relationship of reserpine and breast cancer. *Arch Int Med.*, **145**:1873–1875.

Miettinen O. S. (1988). Striving to deconfound the fundamentals of epidemiologic study design. *J. Clin. Epidemiol.*, **41**:709–713.

Rothman K. J. (1986). *Modern Epidemiology*. Boston: Little, Brown.

Vandenbroucke J. P. and Pardoel V. P. A. M. (1989 An autopsy of epidemiologic methods: the case of 'poppers' in the early epidemic of the acquired immunodeficiency syndrome (AIDS). *Am. J. Epidemiol.*, **129**:455–457.

3 Problems with very low dose risk evaluation: the case of asbestos

Etienne Fournier
Marie-Louise Efthymiou

Summary

In clinical toxicology, it is a constant that, as doses to which individuals are exposed are progressively decreased, the induced effects diminish in frequency and often undergo qualitative changes until, ultimately, a dose is reached where no effect can be detected.

In the case of asbestos, the assumption of 'no threshold' has led to very expensive 'clean-up' operations that may have actually been counterproductive. Even assuming a linear threshold, at environmental exposures of between 200 and 1000 fibres per litre (f/l) (such as might occur in exceptional circumstances in buildings following asbestos removal), the relative risk of lung cancer from asbestos is only 1.05 (and even if such a small effect was actually observed, it is unlikely to be statistically significant). For practical purposes, a threshold of 100 f/l for chrysotile asbestos and 300 f/l for amphiboles is proposed, allowing for a general precaution factor of 10.

A critical study of the methods and models utilized by many epidemiologists to evaluate the effects of low (controlled occupational) and very low (environmental) doses of asbestos on the population, within the framework of 'risk assessment', reveals major distortions in the interpretation of medical observations. This clearly has far-reaching implications for politicians and regulators.

Mathematical modelling may be applied to the development of indices that compare the potential for carcinogenicity of exposure to certain chemicals and other factors as a non-experimental aid to policy making. However, the use of simple linear extrapolation from effects observed at high levels of exposure to predict human risk at low levels of exposure is not a scientifically sound procedure. The cost of employing this false doctrine has been and continues to be enormous. This is not only a matter of public health but a problem for the whole of society, and one that merits the attention of our civic leaders.

Linear extrapolation to zero is an unscientific methodology whose social consequences are so immense that it warrants unconditional elimination.

Introduction

> If a physician were to believe that his reasoning had the value of that of a mathematician, he would have committed the greatest of errors and would be led to the most erroneous conclusions. That is, unfortunately, what has happened and what continues to happen to men whom I will refer to as 'systematics'. In essence, these men start from an idea, founded more or less on observation and which they consider to be an absolute truth. Thus, they reason logically and without experimentation, and arrive, with one conclusion leading to the next, at constructing a system that is logical, but bears no resemblance to scientific reality. Often, superficial persons allow themselves to be swept away by this appearance of logic, and it is thus that discussions worthy of the old school periodically renew themselves today. This excessively profound faith in reasoning, which leads a physiologist to a false simplification of things, is linked in part to an ignorance of the science of which he speaks, and in part to a lack of understanding of the complexity of natural phenomena. *Claude Bernard, Introduction to the Study of Experimental Medicine, 1865.*

Forty years ago, the enactment of the Delaney Clause resulted in a virtually total ban on the use of carcinogens in the USA. At about the same time, the first evaluation of the cancer risk from chemicals was proposed. Both acts were founded on the hypothesis that a single molecule of a carcinogenic chemical would have a single effect on a single cell (one hit, one cancer). Since that time, tremendous progress has been made not only in our comprehension of human cancer, but also in the sensitivity of methods of chemical analysis. The discrepancy that has resulted between old predictions and present reality could nowadays be considered as the most astonishing situation in any applied human science. Nonetheless, biomathematicians still describe models of cancer risk whose figures vary by 5 orders of magnitude. Although medicine is not a very exact science, its estimates are usually accurate to within 50 per cent (for example, an answer of 2 would have a range of between 1 and 3). It is for this reason that we must reconsider our collaboration with the epidemiologists, who apply all manner of mathematical models without sufficient critical analysis.

Medical epidemiology is an accounting method for phenomena observed by physicians. Human toxicological epidemiology is an accounting method for illnesses which have been recognized by physicians as being of toxic aetiology. For more than 20 years, physicians and epidemiologists have proposed limiting on a scientific basis the doses of certain chemicals received by workers in order to prevent occupational diseases (Alies-Patin and Valleron 1985, Ames and Gold 1991). More recently, however, epidemiologists have been called upon to evaluate the impact of exposing the general population to the much lower doses of chemicals present in the environment (especially in towns and cities). To do this, they have attempted to extrapolate from data obtained in studies of acute and chronic toxicological situations and occupational epidemiological evaluations. In doing so, they have typically failed to take into account not only the advances in knowledge of the mechanisms of carcinogenesis, but also the most fundamental elements of toxicology and the consistent observations of clinical medicine in this domain. Epidemiology must adapt itself to the principles of human toxicology and, in general, respect the fundamental biological laws of this discipline, one of which is that for every effect submitted to observation, there is a threshold dose. This is of paramount importance when considering the impact of very low doses of chemicals, such as are frequently found in the modern environment.

Mathematical models and risk evaluation

The 'one hit, one cancer' theory is 40 years old and is correlated to a 40-year-old representation of cancer in biology. We have abandoned the out-dated representation of cancer; now it is time to abandon the out-dated theory of carcinogenesis.

Modified multistage models

There are numerous stages in cancer formation: reaction of a chemical with DNA and enzymatic immediate reversibility (Buss et al. 1990), DNA repair by excision repair, DNA SOS repair, formation through different steps of genetically altered cells, then uncontrolled cells division or death, maintenance of modified cells against physiological elimination of foreign cells, formation of tumours against limiters of tumoural progress (p53 tumour suppresser gene), etc. All these steps may be passed over through a high level of steady-state situations obtained by a sufficient flow of xenobiotics.

The lack of experimental evidence for more than two stages leading to malignant cell transformation has led several authors to modify their multistage models, in order that two or three stages be sufficient to explain observed data. The basic idea is that, as in the classical model of Armitage and Doll (1954), only cells in the final stage, i.e. malignant cells, are allowed to multiply. Modified models consider the possibility of multiplication of cells in earlier stages.

Models with deterministic growth

Armitage and Doll (1957) assumed deterministic exponential growth of intermediate cells, with a defined rate for transitions from stage i to stage $i+1$. The Armitage–Doll model may be seen as a pure jump Markow process with stochastic growth. Neyman and Scott (1967) proposed a theory according to which intermediate cells initiate a subcritical birth and death process, while transformed cells initiate a supercritical one. Thus, intermediate clones may become extinct with positive probability, but transformed clones give rise to tumours with probability one.

Among many mathematical models, those of Moolgavkar et al. are still frequently used. In their two-stage model (Moolgavkar and Venzon 1978, Moolgavkar and Knudson 1981) the authors allow growth and differentiation of both normal and intermediate cells according to 20-year-old concepts of carcinogenesis. According to this model, normal (stem) cells may die or divide*, yielding two normal daughter cells*, or one normal and one intermediate cell*.

In another version of this model, the growth of normal cells is assumed to be deterministic (logistic or Gompertz). Intermediate cells, which thus appear to follow a non-homogeneous Poisson process with intensity proportional to the number of normal cells, then undergo a birth–death–mutation process: they may die or divide*, yielding two intermediate daughter cells*, or one intermediate and one malignant cell*.

The rates depend on time and dose, and any other available covariate of interest. The time between cell transformation and tumour detection is assumed constant. This model was originally devised to explain the age-specific incidence of tumours, both in children and adults. It allows great flexibility with respect to the type of dose–response curves. One of its main advantages is its experimentally founded basis. The price that is paid for these qualities is mathematical complexity (which has led the authors to develop approximations) and the large number of parameters, some of which must, by necessity, be determined from parallel *in vitro* or animal experiments.

The full stochastic version of the model is well adapted to hereditary and non-hereditary embryonal tumours that appear in childhood. The second version, wherein the growth of normal cells is assumed deterministic, is adapted to adult tumours. For this model, an approximation to tumour incidence is valid when the probability of tumour is low. An exact expression for tumour incidence cannot be obtained in closed form, but may be computed by numerical integration. In this model, carcinogens may affect tumour incidence by affecting the mutation rates or the kinetics of cell division and differentiation. Due to the logic of the process, it is a non-threshold model.

Pharmacokinetic models

These are accepted by toxicologists for medium (experimental) and low (occupational) doses; extrapolation to zero, using the sigmoid (S-shaped) curve well known in pharmacology and toxicology, is one of the most exact models. The S-model begins with a representation of type $y = ax^n$ tangential to the origin. Pharmacologists, like toxicologists, only employ the zone of inflexion, which is quasi-linear (using logit, probit, Weibull models, which misinterpret effects of very low doses). The line traced in this manner never passes through the origin. This does not define a validated threshold of toxicity, but gives an approximation which indicates the zone of null pharmaco-toxicological effect.

* N. B. Each step occurs at a specific rate.

The role of pharmacokinetic models is to serve as an interface between exposure and carcinogenesis models, by estimating the effective dose – that is, the dose of carcinogen that reaches the critical target site – as a function of external dose. This may be carried out by using compartmental or physiologically based pharmacokinetic models (Lutz and Dedrick 1987). The absorption, distribution and excretion of chemicals are described by the usual kinetic equations. However, activation, detoxification, and repair are assumed to follow saturable Michaelis–Menten kinetics, while the other reactions are assumed to follow first-order kinetics. The most interesting feature of this model is the existence of saturable steps, impling that effective dose is not proportional to external dose, which may in fact explain, at least in some instances, the non-linearity of some dose–response relationships at low and very low doses.

The importance of modelling the toxicokinetics of carcinogens has been investigated by Whitemore et al. These authors found by simulation that inappropriate kinetic models result in biased risk estimates.

The median effect equation (Stewart and Calabrese 1996) tries to normalize all types of dose–response effects except a threshold (a NOEL, no observable effect level). It uses a steady-state pharmacological method and as reference a dose producing 50 per cent effects (D med) – never observed in chemically induced human cancers – and a very simple formula of the type Fill/F norm = (Dose/D med)m, where m is a Hill-type coefficient which allows any manipulation of the log/log representation; it may be useful for experimental studies on very potent carcinogens (such as those that induce cancers in more than 50 per cent of animals).

The parameters of the chosen model are usually estimated for low doses using the maximum likelihood method, or, equivalently, using iteratively weighted least-squares regression, which is nearly impossible when the model requires the estimation of more than two parameters or when the known effects refer to 2 or 3 doses only. Once the model is identified, it must be assessed against other competing models (Table 3.1, Brown, 1978).

Table 3.1 Behaviour of models at very low dose (near zero point)

Model	Slope of tangent at dose = 0	
Multi-hit	a = 1	if k = 1 (one-hit)°
	a = 0	if k > 1*
	a = inf.	if k < 1
Weibull	a = 2	if m = 1**
	a = 0	if m > 1**
	a = inf	if m < 1
Multistage	a = 3**	
Probit	a = 0	
Logit	a = 1	if b = 1**
	a = 0	if b > 1*
	a = inf	if b < 1

* Slope zero at origin (no extrapolation)

** Slope positive at origin (linear extrapolation to zero) = Infinity (a biological nonsense)

However, it should be noted that, even for those models that have zero or infinite slope at $d = 0$, it has been claimed that the model will in fact be linear if background carcinogenesis mechanisms act additively with chemically dependent mechanisms (Crump et al. 1976). This observation has done much to promote linear methods of extrapolation to low doses, whatever the model. It fits the medical observation in the median range of effective doses, is frequently valid toward smaller doses but cannot be extrapolated to very low doses near the zero point (origin).

Many extrapolation procedures have been described (Brown 1978 ; Krewski et al. 1983). In brief, one can:

- extrapolate linearly from the lowest detectable risk;
- extrapolate the model itself;
- extrapolate linearly from the model-estimated risk corresponding to the dose generating the lowest detectable risk;
- as above, but using the upper 95 per cent confidence limit of the estimated risk;
- use robust and/or non-parametric methods;
- extrapolate linearly with a conservative slope.

Depending on the model, for very low doses (which applies to most environmental pollutants) risk estimates may differ by five or six orders of magnitude. That is not very satisfactory; indeed, it cannot be considered really acceptable either for environmental toxicologists or for the general population. Toxicologists should not submit to mathematics merely in order to support hypotheses for use in political decision-making. The scientific models are normally designed to serve a part of science and should be abandoned if they consistently prove false or inoperative.

Main strategies

Lists of carcinogens

Lists of carcinogens classified with a numerical comparative index (comparative index of carcinogenicity (CIC) on mammals, according to their effects on man) are used in many fields of regulation. In particular, CIC lists are currently used by the US EPA. They are unavoidable when there are no validated observations on humans or other mammals published and are certainly useful – mainly to propose an evaluation of global effects of associations where multiple carcinogens are found and quantitatively analysed (for example the environmental risk of cancer).

The application of toxicological principles

In other cases, clinical toxicologists must apply medical and toxicological principles.

Principle 1. Medical diagnoses of toxic effects must be understood and validated.

The utilization in human epidemiology of data unverified by physicians invariably leads to errors. Data obtained from charts or questionnaires may be very uncertain. Furthermore, the way the results are expressed is often disconcerting and even unethical when imaginary deaths or diseases are added.

When asbestos or air pollution are studied, the terminology 'respiratory disease' is, to say the least, imprecise.

Hazy terms like 'mean reduction of life span' cannot be accepted by physicians. B. L. Cohen (1991) translates some weakly significant figures into reduction of life span attributed to atmospheric pollution (4 days from cancers and 40 days from the 'toxic part' attributed to carbon monoxide and sulphur dioxide – can be compared with 2250 days for a smoker and 207 days for a driver). In French newspapers, one sees the results of some absurd calculations: each day lost is multiplied by 57 million (the population of France), to give 1900 deaths, so that the 40 days lost from CO and SO_2 becomes 76 000 deaths. But for a physician, and probably for a normal citizen, an accidental death is not comparable with the last week of a respiratory insufficiency. Roberts and Abernathy (1996) note that the parameter 'choice of life style' as a personal influence is never or poorly included in global epidemiological studies.

Principle 2. Knowledge regarding the dose of a toxin is indispensable: 'the dose makes the poison'

Toxicology is a strictly medical (human and experimental (veterinary)) discipline whose object is the diagnosis and treatment of acute, subacute, and chronic illnesses caused by chemical products, whenever and wherever they may arise. Its existence relies necessarily on highly competent technical specialists capable of carrying out and validating chemical analyses and estimating the dosage of products in human tissues and in the human environment (analytical toxicology).

Toxicology without dosage or without at least a consideration of dosage, does not exist. We have learned this from the beginning of modern chemistry from Orfila in the nineteenth century – and we should reject any publication that does not indicate the dose.

Principle 3. A threshold dose exists for every toxic action

The law of mass action is the essential principle of chemical reactions and applies to the first stage of genotoxicity (direct reaction of the pure chemical on pure CNA). But a fundamental law of normal or pathological physiology is that a living organism is organized in order to counteract external aggressions, to detoxify most xenobiotics, so that every response of a living organism to a chemical exposure (perception, nociceptive response, irritation, organ injury, immuno-allergic response, therapeutic response, test for pre-cancerous initiation, genotoxicity tests, tests of carnogenicity *in vivo* and *in vitro*) is immediately limited except in extreme (acute) situations. Particularly in the case of very limited possible effects (very low doses), the biological response must be described in a threshold model by a graph beginning at the abcissa and rising toward a plateau or toward a dome-shaped curve for the median doses. This graph never begins at zero*, and it is this curve that the epidemiologist should attempt to construct when writing about toxic illnesses.

For example, the Ames Test, the most elementary of predictive genotoxicity tests, which involves millions of bacterial cells and permits the application of calculations of probability based on a large number of independent cells, has consistently demonstrated the existence of a threshold, regardless of the carcinogen being tested. The Ames biological response is not a proportional linear response beginning at a nil dose. In a living organism, the biological response (damage control) immediately follows the chemical reaction according to the general biological law of homeostasis.

In the two huge experimental studies discussed by Gaylor (1979) and Peto et al. (1991), the conclusions of the authors themselves cannot be accepted if one examines the results. The EDOI test considers the bladder tumours due to acetyl-amino-fluorine (AAF). According to the tables and the graphs, the experimental no observable effect level (NOEL) is 30mg/kg food (6g/day/mouse) (sup. lim.) verified after 33 months (extreme survival time).

The author comments, 'tumour response is clearly non-linear with an apparent no observable effect level' but returns, regardless, to using linear models. Why did he not write 'with an experimental no observable effect level of. . . '?

In Peto et al. (1991), 4080 rats were tested with N-Nitroso-diethylamine and ND dimethylamine (NDEA and NDMA). It was proposed that the dose-related cancer effect was increasing between 1.5 and 16 ppm (w/w food (20g/D/rat), a limited non-linear effect between 0. 06 and 1.5 ppm, and a NOEL for 0. 033 ppm. When autopsies are done, the number of liver cancers (the target disease) was 3 for the 0 dose, 0 for 0.033 ppm and 0 for 0.06 ppm. In the summary, however, the author proposes a risk assessment of 0. 25 per cent bladder cancers with drinking water levels of 0. 01 ppm 'with no indication of any possible threshold'.

In fact, the no-threshold faith seems very strong among epidemiologists. To be precise and to avoid confusion on different possible end points, although cancer is the end-point of carcinogenesis evaluation, one must define the cancer threshold (T1), which corresponds to the observation of clinically or pathologically confirmed maladies. There are other references to significant experimental anomalies, such as enzymatic modifications of certain cellules or abnormal cellular islands demonstrated in experimental chemical liver oncology. The threshold thus defined (T1) is normally

---

*Except in the publications where the design of the graph is artificially made more dense near the zero point (which occurs frequently with very potent carcinogens such as aflatoxin B1) or when the lowest experimental dose is still toxic. No current evidence has disproved this fact and physicians must stick to the facts.

greater than the thresholds (T2, T3,. . .) corresponding to these observed biological phenomena; however, these thresholds (T2, T3) exist. Obviously, when the illness of interest occurs spontaneously in man, the no-effect due to a chemical parameter is defined by the determination of a number of illnesses statistically identical to that of a control population. It is useful to note that a toxicological study may not propose a relative risk less than 1. In such a case, the toxin would become a *medication*. This has been observed in some studies of chronic toxicology, for example following ingestion of low doses of lead or exposure to low levels of ionizing radiation ($<$0. 1 rad) and has led to some models with much improved explanatory power (Van Ewijk and Hoechstrat 1992).

Non-genotoxic materials

When a chemical is not genotoxic, toxicologists commonly accept a threshold of toxicity without any reason. The sporadic appearance of a hepatitis or a nephritis during some medical treatments refers to explanations as complex as that of a cancer. Finally, there is often no evidence of a genotoxic effect despite a clinical demonstration of a causal relation between the material and human cancers. One such case is asbestos.

The case of asbestos

Asbestos is neither toxic over the short-term, nor an allergic sensitizer, nor an irritant; it is simply inert. There is a noxious threshold of action, regardless of the pathological result, viewed from a short- or long-term perspective. The following are the fundamentals of toxicological expertise, as exemplified here by asbestos

There must be precise identification of the compound(s) studied

Asbestos exists in more than one crystalline form, of which at least two are relevant to the toxicologist: amphiboles (pointed at each end) and chrysotile (flexible structure). Even if both may cause cancer, they are not equipotent. The two forms have been examined separately since the beginning of studies performed in the 1960s and 1970s (Wagner et al. 1960) following the observation that they present a different degree of danger (similarly, there are different forms of dangerous crystalline silica).

Until the late 1960s and early 1970s pleural mesotheliomas were not easily identified among pleural tumours (Forster 1960, Choffel and Chretien 1962). They required pathological demonstration. But mesotheliomas were found to be characteristic of massive exposures to amphiboles, and in particular to amosite in very specific occupational exposures (mining, insulating) (Wagner et al. 1960).

Selikoff's study presented 170 cases of amosite mesotheliomas in 1946 deaths among insulators exposed for more than 20 years, giving a rate of approximately

1 in 9, or 250 times the normal rate. MacDonald, studying chrysotile miners in Canada, identified 11 mesotheliomas (8 clinical, 3 by autopsy) among a cohort of 4547 deaths (more than 11 000 employees followed from 1910 to 1975, for up to 53 years each) – that is about 1 in 400, or 4. 8 times the normal rate – despite similar diagnostic procedures in Canada and the United States. MacDonald et al. (1983) demonstrated a two-fold increase in the incidence of lung cancers among workers who were non-smokers and had worked for more than 20 years with dust exposures of the order of 60 000 fibres per litre of air (f/l) or greater. These results should be accepted and compared, not just discarded without examination. However, consideration of asbestos mesothelioma was typically limited to the amphiboles in succeeding discussions.

Since 1976, the same phenomena and the same differences have been observed in the asbestos industries, where dust exposures between the 1940s and 1960s were between 20 000 and 100 000 f/l.

The toxicologist should take into account the noxious structure and its capacities for reaction

Asbestos dusts induce no chemical reactivity and have no notable catalytic capacity. Their action is necessarily linked to their physical structure, which is recognized by the definition of fibres judged dangerous by their dimensions: length, diameter, and aspect ratio (l/d). (Ames and Gold 1996)).

In human beings, inhaled doses have been measured with sufficient accuracy by the asbestos industry for the last 30 years (between 200 000 and 70 000 f/l until 1976, then a rapid decrease to 5 000, then 2 000, and now less than 500 f/l). Occupational surveillance is strict and occupational diseases may be used as a sound human reference.

Experimental confirmation is a poor indicator; the only asbestos dust exposure which has provoked cancers by inhalation in animals was of the order of 10 mg/m^3, is a huge burden compared with typical environmental exposure. As 1 ng/m^3 is equivalent to 0.5 fibres (if length > 5 microns) per litre of air, 10 mg/m^3 is equivalent to 5 000 000 f/l and represents a level of exposure far greater than industrial dust exposures even in earlier times (1940s–1960s). Modern toxicological experimental protocols reject those experiments which distance themselves too much from the real-world situations they are designed to test. In this particular case, a factor of 50 000, in comparison with a maximum concentration of 100 f/l in the non-professional environment, is far too large to validate any extrapolation whatsoever. A common maximum protection factor is 1000, already quite large and scarcely applicable to current chemical compounds. Nonetheless, most of the animals survived an experiment in which they were exposed to 10 mg/m^3 of asbestos which lasted two years (their life expectancy). This implies an almost complete elimination of inhaled fibres, without which the animals would have died of asphyxia in a few weeks. Asbestos inhaled is not asbestos absorbed. Respiratory elimination accounts for almost all dust particles inhaled. The presence of expelled fibres (asbestos bodies) in the sputum, an indicator of the persistence of absorbed fibres in the alveolar region, is only seen, and then inconsistently, in retired workers who were subjected to very high concentrations of asbestos dust.

When inhaled toxins are at issue, the toxicologist should take into account the result of the physiological functions of purification and elimination of foreign substances (inhalation does not imply absorption; absorption does not imply intoxication). Without entering into the details of the calculations of the equivalence between weighted measures (reserved for massive dust exposures), and the evolution of concentration units that are not independent of socio-political campaigns (f/cm^3, million particles per cubic foot of air, f/l, counting classes according to WHO lengths and diameters), breathing air containing one fibre per litre (2 ppt) is equivalent to inhaling 1.6 mg of asbestos in 100 years in 800 000 m^3 of inspired air (or 16 mg for a concentration of 10 fibres per litre). This is minuscule compared with the dozens of grams inhaled by asbestos workers in the 1950s and even the proportional calculation, which we have already refuted, would show that nothing detectable would occur at this level of inhalation.

[Any medically acceptable analysis of the impact of asbestos should begin with data on the incidence of cancers of the pleura and lung and a critical evaluation of their prevalence in the population under study.]

The normal rate of pleural mesotheliomas is approximately 2–3 per million and for old people some historical aetiologies may be found, such as pleuritis and therapeutic pneumothorax (an ancient means of control for lung tuberculosis). The difference between localized (often benign) and generalized mesothelioma was a typical piece of pathological evidence in the 1960s, but it now seems to have been forgotten.

A mesothelioma registry has been maintained in France for many years, and precisely indicates peaks of registration at sites where asbestos was manipulated professionally without precaution or prevention (shipyards, textiles, heating and insulation, during the years 1945 to 1976). It is noteworthy that mesotheliomas are quite rare in certain regions such as the eastern part of France, despite the fact that environmental specialists there are just so capable and the pulmonologists no less competent than elsewhere, and would therefore be likely to detect the disease if it were present. These regions have no asbestos industry, but intensive post-war reconstruction there suggests that a large number of new buildings were likely to have been treated with asbestos to decrease fire risks. The incidence of mesotheliomas in Great Britain (Alies-Patin and Valleron 1985) is very different from that in France (Fournier et al. 1983). The implications for France should thus depend on validated French studies, where they exist. British asbestos or the British themselves may be different from the French. Likewise, the air in Mexico City or Tokyo is not the same as in Paris.

The toxicologist should attempt to explain the mechanism of the toxic effect

How can a mechanical phenomenon provoke the appearance of a cancer, if the whole theory of carcinogenesis, said to be chemical or molecular, occurs through chemical and cellular intermediates, including the effect of ionizing radiation (intermediary hydroxyl radicals), and by identifiable chemical and biological reactions (DNA adducts, genetic mutations)? According to IARC, there is no evidence for such phenomena in the case of asbestos.

A (not strictly personal) hypothesis can be summed up as follows: carcinogenesis by asbestos is an effect of permanent cellular traumas. It is sufficient to follow under the microscope a cellular culture with fibres added, whether involving macrophages or other cell types. The pointed asbestos fibres stick to the cells, skewer them like swords, kill them (giving a positive cytotoxicity test) and thus stimulate the secretion of cytokines or other biomediators (NO fir exaloke (Anderson 1996)) by a mechanical lesion of membranes and organelles. This may lead to death by apoptosis. The fibres also stretch or skewer the nuclei as well, and kill the cell (necrosis). Their effect at this level also simulates what is seen after certain products induce chromosomal breakages (clastogens). Moderate non-lethal cellular lesions facilitate the penetration of classical chemical carcinogens, such as benzopyrenes, radon, or tar from tobacco smoke which provokes a high degree of epithelial dysplasias (associated or facilitated carcinogenesis) and even of viral aggression (Ames and Gold 1991).

The result would be thus:

(a) the traumatic death of cells, liberating cytokines with accumulation of inflammatory cells (experimental granulomas);
(b) for surviving cells, perturbations of ongoing cellular division, while the most resistant cells responsible for scar formation (fibroblasts, for example) arrive to develop (normal and abnormal divisions) in the felt of asbestos fibres accumulated by lymph transport from the deep lung to the pleura.

Modern oncologists insist on one common factor for tumour development: the variations imposed on cellular division. This is only a hypothesis, but we have not found a better explanation in the papers by the WHO and IARC on the carcinogenic effect of asbestos. At an even simpler level, the association of membranous lesions with molecular carcinogens has been clearly shown to augment the frequency of cancers, as evidenced by the tests of cutaneous carcinogenesis utilized at the beginning of this century. Up to this stage, what has been discussed is only a matter of potential toxicity (hazard).

In order to evaluate real effects of environmental doses and their social importance (risk assessment), the toxicologist should record the associated medical facts or their absence

One should first recall the medical facts discovered about asbestos in the 1960s and 1970s. In levels of dust exposure of 20 000 to 100 000 f/l and exposures over several decades, physicians observed particular ailments. This was thanks to several studies dedicated to the accounting of illnesses responsible for deaths in industry (asbestos mines in South Africa and Canada, English and American industries (amosite), among insulation workers and, above all, shipworkers), through the surveillance of specific groups of workers. These studies found (in addition to the classic findings of asbestosis, pleural thickening and pleural plaques already recognized and published):

(a) the appearance of pleural mesotheliomas, in a prevalence completely abnormal among mortality studies; and
(b) an augmentation in the number of lung cancers among smoking workers, all working without protection.

The epidemiologist in these cases should limit his action to the reporting, without interpretation nor commentary, of validated medical data.

Using global epidemiological data, Selikoff (1964) studied a cohort of 17 800 workers with massive amosite exposures (amphibole), some of whom were followed for more than 20 years, and recorded 1577 deaths, among which were 103 mesotheliomas (36 were pleural mesotheliomas and 67 peritoneal mesotheliomas); 321 lung cancers found exclusively among smoking workers, and 119 cases of respiratory insufficiency linked to fibrosis (classic asbestosis). He thus observed the full range of currently recognized occupational illnesses related to asbestos. He also observed at the time fewer cancers of the lung than expected (1 observed for 4 expected) among 1 457 non-smoking workers, but the number of observations was too small to be conclusive. This first study should be accepted as factual, but extrapolation should be prohibited.

Quite curiously, Selikoff co-authored with Hammond in 1979, a text on apparently the same cohort where an account is given of a factor of 5 times the number of lung cancers among non-smoking asbestos workers (51 per 100 000 versus 11 per 100 000 expected, according to certain commentators). In fact, in reading the publication, one notes, in Table 7a, that four cancers were observed among workers who never smoked regularly, versus 0.7 expected.

Two remarks are in order: first, not to smoke at all, and not to smoke regularly are two different notions; secondly, the four cases 'expected' in 1976 became 0.7 in 1979. In both cases, the number of subjects counted is so small and the figures obtained have such a large margin of error that neither has any predictive value. It is, however, from such information that grave conclusions have been linearly extrapolated to 100 000 humans. In any case, the level of 51 per 100 000, postulated without confidence intervals, falls between the classic prevalence of cancers in the population certain to be non-smoking (8 to 15 per 100 000) and that of the general population (60 to 80 per 100 000) and this despite a level of asbestos exposure evaluated at the time to be between 20 000 and 100 000 fibres of amosite per litre. The physician toxicologist would indicate to the epidemiologist that he should pay attention to his counts and certainly should not extrapolate at random.

In dividing by 100 the 'excess mortality' calculated (by linear extrapolation, which we again refute formally), in order to estimate the danger of levels one hundred times lower (200 to 1 000 f/l) the calculation would provide a relative risk of 1.05 (lung cancer as target), which is statistically insignificant, for non-smoking workers. This holds for smoking workers as well.

Extrapolation in medicine is a scientific error. A physician who seeks to account for 20 cases when he has seen only 2 should be prohibited from publishing. We must demand the same loyalty from all who write about diseases, and refuse to admit that 1 case observed in a group of 1 000 is equivalent to 1 000 cases seen among 1 000 000. The violation of scientific ethic is the same. As a physician I would accept a list of indexes of potential carcinogenicity but would reject a citation giving a number of 'real' human cancers.

Currently, although there is a correlation between very high exposures to chrysotile and lung cancers, there is no validated medical data coupling the effects of chrysotile and cancer at a level of 1 000 f/l and less. The epidemiologist should simply state the facts, the physician reserving the right to determine any necessity of prolonged surveillance. These two facets of the issue should, in any case, be clearly separated.

To affirm 'negative' facts is as important in human toxicology as affirmation of 'positive' facts. It is indispensable to indicate, from 1996 forward, that there currently exists no observation of a malady due to asbestos for verified industrial dust exposures maintained uniformly below 1000 fibres per litre. A delay of 20 years (1976 to 1996) is certainly too short to judge the overall impact, but Selikoff, like MacDonald, had published with less than 20 years hindsight. If toxicologists had waited 40 years to speak, preventive measures would not have been taken before 1996. If we must wait 60 years, the medical observation will be blank for lack of survivors (bias of extreme age) (Armitage and Doll 1957).

Therefore a human NOEL of 300 f/l of chrysotile and 100 f/l of amphiboles could be proposed, allowing for a general precaution factor of ten.

The toxicologist should evaluate the current danger of the product (risk) based on current situations

Air concentrations of asbestos affecting the general population are typically less than 100 fibres per litre in existing asbestos-treated buildings. In this instance, the epidemiologist might be quite useful, but with the proviso that the tools employed for reasoning and the calculations be well defined. If he tries to identify a variation in the prevalence of a rare disease (such as mesothelioma or lung cancer among strict non-smokers (even passive)), he should survey at least 10 million people individually. This is nothing but an epidemiologist's dream.

A medical language exists and must be respected. Speaking more generally we must avoid publishing in medicine that which is medical nonsense.

Deaths and illnesses

Using a parallel illustration: it would be medically inadmissible to claim that air pollution in France could provoke early deaths among those patients with respiratory insufficiency and cardiovascular disease who have already arrived at a stage which good sense and classical Latin would designate as 'moribund' (those who will or should be expected to die in the months to come) and who modern nomenclature places in Class IV or V on the scale of gravity (death likely in a very short time or imminent death). Class IV patients require respiratory assistance and oxygen (an oxidative biradial); class V patients (of whom there are 50 000 in France) are receiving intensive care, stimulators, waiting for transplants, etc.

Similarly, to speak of cancer or death in general is medical nonsense. Human toxicology is a chronic discipline in which the location of chemically induced cancers and causes of deaths must be established with precision.

The epidemiology of lung cancer must be approached with at least a minimal degree of prudence. Primary cancer of the lung (mostly squamous cell and small cell bronchial cancers) is indisputably the cancer most often linked to multiple variants in the chemical environment. But this cancer is rare in a non-smoking population. One must thus surround oneself with a modicum of information about the environment of each individual (profession, habitat, leisure activities, lifestyle). Verification of the predominant dominant parameters already known in lung cancer cohorts is essential before any detailed reading of a publication. For example, if tobacco use is not detailed in an article on lung cancer which involves France or other similar countries where tobacco is consumed, such a toxicological study is without any value and should be ignored. Moreover, it should be noted that such pieces of information as 'never smoked', 'stopped smoking' or 'smoking not indicated' that are typical in questionnaire surveys, are not equivalent. The tobacco parameter has a mean 15 times potency (+1500 per cent), which is to be compared with results reported in publications where the relative risks is found between 1.0 and 1.5.

Other investigational procedures are required in order to integrate recently recognized essential factors: for instance in the absence of any indication of asbestos or tobacco smoke exposure, incidence of bronchopulmonary cancer in the general French population varies by a factor of nine depending on the socio-economic status of the individual. Workers do not occupy the most unfavourable rung in France (Damiani 1988).

All oncologists now emphasize the importance of age and of medical histories of old people (Armitage and Doll 1957, Crump and Howe 1984), and so the epidemiology of humans older than 65 must be reconsidered, taking into account such factors as profession, chronic diseases, drugs, and lifestyle. Recently, over-consumption of fats, and obesity have been found to be significant.

Primary lung cancers (exclusive of tobacco use) and other rare spontaneous cancers are currently the subject of genetic studies. This is certainly a difficult task – due to human genetic polymorphism – but it is nevertheless promising (Forster and Ackermann 1960).

For mesotheliomas, too, which are not linked to tobacco, a genetic predisposition should at least be considered, as they are exceptionally rare as spontaneous cancers (about 120–150 cases of subjects not exposed to asbestos for 520,000 deaths, or 2 to 3 per 10,000 deaths, as a baseline).

The enumeration of errors and biases to be expected in serious studies of primary lung cancer in non-smokers (active and passive), even for mesotheliomas, far exceeds the scope of this study.

Medical languages and publicity

Medical professionals should avoid slogans. The current state of biology, including indices which summarize the extent of human and experimental knowledge is preferable, despite its imperfections. And the current knowledge of formation of cancers strays far beyond the biomathematical hypotheses considered to be the better modern representations of the medical facts. There is a considerable gap between the pre-mathematical schemes and the reality.

Any extrapolation to zero is either measuring a placebo effect or else it is a toxicological nonsense made doubly offensive by a sophism of extremely grave social consequences: extrapolation to zero implies the absence of nil risk from any particular substance, such that no matter how small the dose, something will occur. Whilst a coma is almost certain to be induced following ingestion of ten grams of gardenal for example, should we really expect a coma to be provoked by a milligram of Phenobarbital, which supposedly might occur in 1 in 10,000 cases if we made a linear extrapolation? Unprotected sandblasters may develop fatal silicosis, but has a single case of silicosis been observed among the millions of children who spend a month at the beach playing on fine sand?

Absurd though these examples may appear, they are representative of the regular conclusions of certain recent epidemiological works. The social costs of such claims are enormous and increasing, as we live through a period during which the digressions of ecologists are becoming more rampant in all industrialized countries.

A further consequence of the above-mentioned sophism, and one no less catastrophic, is the recurrent calculation applied to whole populations and frequently published in the media. A figure appearing in a text, classified as scientific, of the form 'one cancer per 10 000 expected' becomes for France, according to the usual journalistic interpretation (the journalist is, of course, paid to dramatize the facts) either 50 of every 500 000 deaths, or 5 700 deaths for the 57 million inhabitants every year, one figure being equally as ludicrous as the other. These are imaginary deaths, not those observed by the physician.

Linear extrapolation has never been justified, except for administrative reasons in the USA, where a veritable witch hunt was instituted against all carcinogens (Delaney Clause). Topping this decision, the oncologists and toxicologists in the USA said, and wrote, that because there are carcinogens everywhere (Cohen 1991), it was 'tolerable and legally tolerated' to accept a figure linearly extrapolated to zero, if the risk was less than one in a million, which in reality means a zero risk, but some French and other European media have stupidly translated this into 57 or 570 (depending on the journalist) real supplementary deaths in France.

This model has other important consequences. Biomathematicians do not hesitate to propose an evaluation of current danger (risk) using the formula for 'supplemental illnesses per unit dose'. They speak of illness rather than a risk index: for them one dose unit (1 fibre per litre for asbestos) would provoke real supplemental cancers for a million inhabitants, a rationale which provokes unwarranted fear and allows journalists to terrorize the whole population. In adding the cancers thus calculated for each chemical product on the planet, the calculators would arrive at a figure of 100 times the number of cancers observed in the human population (and representing 95 per cent of all cancers), a completely nonsensical conclusion but one which seems to disturb only a limited number of physicians and toxicologists. Many toxicologists estimate the number of actual cancers due to chemical products at less than 5 per cent of the total (tobacco and alcohol excluded) thanks (in part) to current regulations in all developed countries. No law has ever imposed the model of linear extrapolation to zero (to the origin) in France or in Europe but the American way of thinking is so pregnant in young European brains that it has begun to appear in European directives – promoted by ecology parties and their partisans. Currently, a group of US scientists (Crump et al. 1976–1996) is pressing the American legislature to review its attitudes of the past. Clinical toxicologists might ask the biomathematicians to help them better to approximate the threshold dose for a life-long exposure (or for 100 years), for each product identified (i) and for each illness (or biological variation) targeted (j), known as 'Thij'.

Medico-scientific understanding of chronic very low doses exposures is limited to the zero effect level (or NOEL). In order to reflect this, the results of consensus meetings on the carcinogenic potential of chemicals and complex mixtures should be presented as validated studies of experimental or human carcinogenesis and a general comparative index of carcinogenicity referring to quantities of substances (exposed or ingested) and identity of species as non-interpreted figures. This could lead to an international scientific consensus.

Rejection of 'everything goes' is a pressing necessity in chronic human very low dose toxicology

It is always reassuring to boast of having thousands of references, but experience shows that essential scientific facts are established with very few publications, sometimes only one. Owing to the proliferation of 're-writes' and the considerable number of meaningless papers indefinitely reproduced, toxicologists should stick to their principles and reject, purely and simply, all those studies which do not follow the principles outlined above. In chronic epidemiology, the facts registered at any one time reflect the previous 20 to 40 years research, so a study which is in disagreement with older references may validly follow a bibliographic habit that rejects documents, other than fundamental ones, that are more than 20 years old. Either the observers go back to the past (to establish occupations, lifestyles, and pollution) and conclude for the past, or the conclusion reached in the past is unverifiable and requires further studies.

A mixture of good and bad analysis certainly makes up part of the history of medicine, as it does the history of all sciences. Nonetheless, we are not constrained to treat the good and the bad indiscriminately as equals. To dwell on reports that no toxicologist would judge acceptable would be abnormal, at the very least. Nobody is obliged to attempt to reproduce, even indirectly by the game of meta-analysis, those studies which have been invalidated; meta-analysis being followed by reanalysis, by intuition, by conviction and by other more or less literary procedures.

Only judges should hold themselves to their intimate convictions, and even they are faced with the problem of seeking advice from competent experts and accepting the proofs with which they are furnished.

Under the rules of medical ethics and common law, one is obliged to refuse to offer to the media terrifying and simplistic information based on ancient and outdated studies and elevated to a form of science fiction. On the contrary, toxicologists must confirm the absolute inocuousness of levels of exposure below the threshold – for example, a level equal to 5–10 fibres per litre in the case of asbestos, which linear extrapolation to zero (to the origin) cannot explain, but which clinical toxicology explains quite simply.

Human toxicology is not a forum for prestige or media sound-bites but a part of medicine and the life sciences.

Acknowledgements

The authors thank the Fondation de France who permitted the Centre de Toxicovigilance to develop its informatics capability, Dr Stephen Borron, toxicologist, for his friendly assistance and Prof. G. Thomas for his advice on mathematical models.

References and bibliography

Alies-Patin A. M., and Valleron A. J. (1985). Mortality of workers in a French asbestos cement factory (1940–1982). *Brit.J.Ind.Med.*, **42**, 219–225.

Ames B. N., and Gold L. S. (1991). Carcinogens and human health (Letter). *Science*, **251**, 607–608.

Ames B. N., and Gold L. S. (1996). The rodent high-dose cancer test is limited at best: When cell division is ignored, then risk assessment will be flawed. *Belle Newsletter*, **5** (2/3) 4–7.

Anderson L. M. (1996). The cancer risk assessment paradigm : An experimentalist's viewpoint. *Belle Newsletter*, **5** (2/3),7–10.

Armitage P., and Doll R. (1954). The age distribution of cancer and a multi-stage theory of carcinogenesis. *Br.J.Cancer*, **8**, 1–12.

Armitage P., and Doll R. (1957). A two-stage theory of carcinogenesis in relation to the age distribution of human cancer. *Br.J.Cancer*, **11**, 161–169.

Brown C. C. (1978). Statistical aspects of extrapolation of dichotomous dose–response data. *J.Natl.CancerInst.*, **60**, 101108.

Buss P., Caviezel, M. and Lutz W. K. (1990). Linear dose–response relationships for DNA adducts in rat liver for chronic exposure to aflatoxin B1. *Carcinogenesis*, **11**, 2133–2135.

Choffel C. l. and Chrétien J. (1962). Etude critique de 250 ponctions biopsies de la plevre parietale à l'aiguille d'Abrams. *J.Fr.Med.Ch.Thor.*, **16**, 571–580.

Chrétien J. (1979). Mesotheliomes malins de la plevre. *Rev.Fr.Mal.Resp.*, **7**, 221–222.

Cohen B. L. (1991). Catalog of risks extended and updated, *Health Phys*, **36**, 707–722).

Crump K. S., Clewell, H. J., and Anderson M. E. (1996). Cancer and non-cancer risk assessment should be harmonized. *Belle Newsletter*, **5** (2/3), 2–4.

Crump K. S., Hoel, D. G., Langley, C. H., and Peto R. (1976). Fundamental carcinogenic processes and their applications for low dose risk assessment. *Cancer Res*. **36**, 2973–2979.

Crump K. S., and Howe R. B. (1984). The multistage model with a time-dependent

dose pattern: applications to carcinogenic risk assessment. *Risk Anal.* **4**, 163–176.
Crump K. S. (1982). Designs for discriminating between binary dose–response models with application to animal carcinogenicity experiments. *Commun. Statist. Theor. Meth.*, **11**, 375–393.
Damiani P. (1988). Liaison de la mortalité par cause avec l'urbanisation et la catégorie socio-professionnelle. *J. Soc. Stat.*, **129** (4),169–276.
Forster E. A., and Ackermann L. V. (1960). Localized mesotheliomas of pleura; pathologic evaluation of 18 cases. *Am. J. Clin. Path.*, **34**, 349–360.
Fournier E., Zivy P., Diamant-Berger, O., Valleron A. J. et al. (1983). Parametres biocliniques utilisables pour la surveillance des ouvriers exposés à l'amiante. *VDI-Ber*, **475**, 197–201.
Gaylor D. W. (1980). The EDOI study summary and conclusions. *J.Environ.Pathol.Toxicol.*, **3**, 179–83.
Hammond E. C., Selikoff I. J., and Seidman H. (1979). Asbestos exposure, cigarette smoking and death rates. *Ann. NY Acad. Sci.*, **330**, 473–490.
Hart RW. (1996). Is a new cancer risk assessment paradigm needed? *Belle Newsletter*, **5** (2/3),14–18.
Hartley H. O., and Sielken J. R. (1977). Estimation of 'safe doses' in carcinogenic experiments. *Biometrics*, **33**, 1–30.
Hémon D. (1994). Interactions entre caractéristiques génétiques et expositions professionnelles dans le risque de cancer du poumon : Approche épidémiologique. In Séminaire INRS. Nancy, France: INRS.
International Program for Chemical Safety, (1993). Principles for evaluating chemical effects on the aged population. Geneva: World Health Organisation, *Environmental Health Criteria*, **144**.
Johannsen F. R., (1990). Risk assessment of carcinogenic and non-carcinogenic chemicals. *Crit. Rev. Toxicol.*, **20**, 341–367.
Krewski D., Brown C., and Murdoch D. (1984). Determining 'safe' levels of exposure: safety factors or mathematical models? *Fund Appl. Toxicol.* **4**, S383–S394.
Krewski D, Crump K. S., Farmer J., Gaylor D. W. et al. (1983). Comparison of statistical methods for low dose extrapolation utilizing time-to-tumor data. *Fund Appl. Toxicol.*, **30**,: 140–160.
Lutz R. J., and Dedrick R. L. (1987). Implications of pharmacokinetic modelling in risk assessment analysis. *Environ. Health Persp.*, **76**, 97–106.
MacDonald J. C., Liddel F. D. K., Gibbs G. W. et al. (1983). Dust exposure and mortality in chrysotile mining 1910–1975.
Moolgavkar S. H., and Knudson A. G. (1981). Mutation and cancer: a model for human carcinogenesis. *JNCI*, **66**, 1037–1052,.
Moolgavkar S. H., and Venzon D. J. (1979). Two-event models for carcinogenesis : Incidence curves for childhood and adult tumours. *Math. Biosci.*, **47**, 55–77.
Neyman J., Scott E. L. (1967). *Statistical aspects of the problem of carcinogenesis.* 5th Berkeley Symposium on Math Stat and Prob, University of California Press, pp. 745–776.
Pennisi E. (1997). Monkey virus DNA found in rare human cancers. *Science* (News) 3275, 748–749.
Peto R., Gray R., Brantom P. and Grasso P. (1991). Effects of chronic ingestion of N-nitrosodimethylamine : a detailed dose–response study. *Cancer Res.*, **51**, 6415–51.
Prentice R. L. (1976). A generalization of the probit and logit methods for dose–

response curves. *Biometrics*, **32**, 761–768.

Rai K., and Van Ryzin J. (1981). A generalized multihit dose–response model for low-dose extrapolation. *Biometrics*, **37**, 341–352.

Roberts W. C. Abernathy C. O. (1996). In *Risk Assessment: Principles and Methodologies in Toxicology and Risk Assessment*. Fan A. M. and Chang L. W. (eds). Marcel Decker.

Selikoff I. J., Hammond E. C., and Seidmann H. (1979). Mortality experience of insulation workers in the United States and Canada 1943–1976. *Ann N. Y. Acad Sc.*, **330**, 91–117.

Selikoff I. J., Hammond E. C., and Chong J. (1964). Asbestos exposure and neoplasia. *JAMA*, **188**, 22–26.

Stewart J. and Calabrese E. J. (1996). The median effect equation. A useful mathematical model for assessing interaction of carcinogens and low dose cancer quantitative risk assessment. In *Toxicology and Risk Assessment*. Fan A. M. and Chang L. W. (eds). Marcel Decker.

Thomas G. (1994). Asbestos induced nitric oxide production: synergistic effect with interferon–gamma. *Ann. NY Acad. Sci.*, **725**, 207–212.

Valleron A. J. Bignon J., Hughes J. M. et al. (1992). Low dose exposure to natural and man made fibres and the risk of cancer: towards a collaborative European epidemiology. *Br.J.Indust.Med.*, **49**, 606–614.

Van Ewijk P. H. and J. A Hoechstrat. (1993). Calculation of the ECEO and its confidence interval when subtoxic stimulus is present. *Ecotoxicology and Environmental Safety*, **25**, 25–32.

Vijg J., and Papconstantinou J. (1990). Ageing and longevity genes strategies for identifying DN sequence controlling life span. *J. Gerentol.*, **45**(5), 179–182.

Wagner J. C., Slegg C. A., and Marchand P. (1960). Diffuse pleural mesothelioma and asbestos exposure in the North Western Cape Province. *Brit. J. Indust. Med.*, **17**, 260.

Whitemore A., and Keller J. (1978). Quantitative theories of carcinogenesis. *SIAM*, **20**, 1–30.

Zenser T. V., and Coe R. M., (eds) (1989). *Cancer and Ageing*. Springer Verlag.

II Science

4 Benzene and Leukaemia

Joan Munby and Donald Weetman

Summary

Benzene is a common trace gas originating primarily from the use of motor vehicles. This paper considers the risk to man of leukaemia from benzene, and what may be done to mitigate the risk.

Exposure to benzene fumes in the air was linked to leukaemia in studies of highly exposed shoe workers in Turkey and Italy in the 1970s. It is now generally accepted that the association is causal, but the mechanism is not known. Workers are now protected from the leukaemic risk by the phasing out of benzene wherever possible, by industrial hygiene procedures and by regulation of maximal exposure.

Non-industrial exposure to benzene is virtually unavoidable, and the risk of leukaemia has been postulated by extrapolation from workers in the US Goodyear tyre factories, using the process called quantitative risk assessment. It is argued that the risk has been exaggerated because of inadequate exposure assessment in the affected workers and improbable assumptions of the extrapolation.

In particular, this classic study cited by US regulators almost certainly underestimated the exposure of people who subsequently suffered from leukaemia. As a result, the slope of the hypothetical linear dose–effect curve was increased and the effect at low doses was overestimated. Moreover, hypothetical linear extrapolation to zero is probably not justified; physiological control mechanisms mean that a threshold for human exposure to benzene is likely.

It is concluded that the risk of leukaemia from non-industrial exposure to benzene is probably zero or so close to zero as to be undetectable in any possible investigation.

Introduction

Airborne pollution is obvious to all who spend time in towns. Clothes become soiled by the dirt in the air, and some people suffer respiratory problems. Equally obvious is the source of much of the pollution: the exhaust emissions from petrol and diesel-powered vehicles. However, within this complex mixture of chemical pollutants, benzene is of special interest, because benzene causes leukaemia. This chapter considers the risk from benzene of leukaemia in humans, with particular emphasis on the levels of exposure likely to be experienced by the general population living in an urban environment.

One way to tackle the problem of air pollution from vehicular exhaust emissions is to set up air quality standards. Of course, the standards themselves achieve nothing, but if they are linked with measures to reduce traffic density, and to modify exhaust emissions, some progress may be possible. In this chapter, the scientific basis of the regulation of air quality is reviewed, at least as far as the serious health effect of leukaemia is concerned. It is not our intention to cover all the adverse health effects from airborne benzene, but to concentrate on the risk perceived to be of singular importance, the cancer risk.

Industrial use of benzene

Benzene is a cheap and readily available by-product the fractionation of of crude oil in the production of petrol, paraffin, domestic heating oil, etc. In Western Europe about 90 per cent of the annual production of about 5 million tonnes is derived from oil (Neumeier 1992). The world-wide production of benzene is about 14.8 million tonnes per year (WHO 1993), which rises to 30 million tonnes if the benzene in fossil fuels is included (Neumeier 1992). For these compelling reasons, benzene used to be the industrial solvent of first choice in many manufacturing processes such as shoe making, printing, dyeing, painting, and in the production of artificial leather, linoleum and oil cloth.

Today benzene is a starting material in industrial synthetic chemistry; for example, it is employed in the production of some medicines, pesticides and synthetic plastics. European petrol contains about 2 per cent benzene (1 per cent in USA), which adds to the octane value of the fuel. Much of the benzene burden on the general population is derived from motor fuel (87 per cent in Britain (Anon 1994a). However, even with air unpolluted by the evaporation of fuel and exhaust emissions, there are measurable levels of benzene: for instance 14 ppb in Amazon air, which originates mainly from the combustion of natural vegetation (Greenberg 1984).

Food and drinking water contribute to the daily dose of benzene. Smokers receive an additional dose, because tobacco smoke also contains appreciable concentrations of benzene. Typical daily doses of benzene for smokers and non-smokers are summarized in Figure 4.1. It should be noted that where one lives and whether or not one smokes provide the main avoidable sources of benzene. Nevertheless, an unavoidable dose of about 300 μg per day will be received from the diet and ambient air, which in the least favourable circumstances may be increased approximately fourfold; under current regulations, the dose from occupational exposure may be as much as sixty-fold higher.

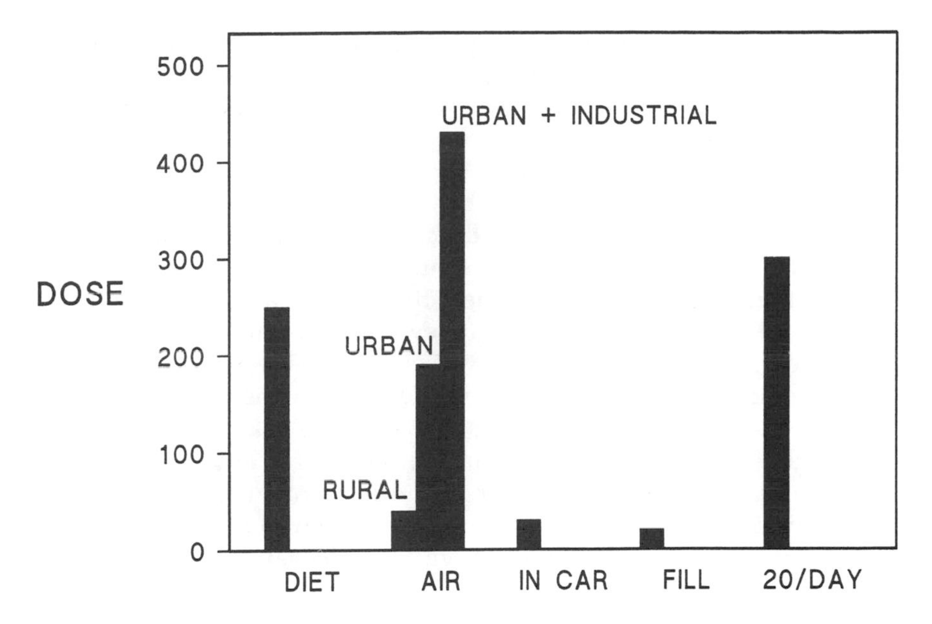

Figure 4.1 Typical daily doses of benzene (μg/day). DIET = the dose received from food and water, AIR = the dose from air, RURAL, URBAN and URBAN with INDUSTRIAL emissions respectively: IN CAR = the dose from travelling in a car: FILL = the dose from filling the car with fuel at a rate of once per week: 20/DAY = the dose from smoking 20 cigarettes per day.

Health effects: acute exposure

Very high concentrations of airborne benzene exert acute effects on the central nervous system, seen in the following successive stages: headache, sometimes convulsions, paralysis, loss of consciousness, coma and death. Non-fatal exposure is sometimes characterized by central nervous system excitation, with initial euphoria and nervous exhilaration, followed by headaches, insomnia, nausea and vertigo. All these acute effects result from inhalation of benzene.

Acute exposure to benzene, as may arise in close proximity to the spillage of the solvent, resulting in air concentrations of over 1 000 ppm, may be lethal (Stenstam 1942). In general, benzene exposure to 3 000 ppm will be survived if it is for only about 30 min, whilst air levels of 7 500 ppm are often lethal after 30–60 min (Browning 1965).

Reduction of the risk to health from hazardous chemicals

When it became clear that benzene was a potential threat to health, attempts were made to protect workers. There are three ways that the damage to health from an industrial chemical can be reduced or eliminated.

1. The use of the chemical can be phased out by replacing it with less toxic substances; on this basis, benzene is no longer a constituent of domestic paints. However, the main current uses of the hydrocarbon are in petrol and as a starting material in synthetic chemistry, for both of which there is no obvious, risk-free, and readily available alternative to benzene.
2. Workers can be protected from the chemical by improved industrial hygiene procedures. Industrial synthetic chemistry often involves reactions in closed systems, where the workers are not exposed to the hazardous substances, at least when the reactions are proceeding, although retort vessels still have to be repaired periodically and cleaned after use. On such occasions hoods and other protective wear can be used to mitigate risk.
3. Rules and regulations can be introduced so that any unavoidable contact with the chemical, or any other potentially hazardous industrial chemical, will keep exposure down to a minimal level. Regulations should be derived from an understanding of the likely mechanism of toxicity.

Health effects: chronic exposure

It is now generally accepted that the main risk to health arises from chronic exposure to airborne benzene, as may be encountered in some industrial processes. In the 1920s it was established that workplace levels of benzene between 100–250 ppm could be tolerated by the body for many years (Greenburg 1926). When toxicity to benzene occurs, it affects blood cells, resulting in low circulating levels of the different types of white and red blood cells (Jacobs 1989). These adverse effects are usually reversible if the subject is removed from the exposure, so it is worthwhile to perform regular haematological examination on workers exposed to benzene. But with leukaemia, the condition is not detectable by such routine blood tests, because the clinical signs of the disease are not apparent until ten or more years after exposure.

Leukaemia

Leukaemia is a disease characterized by an excess of white blood cells in the blood. Patients usually consult their doctor because they notice one or more of the following signs: an increased ease of bruising, bleeding from the gums, pallor and fever. Diagnosis requires a blood test of the number and type of white blood cells. The fundamental change caused by the disease occurs in the blood cell-forming tissue in the marrow of the long bones, from which the blood cells are derived by cell division and maturation. In effect an immature cell is changed in some way, so that an abnormal clone develops, which in time displaces the normal white blood cells. Blood cells are considered to be derived from undifferentiated stem cells, which may give rise to more than one type of cell. The normal progression is from stem cell to differentiated blast cells and then to white blood cells. In healthy people, stem cells and blast cells are found only in the bone marrow. Thus in medical terms, leukaemia is the unregulated malignant proliferation of haematopoietic tissue that progressively displaces normal blood cells.

Leukaemia is sub-classified into acute and chronic conditions, depending upon the degree of maturity of the affected white blood cells. The acute form of the disease is usually diagnosed on the basis of the signs and symptoms of the patient, whereas it is not unusual for the chronic diseases to be discovered in routine medical examinations for other purposes. A further division arises from the type of white cell involved. In lymphoblastic leukaemia, which is the predominant form of the disease in children (Rowan 1992), it is the lymphocytes that are affected. When other white cells are involved, the conditions are classified together as non-lymphoblastic leukaemia (e.g. acute myelocytic, myelogenous, myeloblastic and myelomonoblastic). However, this degree of subdivision represents a superficial approach to complexity: there are 116 categories of leukaemia described in the 1976 *International Classification of Diseases in Oncology* (Anon 1976).

Leukaemia is an uncommon disease, which in Britain affects about 1 in 6 000 adults per year (Anon 1994a), and is responsible for less than 1 per cent of all deaths: 1 806 male (of 254 212 total deaths) and 1 521 female deaths (of 268 444 total deaths) in England and Wales in 1992 (Anon 1994b). As with most conditions, the morbidity rates increase with age (69 per 100 000 per year in those over 65, whilst the overall rate is 9 per 100 000 per year (Rowan 1992)). There was little change in either the cumulative, or age-adjusted, death rates in England and Wales between 1954 and 1992, the ratios of the rates being between 1.0 and 1.1 (Anon 1994b; Segi 1981).

The putative causes of leukaemia include some genetic defects, viral infections, exposure to ionizing radiation, non-ionizing radiation such as the EMF fields from power lines, certain chemicals including benzene, and some anti-cancer drugs (Alderson 1990; Berkow and Fletcher 1987).

The four major classes of leukaemias are generally regarded as distinct cancerous states (Lamm et al. 1989), often affecting different groups of people, with different causations, and different progressions of the diseases (Cline 1994). The number of classes of leukaemic disease reported in the medical literature depends upon the availability to the haematologist of 'cell markers' to distinguish between different types. Only acute myeloid leukaemia, a type of acute non-lymphoblastic leukaemia (ANLL), has been consistently associated with benzene exposure. Table 4.1 summarizes the main features of the different diseases.

Table 4.1 Types of leukaemia

Type of leukaemia	*Types and number of cells in the blood*	*Peak in age first manifest*	*Time to death in the untreated*
Acute non-lymphoblastic (ANLL)	Blasts and other early precursors	All ages	4 months
Acute lymphoblastic (ALL)	Immature lymphocytes	3-7 years	4 months
Chronic non-lymphoblastic (CNLL)	Very high leucocyte counts	Young adulthood	1 to 4 years
Chronic lymphoblastic (CLL)	Small lymphocytes, initially in the bone marrow, lymph nodes and spleen, then in the blood	Middle and old age	10 years

The nomenclature in Table 4.1 describes the four major types of leukaemia. About 85% of childhood leukaemia is ALL (Rowan 1992).

The link between benzene exposure and leukaemia

Benzene was first suspected to cause leukaemia in a French case report in 1897. The medical literature since that time contains repeated case reports linking leukaemia with exposure to benzene (Quadland 1943). However, the medical profession were first persuaded that there was a real problem by findings from Turkey (Aksoy et al. 1972; Aksoy et al. 1974; Aksoy et al. 1976) and Italy (Vigliani and Forni 1969; Vigliani and Saita 1964) in the 1970s. In both countries, cases of leukaemia were noted, particularly in workers from the shoe industry, where benzene was employed as a solvent for glues and dyes. Shoe production in Turkey was a cottage industry, with most of the exposure to benzene occurring in the home, or attached small workshop, which resulted in cases of disease, but without reliable exposure information. Most of the workshops were poorly ventilated. On the basis of suspicions raised from the published case studies and the high exposure levels of benzene in the workshops, it was concluded that benzene was responsible for the excess cases of disease relative to the general Turkish population. So it became accepted that there were excess deaths from leukaemia in certain industries. From that time it has become important to try to find the level of benzene that would not cause adverse health effects in general and leukaemia in particular.

Mechanism of the leukaemic effect of benzene

Dosing experimental animals with benzene does not cause leukaemia (see below), so the mechanism of this effect must be investigated in man. Several sets of workers who were affected by overt disease after chronic exposure to high levels of benzene in various industrial processes are also described in a separate section below.

As benzene is volatile (boiling point = 80.1°C (Cheremisinoff and Morresi 1979)), the most important route of entry into the body is via the lungs. Absorption through the skin is of little significance for airborne benzene, but direct contact with the liquid form may give rise to an appreciable dose, as well as resulting in a high local concentration in the breathing zone from evaporation of the liquid.

Up to half the dose of inhaled benzene is unabsorbed and eliminated from the body in the exhaled air. The retained dose is distributed to tissues, especially those rich in fat, including the bone marrow. Within the body, a proportion of the dose of benzene is metabolized to products, mainly phenol, catechol and hydroxyquinone derivatives, which are thought to be responsible for the haematotoxicity (Sato 1988). The mechanism by which benzene, or more likely its metabolites, causes leukaemia is not known.

Over the last 20 years or so it has become possible to investigate the effects of much lower exposures to benzene, which do not cause overt disease. This is achieved by identification of a chemical that has combined covalently (in effect irreversibly) in an exposed person with a macromolecule, usually a protein, to form what has become known as an adduct. In effect, the adduct is a chemical signal of an addition reaction, which of course will differ with different pollutants. The phenomenon is highly relevant because many chemical carcinogens are thought to act directly in initiating cancer by forming an adduct with DNA.

In practice, the human tissue selected for investigation is often chosen on the grounds of convenience rather than relevance (Farmer 1995). Thus toxicology literature is filled with details of adducts to haemoglobin, or some other readily available material, which are not involved in cancer causation, but do provide evidence of exposure. In virtually all individuals there are many adducts to macromolecules that can be identified by the sophisticated methods of analytical chemistry, so there is a problem in deciding which ones are pathologically relevant. Unlike cancer-related adducts to DNA, the convenience samples are devoid of repair mechanisms, so the number that can be identified effectively measures exposure throughout the lifetime of the macromolecule (the normal life-span of human erythrocytes is about 120 days).

The problem of adduct determination is that the investigator might be misled into the incorrect attribution of an adduct to the cause of a disease (Boffetta 1995), by what Skrabanek and McCormick (1989) have called the 'surrogate-outcome fallacy'. In other words, a pathologist-confirmed diagnosis of the target disease is the only utterly reliable endpoint in the study of disease causation.

Benzene is a carcinogen

For non-cancer-causing agents, all that is necessary is to regulate the maximal permitted level to one below the level causing toxicity. Thus a no observable effect level (NOEL) of the chemical can be determined, for example in animal tests, or in volunteers in exposure chambers, and the regulatory level set from this by incorporating a safety factor (usually 100 times lower than the NOEL) (Rodricks 1992). However, as leukaemia is a form of cancer, it is necessary to regulate any substance thought to cause an excess of the disease according to a quite different set of principles.

One mechanism by which cancer is thought to be caused by chemicals requires an initial action to be exerted on the replication systems of the host animal by the chemical, or one of its products in the body. According to this mechanism, the DNA of the affected animal, or person, is transformed, and the resulting abnormality will be perpetuated through subsequent generations of cell division. Such chemicals are described as genotoxic carcinogens, and the effect they produce can be described as a mutation.

There are very many tests for mutagenicity, with the most used being performed *in vitro* (literally in glassware) on comparatively simple biological systems, such as bacteria (usually *Salmonella typhimurium*), because of their high rate of replication. The microbiologist is able to harvest bacteria that have spontaneously mutated in such a way that they can only grow in a special medium. The 'mutated' bacteria cannot synthesize the amino acid, histidine, so colonies will not grow unless histidine is added to the medium. Of course, the 'wild type' bacterium will grow in the absence of histidine. Thus the test measures the ability of a chemical, or mixture of chemicals, to back-mutate the organism (i.e. from 'mutated' to 'wild type') by growing it in a normal medium, with the result measured as 'reverents', which is determined by the size of the colonies.

Many mutagenic chemicals are not active themselves, but require to be metabolized into an active derivative by the normal bodily reactions. This problem can be overcome by adding extracts of liver, the organ responsible for most such metabolic reactions. With this modification, the Ames test, as it is called, is a powerful tool for the detection of mutagenicity, which takes only two days to perform. In practice, most substances are evaluated in a battery of tests of this type. Benzene is not mutagenic in the Ames test.

The problem with any *in vitro* mutagenicity test, such as the one described above, lies in its interpretation. Mutagenicity is not the same as carcinogenicity, which can only be determined in intact animals, or inferred from epidemiological studies in man, and both of these techniques are surrounded by their own sets of problems.

In many cases of cancer causation in man there is an initial abnormality in DNA, much as described above. However, it has now been established that there are very many errors in replication, occurring each time a cell divides and throughout life. Some of these spontaneous mutations may even trigger a cancer and, considering the number of mutations, it is possible that malignant cells could be found in man at any age. However, because of effective defence by biological systems, these affected cells develop into tumours relatively rarely. Part of the defence is provided by the immune system, the importance of which can be seen by the frequent occurrence of cancers, such as Kaposi sarcoma, in immuno-deficient patients. Additionally, there are cellular defences that are effective much earlier in the replication cycle.

It has been claimed that each of our genes is likely to mutate spontaneously more than 10 milliard (10^{10}) times in a human lifetime, and there are about 100 000 genes (Alberts et al. 1983). Clearly errors in DNA must be correctable, and systems have been discovered that effect the correction. At least one gene, p53, which has been variously described as a tumour suppresser gene, anti-oncogene, or recessive oncogene, is involved in the integrity of DNA replication. Under normal circumstances, gene p53 will arrest DNA replication at the error point, allowing other systems to repair it. Should this correction fail, p53 usually will trigger the death of the cell by a process known as apoptosis (cell suicide). Thus although the processes of cancer causation are complex and far from fully understood, there are known mechanisms to maintain DNA integrity, and probably others awaiting discovery.

It is also quite clear that there are well-established mechanisms of cancer causation that do not involve the combination of a chemical with DNA as described above. It has been known for many years that over-eating is associated with excess cancer cases in man, and this finding is now supported by reliable data from animal experiments (Tucker 1979). If the quantity of food fed to mice is restricted to 80per cent of that consumed *ad libitum*, the incidence of tumours is reduced, in some cases six-fold (Conybeare 1980). Furthermore, disturbance in the balance of hormones may lead to cancer (Roe 1988), as may chronic irritation (Furst 1992).

The process of malignant transformation requires expression of genes (oncogenes) which are 'normally' inactive. The action of various potential carcinogens may simply involve direct or indirect activation of oncogenes. Oncogenic viruses already carry the code for malignant transformation, so it is only of importance to know where in the DNA this genetic material will be inserted. If the oncogenic virus is incorporated into the active region of the genome, the host cell may become transformed into a cancer cell. This may also occur if previously inactive chromatin, containing an oncogene or oncogenes, is activated in the process of development and ageing. Nevertheless, at least for those responsible for ensuring a safe environment, most attention is paid to a single combination between a molecule of a chemical agent and DNA.

Perhaps the most important unknown feature of cancer causation by chemicals, especially genotoxic carcinogens, concerns the extent of damage to DNA necessary for the process to be initiated and proceed to completion. If a 'single hit' on DNA is all that is needed, then it is impossible to envisage a safe level for any such agent. Alternatively, if a distinct level of damage must be achieved to overcome the gene p53 and other defences, in effect a threshold for carcinogenesis must exist. The correct answer to this fundamental question of how much damage to DNA is needed for cancer causation is unknown, which presents the regulator with a dilemma. If there is indeed a threshold for cancer causation, all those exposed below this level run no risk whatsoever, and over-regulation is possible. But what body of evidence would be needed to convince a public servant of this? The prime obligation of the regulator is to be prudent, so it would be reckless, and damaging to his or her career prospects, to accept that there is a threshold for carcinogenesis, unless the evidence was utterly overwhelming. In the words of the author of the guidelines used by the Department of Health:

> ... it is prudent to assume that genotoxic carcinogens have the potential to damage DNA at any level of exposure and that such damage may lead to tumour development. Thus for genotoxic carcinogens it is assumed that there is no discernible threshold and that any level of exposure carries a carcinogenic risk ... *(Anon 1991).*

Although the evidence is by no means unequivocal (see below), benzene has been treated as a genotoxic carcinogen, which has led to statements such as the following:

> ...If we are serious about controlling benzene exposure for health reasons, should not we view each benzene molecule as our enemy and work to stop as many as possible from reaching people? (Smith 1993).

There are profound implications when a regulator rejects the possibility of a threshold for carcinogenesis, because, with this decision, he has accepted that there can be no safe level for a genotoxic carcinogen. The regulation of the quality of food in the USA has taken this path, at least with respect to food additives and agrochemical residues, under the so-called Delaney Clause of the Pure Food and Drug Act (Albert et al. 1977; Aldrich and Griffith 1993), where it is prohibited to use any chemical shown to cause cancer in laboratory animals in human food. The absurdity of this is that as the techniques of chemical analysis develop, lower and lower concentrations of chemicals can be detected, resulting in less and less food passing the quality standard, in a sequence that presumably has no end. Ironically, many naturally occurring carcinogens, such as the fungal aflatoxins that often contaminate peanuts, are less severely regulated.

It would appear that the assumptions of regulators in dealing with genotoxic carcinogens are scientifically unjustified, and run counter to one of the fundamental tenets of toxicology, that dose is important, first enunciated over 400 years ago by Paracelsus:

> All substances are poisons: there is none which is not a poison. The right dose differentiates a poison and a remedy (Poole and Leslie 1989).

Regulation of benzene

The first regulations concerning benzene date from 1927 when Winslow first recommended a workplace exposure limit for benzene of 100 ppm (1 ppm = 3.25 mg/m^3) (Cheremisinoff and Morresi 1979), based on an examination of the effects in exposed workers and animal inhalation data. Since that time, as the true nature of the hazard to health from benzene has become clear, there has been progressive reduction in the permitted maximum exposure level. In some situations, *regulation* has been effected, with procedures to test compliance and legal sanctions attached, whereas in others the maximum exposure is subject to a *guideline*. The general population is also exposed to benzene, especially benzene from petrol. Although the issue is not entirely argued from a specific health point of view, the quality of outdoor air is now an important topic and has been subjected to scrutiny within the European Community. By the end of 1997 there will be a recommended maximal level of benzene in air under the Framework Directive: these levels should be seen as targets to be achieved, with penalties for non-compliance to be administered by the member state. Table 4.2 contains some current exposure limits.

The level a government agency sets as the exposure limit for workers can be quite misleading. In the former Soviet Union and its satellites, many industrial substances appeared to be very tightly regulated, because the maximum permitted levels were much lower than those in the west (Levy 1990, Lippmann and Schlesinger 1979). For example, the USSR limit for benzene established in 1972 was 1.6 ppm, whereas in Britain the HSE reduced the exposure limit for benzene from 10 to 5 ppm in 1991. However, the Soviet workers were not better protected than their western equivalents, because of low compliance with the regulations.

How should one seek to regulate effectively? Obviously, a regulator must be appointed and the legal framework and other apparatus of state set up. Then regulators will regulate. As hazards become known, the public, or more likely a self-appointed pressure group, will conduct a media campaign for action. Ideally, the regulator should order the priorities of substances to be regulated, so that the most hazardous will be controlled first.

Table 4.2 Long-term exposure limits for benzene (from Krstic 1994b)

Country	*Year set*	*Protected group*	*Authority*	*Status*	*Level (ppm)*
USA	1977	Workers	ACGIH	Guideline	10
USA	1989	Workers	OSHA	Regulation	1
Europe	1992	Workers	EC	Guideline	0.5
UK	1991	Workers	HSE	Guideline	5
Denmark	1991	Workers	Government	Regulation	5
England and Wales	1997	General public	Department of the Environment	Air Quality Standard	0.005

ACGIH is the *American Conference of Governmental Industrial Hygienists* (USA); OSHA is the *Occupational Safety and Health Administration* (USA); EC is the *European Community*; and the HSE is the *Health and Safety Executive* (UK).
(1 ppm = 3.25 mg/m^3; 1 ppb = 3.25 μg/m^3; 1 ppt = 3.25 ng/m^3)
NB the odour threshold for airborne benzene is about 1ppm (Holmberg and Lundberg 1985).

Quantitative risk assessment (QRA)

Quantitative risk assessment (QRA) is a stepped process by which it is possible to order regulatory priorities. In effect, all QRA comes down to estimating the number of people affected by a hazard. However, the reason that QRA has become the cornerstone of US regulation of carcinogens originated from a decision in the Supreme Court. The Occupational Safety and Health Administration (OSHA) attempted to lower the benzene standard from 10 ppm to 1 ppm in 1977, using a qualitative assessment of the risk of leukaemia. The proposed revision was defeated on the grounds that OSHA had not demonstrated that a significant risk existed at the existing standard.

In the USA, QRA for carcinogens is now mandatory (Anon 1984), but the results of QRA are not always reliable. In the 1979 United States Environmental Protection Agency (US EPA) QRA of benzene reads, '... approximately a total of 90 cases of leukaemia per year could be expected due to benzene exposure' (Albert 1979).

It should be noted that these 90 cases are not *additional* to the normal US leukaemia rate, but could explain the cause of a proportion of the existing disease. The 90 cases should be considered with the 95 per cent confidence limits of the estimate, viz. 34–235. In other words, a value between 34 and 235 would be obtained in 19 of every 20 such calculations, considering the quality of the data.

Thus QRA is an useful tool for the regulator in setting priorities for action, and in building public expectation of the number of lives saved by prudent regulation. All the estimates appear to possess scientific reliability. Unfortunately, real life is not so simple, and regulation is not a scientific process, even though it uses scientific information and employs people qualified in science. The fundamental problem is that there will be no evidence of the lives saved, because it is impossible to identify those who avoid leukaemia, if indeed this followed from occupational and public health initiatives to reduce exposure to benzene. This specific example from regulatory toxicology is a subset of the general case in logic, sometimes expressed as: 'it is impossible to prove a negative'.

Whilst the claim of any deaths prevented by regulatory action is a hypothetical construction of QRA, the 'scientific' information is often communicated to the lay public via journalists briefed in press conferences or from press releases. The result is disturbing headlines in the newspapers, and special items on the radio and television news broadcasts. The public may be alarmed and anxious, and some will become unhealthily preoccupied with environmental threats. All this anxiety, while life expectancy continues to rise each year!

The best way to understand how such misrepresentation of environmental threats arises is to look more closely at the process of QRA. In the USA, the use of QRA is mandatory in the regulatory process (see above), but it is not required in Britain, although use may be made of it from time to time (Anon 1991). However, the US regulators are very skilled at setting the tone of public health arguments, and most other countries just follow their lead. The result of this is that QRA is important world-wide.

There are four stages in QRA.

Hazard identification The evidence for potential risk from an industrial chemical, such as benzene, could be derived from toxicity testing in animals or epidemiology in man. When epidemiological evidence in man is available, this is generally treated as being more relevant than results from animal tests, because it is always problematic to transfer effects between species. Of course, with epidemiology, the damage has already been suffered by individuals, so with new products, like drugs and pesticides, there is a high reliance on the results of animal experimentation. In epidemiology, the main difficulty is associating the disease causation with just one of the complex mixture of chemicals normally encountered by those exposed.

Hazard evaluation In this phase, the possible mechanisms involved in the adverse effect are considered, including whether or not the hazard is a genotoxic carcinogen. It is important to relate the probable concentration of the hazardous substance to its effect in workers, including the relationships between different exposure levels, resultant doses, and the adverse effects produced. There is a special difficulty with genotoxic carcinogens, because owing to the long latency of the adverse effect (at least several years) the past exposure has to be estimated and cannot be checked by direct measurements.

Exposure evaluation At this stage it is essential to decide how to make the extrapolation from the high doses experienced in affected workers to other, target populations. In most cases, a mathematical model relating exposure to effect has to be selected from many possible such models.

Risk estimation The results are determined by extrapolation of risk from that observed to occur in highly exposed workers to a value that corresponds to the low concentration experienced by the general public. A mathematical model is used for the extrapolation. The answer is in the form of an estimated rate of disease in the target population, together with all the uncertainties that surround this rate.

The target population will be quite different from the industrial workers in that it will contain: about 50per cent females, a proportion of people too sick to work, some with other diseases, and individuals at both extremes of the age range. It is well known that workers are generally healthier than the general population, which epidemiologists usually describe as the 'healthy worker effect' (McMichael 1976; Monson 1990).

There are no established rules to help decide how to deal with the considerable uncertainty described above. In practice, it is normal for a panel of independent experts drawn from the different disciplines to attempt to reach a consensus of what is likely to occur. Wherever QRA operates, there are published guidelines to help focus the attention of the regulator on the factors that should be considered.

In Britain, where QRA is not a step required by law, much of the deliberation proceeds by consent and consensus, but is conducted behind closed doors. The conclusions are published in a report, but there is no communication to the public of the debate prior to the writing of the report. In the USA, a quasi-jurisprudential system is in force, where each of the conclusions can be challenged. It is worthwhile to quote one of the conclusions of the Expert Panel on Air Quality Standards, set up by the Department of the Environment.

> Consideration of the evidence has led the Panel to conclude that the increased risk of leukaemia in cohorts of workers exposed to 500 ppb of benzene over a working lifetime would be too small to detect in any feasible study. In order to take account of the difference between working lifetime (approximately 77 000 hours) and chronological life (about 660 000 hours) the figure of 500 ppb has been divided by 10. Further, since it is reasonable to suppose that the population includes people, such as those exposed to other causes of leukaemia, young children and those with impaired defence mechanisms, who might be unduly sensitive, the Panel recommend that a further safety factor be applied. In the absence of any scientific evidence on individual differences we have chosen a factor of 10, analogous to factors used in regulatory toxicology for non-carcinogens. We thus arrive at a recommendation for an *Air Quality Standard* of 5 ppb, measured as a running annual average *(Anon 1994a).*

The US process of QRA

The US QRA process for the regulation of environmental pollutants, such as benzene, usually involves the preparation of an external review draft report, often by environmental consultancy companies, under the management of a team of staff from the Environmental Protection Agency (EPA). The draft report is made available to interested parties, and comments and other contributions invited. The draft report is also circulated to independent experts drawn from the universities and institutions. After sessions of public hearings, comments from interested parties and the independent science review board, a final report is produced. Only at this stage does the report become an official statement by the EPA. Regulation may follow this protracted period of consideration, although it is sometimes the responsibility of another agency, such as (OSHA), when the objective is to protect workers. It is on the publication of the official report that one sees the unqualified death toll for a pollutant in the media. However, the US *Guidelines for Carcinogen Risk Assessment* require the following action,'... it is critical that the numerical estimates not be allowed to stand alone, separated from the various assumptions and uncertainties upon which they are based ...' (Anon 1984). By the time the conclusions are reported in the press and in radio and television news bulletins, all degree of uncertainty becomes uncoupled from the simple message, leaving anxious citizens to cope with a what appears to be a certain and frightening environmental threat.

QRA: benzene

Carcinogenic tests in animals

Repeated doses of benzene have been applied to a variety of animal species for long periods of time in many toxicological studies (e.g. in rats 104 weeks; untreated rats normally live about 130 weeks). Administration was initially by injection or the oral route, and subsequently by inhalation. Oral dosing of benzene may reveal its carcinogenic potentiality, but does not simulate the occupational exposure. In most studies, there was an increased number of cancers detected in treated rats compared with controls, depending upon the severity of the treatment. The only tissue similarly affected in both rats and mice was the Zymbal gland, which lubricates the ear, and for which there is no equivalent structure in man. A raised rate of leukaemia in the benzene-treated group was not found in these studies.

This type of study provides detail about the incidence of cancers in the rats, but tells us nothing about mortality from the cancers, because the animals have to be killed before the necropsy. In man, the opposite is true: we know much about the cause of death, but little about the prevalence of cancers. The only way to be sure of the incidence of cancer is to complete a whole body post-mortem examination, which is both time-consuming and expensive. Most people are over 70 when they die and a post-mortem is not routinely performed. There must be much undetected cancer, especially when some other disease is considered to be the cause of death. In Britain, the main reason for performing a post-mortem is to exclude foul play (Roe 1992).

After review of all the evidence in 1988, the International Agency for Research on Cancer (IARC) concluded that there was '... only limited evidence of benzene carcinogenicity in experimental animals ...' (IARC 1988).

Epidemiology of industrially exposed workers

As described above, there is evidence that exposure to benzene is associated with increased incidence of leukaemia in industrial workers. The consensus view of the panel of experts assembled by IARC in 1982 was that 'benzene was an agent carcinogenic to humans' (IARC 1982), and this opinion was not revised in 1988 (IARC 1988). However, most of the studies in workers contained little or no information about the levels of benzene to which the affected workers were exposed.

If all the cohorts of industrially exposed workers are considered, it is apparent that acute myeloid leukaemia, classified as ANLL, is the prominent form of leukaemia that can be linked to benzene (Lamm et al. 1989) (Table 4.3).

Not all the studies reported an increased risk of leukaemia (see the

Table 4.3 Leukaemia in worker cohorts with some benzene exposure information

Country	*Reference*	*Relative risk*	*Year*	*Type of work*	*AML*	*Total*	*% AML*
Turkey	(Aksoy et al. 1972, Aksoy et al. 1974, Aksoy et al, 1976)	2.25**	1974	Shoe fabrication	43	51	84
Italy	(Vigliani and Forni 1969, Vigliani and Saita 1964)	20	1976	Shoe fabrication Printing	24	24	100
USA	(Rinsky 1989b, Rinsky et al. 1987, Rinsky et al. 1981), updated to 1992 from (Paustenbach et al. 1993): see below	3.37*	1986	Synthetic rubber production	6	14	43
China	(Yin et al. 1987, Yin et al. 1989)	5.74	1987	233 factories of various types	20	30	67
USA	(Wong 1987a, Wong 1987b, Wong et al. 1986, Wong and Raabe 1989)	0.75	1987	Chemical industry	0	7	0
USA	(Tsai et al. 1983)		1983	Refining	0	0	
UK	(Hurley et al. 1991)	0.4	1991	Coke over and coal	?	5	
				Totals	93	131	

* = Statistically significant risk ($P<0.05$)
** = Statistically significant risk ($P<0.001$)
AML = Acute myeloid leukaemia, a type of non-lymphoblastic leukaemia (ANLL)

reports by Wong et al. of the US chemical industry workers; Hurley et al. of the UK coke oven operatives; risk could not be determined in the study of USA oil refinery workers because of an absence of cases). There are additional studies that report the rate of leukaemia in workers, but do not contain any exposure information (see Table 1 in Kristic (1994) which reports seven such studies).

From all these reports the study of Rinsky et al. is generally considered to be the most suitable for QRA. These Goodyear factory employees are possibly the most intensively and carefully studied set of workers in occupational health (Kipen et al. 1989).

Problems in the Rinsky et al. QRA

The workers exposed to benzene from 1940 until 1965 in the Goodyear factories that produced rubber hydrochloride (sometimes called the Pliofilm cohort) were considered the best for QRA because '...of size, the relative purity of the exposure, and the exceptional record system that existed chronicling the changes in measured area benzene concentrations over 35 years of operations ...' (Rinsky 1989a).

Goodyear operated three facilities in two locations (hereafter Akron 1 and 2 and St Mary's) in Ohio. From the cohort of 1,800 white male workers, eventually 14 cases of leukaemia were identified (1 acute lymphoblastic, 2 chronic non-lymphoblastic, 1 monocytic, 1 acute granulocytic, 1 myelogenous, 2 unspecified, and 6 acute myeloid; which represents 7 cases classified as ANLL). At the time of the Rinsky et al. risk assessment, only 9 cases of leukaemia had been diagnosed. Six cases of ANLL arose from Akron 1, and for which there was virtually no benzene monitoring data (not even a single sample taken in 27 of the 29 years the plant operated), whilst the seventh case occurred in a worker from St Mary's. The likely exposure to benzene of workers in Akron 1 was 'estimated' from data relating to the other sites, although other industrial processes, also concerning benzene, were confined to the Akron 1 site (Brett et al. 1989).

Rinsky et al. assumed that the workers were exposed to between 1.2 and 32 ppm benzene. This is probably an under-estimate, because it is known that blood dyscrasia occurred quite frequently in the Goodyear cohort, and some workers even had to be admitted to hospital with aplastic anaemia (a condition in which little or no cellular bone marrow exists, often resulting in pancytopenia – dangerously low levels of all blood cell types), which is not normally seen with exposure levels of benzene below the upper limit of Rinsky's assumption (Goldstein 1977).

The QRA assumed that inhalation was the only means by which benzene entered the bodies of workers. However, it is well established that benzene can also be absorbed through the skin. In the 1977 OSHA hearings on benzene it became clear that some workers had been accidentally drenched in liquid benzene, which must have resulted in some absorption through the skin and must also have elevated substantially the airborne concentration in the breathing zone. Furthermore, the exposure information came from fixed position monitors for airborne benzene, which were set up to determine compliance with the regulatory level, and which would markedly underestimate any exposure of this type (Brett et al. 1989).

Another problem concerns how the total dose of benzene should be calculated. In QRA it is normal practice to aggregate likely dose, which is related to exposure, to arrive at a life-time dose. Without a clear understanding of the mechanism of toxicity, this is an inadequate procedure. It is quite possible that the initial lesion that eventually leads to leukaemia could arise from an isolated episode of exposure, as may result from an accidental spillage. Should this be so, then the calculation of an aggregate life-time dose from chronic exposure is irrelevant.

Although the QRA by Rinsky et al. was sponsored by NIOSH, others have used the same data for their own independent analyses: by 1989 at least 14 risk assessments had been published (Byrd and Barfield 1989). The differences in approach focus particularly on the 'estimation' of exposure levels in Akron 1, where most of the cases occurred. Rinsky et al. classified workers by job, and assumed generally that exposure was constant for a given job throughout the 40 or so years the plant operated. Crump and Allen (in a QRA performed for OSHA and cited in Paustenbach et al. 1993) calculated exposure as a function of the prevailing regulatory level for benzene. Thus if data existed in the 1960s to show that average exposure was, say, half the regulatory level (12.5 ppm, i.e. 50per cent of 25 ppm), they then assumed that exposure in 1945 would also be 50per cent of the prevailing regulatory level (viz. 50 ppm, because the regulatory level was 100 ppm at this time). The benzene exposure limit was revised downwards in 1941, 1947, 1948, 1957, 1963 and 1969 (Kipen et al. 1989).

Yet another approach was taken by Paustenbach and others (Paustenbach et al. 1992). They attempted to reconstruct possible benzene exposure in the 1940s from the 1960s monitoring data that were available, making allowances for contributions from benzene spillages, dermal exposure, the large number of hours worked each week (it was in wartime), and the morbidity and mortality records of the workers for all conditions.

Of the three approaches to the 'calculated' individual exposure for each of 7 cases (6 from Akron 1 and 1 from St Mary's) of ANLL, Rinsky et al. estimated the lowest level for 6 (Crump and Allen claimed the other), and Paustenbach et al. postulated the highest exposure for 5 cases (with 2 cases for Crump and Allen). The average exposure calculated by Crump and Allen and Paustenbach et al. were both in excess of the value employed by Rinsky et al., by factors of 2.1 and 3.35 times, respectively.

Kipen et al. analysed haematological data of the workers from the St Mary's plant, which was obtained under the freedom of information rules that pertain in the USA (Kipen et al. 1989). On the basis that high exposure to benzene would be expected to reduce both the white and red blood counts, they found no significant correlations between the calculated exposure levels of Rinsky et al. and counts. Under the assumptions of Crump and Allen, there was a statistically significant correlation between estimated exposure and white cell counts, but not for red blood cells.

The exposure–leukaemia equation selected by Rinsky and co-workers (Rinsky et al. 1987) that formed the basis for the extrapolation of risk from the high industrial levels of benzene to the low levels encountered non-industrially, may also have been quite incorrect. Ideally there should be a simple relationship between exposure to benzene (level multiplied by time to give cumulative exposure in ppm-years) and the increased risk (the rate for cases relative to that for population controls). The extrapolation of risk depends upon the shape of the curve (presumed to be a straight line by Rinsky et al.) and its slope, (i.e. the regression of increased risk against cumulative exposure). Alternative equations will predict different degrees of risk. Paustenbach et al. (1993) using the Goodyear data cite the level of exposure to benzene required to cause 1 excess case of leukaemia in a million. The values calculated in four models with different assumptions are as follows: 25, 27, 100 and within the range 1 000–10 000 ppt.

Close examination of the calculated cumulative exposure, using Rinsky's possibly flawed method (see above), for each of the nine cases of leukaemia identified by Rinsky et al., shows that they fall into four groups: 3 cases between 470 and 640 ppm-years with an increased risk of 66.4; 2 cases at 250 and 260 ppm-years, increased risk of 11.86; 2 cases at 99 and 50 ppm-years, increased risk of 3.22; and the remaining 2 cases at 10 and 0.1 ppm-years, increased risk of 1.09) (Rinsky 1989a; Rinsky et al. 1987). If the group consisting of the 3 most highly-exposed cases is removed from the analysis, the model ceases to predict a statistically significant risk of leukaemia (Yardley-Jones et al. 1991), so in effect these three cases drive the model.

Extent of extrapolation

Figure 4.2 illustrates the range of airborne benzene concentrations used in the various types of study and typical exposures described in this chapter. The degree of extrapolation necessary to project the risk of leukaemia from animal toxicity tests and the Rinsky et al. study (using Rinsky's low estimate of exposure: see above) to the general public is apparent from the figure.

Whatever may have been the true level of benzene in the Goodyear plants, and whatever may be the real leukaemic risk, speculative and highly disputed values have to be entered into the risk equation. This is so with all QRA. Thus the number of deaths predicted from QRA is always uncertain, frequently unknown and in some instances, unknowable. It is beyond doubt that some cases of ANLL were caused by industrial exposure to benzene, under the occupational health and safety procedures that prevailed in the 1940s. However, no new cases of leukaemia have been identified in the Goodyear Pliofilm workers who commenced employment after 1950 (Paustenbach et al. 1993). It seems incredible that the risk of leukaemia from non-industrial exposure to benzene should be seen as a topic of concern to the general population.

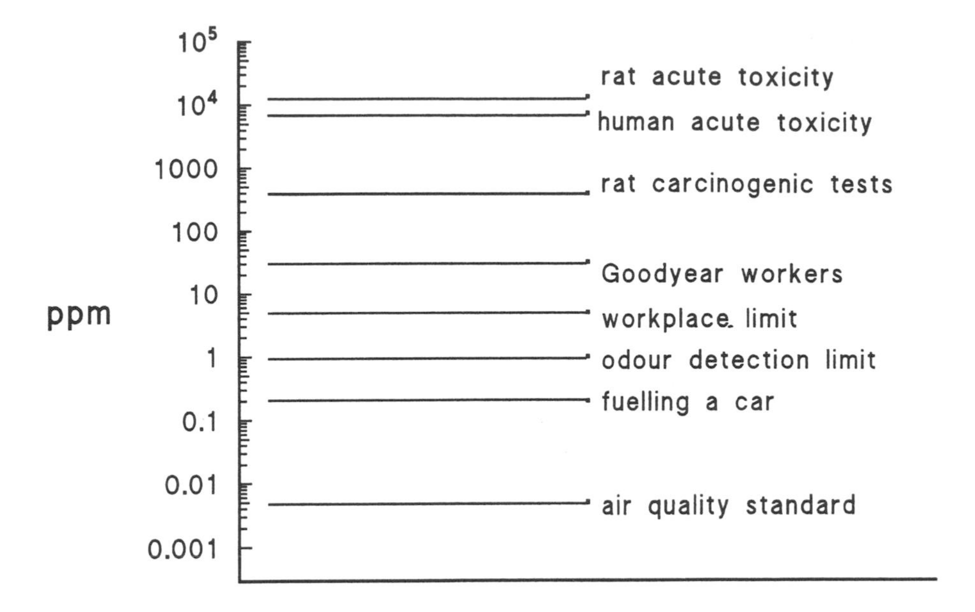

Figure 4.2 The range of concentrations of airborne benzene used in the various studies described in this chapter. The ordinate axis is in μg/m³, on a log_{10} scale

Conclusion

Benzene has undoubtedly caused ANLL in highly exposed workers in the past. It is also necessary to protect the workers of today from benzene by regulation. But what is the risk to the general public? If there is any risk at all, it must be very low, because of the immense degree of extrapolation necessary to reach current ambient levels from the occupational ones of the 1940s. With the quality of the information available, it is impossible to decide between two scenarios: low risk and no risk, which is exactly the conclusion reached by an expert panel of the World Health Organisation in 1993 (WHO 1993). Neither possibility should be a cause for alarm.

References

Aksoy, M., Dincol, K. Erdem, S. and Dincol. G. (1972). Acute leukemia due to chronic exposure to benzene. *American Journal of Medicine* **52**,160–166.

Aksoy, M., S. Erdem, and G. Dincol. (1974). Leukemia in shoe-workers exposed chronically to benzene. *Blood* **44**, 837–841.

Aksoy, M., S. Erdem, and G. Dincol. (1976). Types of leukemia in chronic benzene poisoning. A study in thirty-four patients. *Acta Haematologica* **55**, 65–72.

Albert, R. E. (1979). Carcinogen Assessment Group's Final Report on Population Risk to Ambient Benzene Exposure. United States Environmental Protection Agency. *NTIS, EPA*-450/5–80–004.

Albert, R. E., Train, R. E., and. Anderson, E. (1977). Rationale developed by the

Environmental Protection Agency for the assessment of carcinogenic risks. *Journal of the National Cancer Institute* **58**, 1537–1541.
Alberts, B., Bray, D. and Lewis, J. (1983). *Molecular Biology of the Cell.* New York: Garland Publishing, Inc.
Alderson, M. (1990). The epidemiology of leukaemia. *Advances in Cancer Research* **31**, 1–77.
Aldrich, T. E., and Griffith, J. (1993). *Environmental Epidemiology and Risk Assessment.* New York: Van Nostrand Reinhold.
Anon. (1976). *International Classification of Diseases for Oncology,* first edition. Geneva, Switzerland: World Health Organisation.
Anon. (1984). Proposed guidelines for carcinogen risk assessment. US Environmental Protection Agency. *NTIS, Federal Register* **49**:46294.
Anon. (1991). *Guidelines for the Evaluation of Chemicals for Carcinogenicity.* London: HMSO.
Anon. (1994a). *Expert Panel on Air Quality Standards: Benzene.* London: HMSO.
Anon. (1994b). *Health and Personal Social Services Statistics for England.* London: HMSO.
Berkow, R., and Fletcher, A. J. (1987). *The Merck Manual of Diagnosis and Therapy.* 15th edn. Rahway, N. J: (ed.). Merck & Co, Inc.
Boffetta, P. (1995). Sources of bias, effect of confounding in the application of biomarkers to epidemiological studies. *Toxicology Letters* **77**, 235–238.
Brett, S. M., Rodricks, J. V. and Chinchilli, V. M. (1989). Review and update of leukemia risk potentially associated with occupational exposure to benzene. *Environmental Health Perspectives* **82**, 267–281.
Browning, E. (1965). *Toxicity and Metabolism of Industrial Solvents.* Amsterdam: Elsevier.
Byrd, D. M., and E. T. Barfield. (1989). Uncertainty in the estimation of benzene risks: Application of an uncertainty taxonomy to risk assessments based on epidemiology study of rubber hydrochloride workers. *Environmental Health Perspectives* **82**, 283–287.
Cheremisinoff, P. N., and A. C. Morresi. (1979). *Benzene: Basic and Hazardous Properties.* Edited by R. A. Young and P. N. Cheremisinoff. Pollution Engineering and Technology. New York: Marcel Dekker, Inc.
Cline, M. J. (1994). The molecular basis of leukemia. *New England Journal of Medicine* **330**, 328–336.
Conybeare, G. (1980). Effect of quality and quantity of diet on survival and tumour incidence in outbred Swiss mice. *Food and Cosmetics Toxicology* **18**, 65–75.
Farmer, P. B. (1995). Monitoring of human exposure to carcinogens through DNA and protein adduct determination. *Toxicology Letters* **82/83**, 757–762.
Furst, A. (1992). Why not use sensible criteria for assessing the carcinogenicity for indoor air pollutants? *Indoor Environment* **1**, 119–122.
Goldstein, B. D. (1977). Hematotoxicity in humans. *Journal of Toxicology and Environmental Health* **2** (Suppl),69–105.
Greenburg, J. P. and Zimmerman, P. (1984). Non-methane hydrocarbons in the remote tropics, continental and marine atmospheres. *Journal of Geophysical Reviews,* **89**, 4767-4778.
Greenburg, L. (1926). Benzol poisoning as an industrial hazard. VI. Intensive study of selected industries with respect to factory conditions and pollution of the atmosphere by benzol. *Public Health Reports* 41, 1516–1539.
Holmberg, B., and P. Lundberg. (1985). Benzene: Standards, occurrence, and exposure. *American Journal of Industrial Medicine* **7**, 375–383.
Hurley, J. F., J. W. Cherrie, and W. Maclaren. (1991). Exposure to benzene and

mortality from leukaemia: results from coke oven and other coal product workers. *British Journal of Industrial Medicine* **48**, 502–504.

IARC. (1982). Some industrial chemicals and dyestuffs; Benzene. International Agency for Research on Cancer Monographs. Lyon.

IARC. (1988). Environmental carcinogens – methods of analysis and exposure measurement. Benzene and Alkylated Benzenes. International Agency for Research on Cancer Scientific Publications No. 85. Lyon:

Jacobs, A. (1989). Annotation: Benzene and Leukaemia. *British Journal of Haematology* **72**, 119–121.

Kipen, H. M., Cody, R. P. and Goldstein, B. D. (1989). Use of longitudinal analysis of peripheral blood counts to validate historical reconstructions of benzene exposure. *Environmental Health Perspectives* **82**, 199–206.

Krstic, G. (1994a). Benzene: Chains of influence in deciding exposure limits. *Indoor Environment* **3**, 22–34.

Krstic, G. (1994b). The Rationale of Regulation of Environmental Pollutants: Role of Expert Committees. PhD thesis, University of Sunderland.

Lamm, S. H., Walters, A. S., Wilson R. et al. (1989). Consistencies and inconsistencies underlying the quantitative assessment of leukemia risk from benzene exposure. *Environmental Health Perspectives* **82**, 289–297.

Levy, L. S. (1990). The setting of occupational exposure limits in the UK and their interpretation under the COSHH 1988 regulations. *Institute of Chemical Engineers, Symposium series No. 117*, 4.1–4.11.

Lippmann, M., and R. B. Schlesinger. (1979). *Chemical Contamination in the Human Environment*. Oxford: Oxford University Press.

McMichael, A. J. (1976). Standardized mortality ratios and the 'healthy worker effect': Scratching beneath the surface. *Journal of Occupational Medicine* **18**, 165–168.

Monson, R. R. (1990). *Occupational Epidemiology*. Boca Raton, Florida: CRC Press, Inc.

Neumeier, G. (1992). Occupational Exposure Limits Criteria Document for Benzene. Luxembourg: Office for Official Publications of the European Communities.

Paustenbach, D. J., Bass, R. D. and Price, P. (1993). Benzene toxicity and risk assessment 1972–92: Implications for future regulation. *Environmental Health Perspectives* **101** (Suppl 6),177–200.

Paustenbach, D. J., Price, P. S, Ollison, W. et al. (1992). A re-evaluation of benzene exposure for the Pliofilm (rubberworker) cohort (1936–1976). *Journal of Toxicology and Environmental Health* **36**, 177–231.

Poole, A. and Leslie, G. B. (1989). *A Practical Approach to Toxicological Investigations.* Cambridge: Cambridge University Press.

Quadland, H. P. (1943). Reports of occupational injuries attributed to volatile solvents. *Industrial Medicine* **12**, 734–737.

Rinsky, R. A. (1989a). Benzene and leukemia: An epidemiologic risk assessment. *Environmental Health Perspectives* **82**, 189–191.

Rinsky, R. A. (1989b). RE: Benzene and leukaemia: a review of the literature and a risk assessment. *American Journal of Epidemiology* **129**, 1084–1086.

Rinsky, R. A., Smith, A. B., Hornung, R. et al. (1987). Benzene and leukemia: An epidemiologic risk assessment. *New England Journal of Medicine* **316**, 1044–1050.

Rinsky, R. A., Young, R. J. and Smith. A. B. (1981). Leukemia in benzene workers. *American Journal of Industrial Medicine* **2**, 217–245.

Rodricks, J. V. (1992). *Calculated Risks. Understanding the Toxicity and Human Health Risks of Chemicals in our Environment.* Cambridge: Cambridge University Press.

Roe, F. J. C. (1988). How do hormones cause cancer? In O. H. Iversen (ed.), *Theories of*

Carcinogenesis. pp. 259–272. Washington: Hemisphere Publishing Corporation.
Roe, F.J.C. (1992). Let's have a post-mortem. *Indoor Environment* **1**, 179–181.
Rowan, R. P. (1992). The blood and bone marrow. In R. N. M. MacSween and K. Whaley, (eds) *Muir's Textbook of Pathology*, pp. 585–644. London: Edward Arnold.
Sato, A. (1988). Toxicokinetics of benzene, toluene and xylenes. In L. Fishbein and I. K. O'Neill, (eds) *Environmental Carcinogens: Methods of Analysis and Exposure Measurement: Volume 10 – Benzene and Alkylated Benzenes*, pp. 47–64. Lyon, France: International Agency for Research on Cancer.
Segi, M. (1981). *Age-adjusted death rates for cancer for selected sites (A-classification) in 40 countries in 1976*. Segi Institute for Cancer Epidemiology.
Skrabanek, P., and McCormick, J. (1989). *Follies and Fallacies in Medicine*. Glasgow: The Tarragon Press.
Smith, K. R. (1993). Taking the true measure of air pollution: We have to look where the people are. *Environmental Protection Agency Journal* **19**, 6–8.
Stenstam, T. (1942). Benzol and the blood picture. *Acta Medica Scandinavia* **112**, 111–138.
Tsai, S. P., Wen, C. P. Weiss, N. S. et al. (1983). Retrospective mortality and medical surveillance studies of workers in benzene areas of refineries. *Journal of Occupational Medicine* **25**, 685–692.
Tucker, M. J. (1979). Effects of long term diet restriction on tumours in rodents. *International Journal of Cancer* **23**, 803–807.
Vigliani, E. C., and Forni, A. (1969). Benzene, chromosome changes and leukemia. *Journal of Occupational Medicine* **11**, 148.
Vigliani, E. C., and G. Saita. (1964). Benzene and leukemia. *New England Journal of Medicine* **271**, 872–876.
WHO. (1993). Environmental Health Criteria 150: Benzene. International Programme on Chemical Safety.
Wong, O. (1987a). An industry wide mortality study of chemical workers occupationally exposed to benzene. I General results. *British Journal of Industrial Medicine* **44**, 365–381.
Wong, O. (1987b). An industry wide mortality study of chemical workers occupationally exposed to benzene. II Dose response analyses. *British Journal of Industrial Medicine* **44**, 382–395.
Wong, O, Morgan, R. W., Bailey, W. J. et al. (1986). An epidemiological study of petroleum refinery employees. *British Journal of Industrial Medicine* **43**, 6–17.
Wong, O., and. Raabe, G. K. (1989). Critical review of cancer epidemiology in petroleum industry employees, with a quantitative meta-analysis by cancer site. *American Journal of Industrial Medicine* **15**, 283–310.
Yardley-Jones, A., Anderson, D. and Parke, D. V. (1991). The toxicity of benzene and its metabolism and molecular pathology in human risk assessment. *British Journal of Industrial Medicine* **48**, 437–444.
Yin, S. N., Li, G. L. Tain, F. D. et al. (1987). Leukaemia in benzene workers: a prospective cohort study. *British Journal of Industrial Medicine* **44**, 124–128.
Yin, S. N., Li, G. L., Tain, F. D. et al. (1989). A retrospective cohort study of leukemia and other cancers in benzene workers. *Environmental Health Perspectives* **82**, 207–213.

5 Is environmental tobacco smoke a risk factor for lung cancer?

Robert Nilsson

Summary

It is biologically plausible that environmental tobacco smoke (ETS) has a contributory role in the induction of lung cancer in non-smoking individuals. However, recent findings suggest that a major part of the observed increase in lung cancer risk reported from epidemiological studies on ETS-exposed non-smokers is the result of misclassification of smoking status and inappropriate selection of controls, as well as certain confounding factors related to lifestyle, and possibly also to hereditary disposition. Linear extrapolation from high dose (active smokers) to low dose indicates that the increase in lung cancer that might be expected from ETS is at least one order of magnitude lower than that indicated by the epidemiological studies used to support regulatory action in the US. The epidemiological studies on ETS conducted so far lack the required power of resolution to confirm increases in risk of such low magnitudes. All epidemiological studies of ETS suffer from numerous biases, obviating their use as guides to regulatory policy making.

Self-reported information on exposure to tobacco smoke has been found to be unreliable, and data from interviews with proxy respondents even more so.

In addition, the presence of cotinine (a biological marker of tobacco use) is inadequate as a means of ensuring that misclassification does not occur, since genetically based differences in the rate of nicotine metabolism mean that some active smokers will not be detected. Furthermore, due the short half-life of cotinine in the organism, a self-reported non-smoker may, in principle, have been a life-long heavy smoker until just before the sampling took place.

For some of the major studies, an excessive proportion of disease-prone individuals, mostly of low socio-economic status, seems to have been included as cases. Due to incorporation of this group, lifestyle as well as hereditary disposition may result in a disproportionately large impact on the recorded overall lung cancer rate.

Such factors apparently include *inter alia* a high intake of saturated fat, which to a varying extent is coupled with inadequate intake of anti-carcinogens present in fruits and vegetables. In comparison with background exposure of the general population, for example, to arsenic in drinking water and in seafood, or to the multitude of other natural carcinogens present in foods, the levels of cancer risk associated with ETS hardly merit regulatory action. The one-sided preoccupation with ETS as a causative factor of lung cancer in non-smokers may be seriously hindering the elucidation of the true causes of these tumours.

Introduction

General

The scientific literature on the potential association between environmental tobacco smoke (ETS) and lung cancer is extensive. The epidemiological analysis has largely been confined to 'non-smoking' women exposed to ETS, since available studies for non-smoking men are relatively few, and based on a small number of cases. Notable attempts by government bodies and institutions to review comprehensively this complex topic have been made by the Surgeon General (USDHHS, 1982, 1986), the National Research Council (NRC, 1986), as well as by the US EPA (1992), but the debate on the merits and shortcomings of these analyses has been acrimonious and long-lasting. A recent critical expert analysis that mainly focused on the statistical shortcomings of the published epidemiological studies has been conducted by the Science Policy Research Division of the US Congressional Research Service (CRS, 1995).

The following arguments have been advanced in support of a causal relationship between ETS exposure and lung cancer:

- Since ETS is similar in chemical composition to mainstream smoke (MS), and contains several established human and experimental carcinogens, it is biologically plausible that exposure to ETS may induce lung cancer.
- A number of epidemiological studies carried out in several countries have demonstrated small, but consistent increases of lung cancer in non-smoking women exposed to ETS.
- In many studies there is a positive trend for increased incidence of lung cancer in non-smoking women with increasing exposure to ETS as judged by the number of cigarettes smoked by the spouse, i.e. there is a dose–response relationship.
- Although misclassification bias may exist, the extent of the induced upward bias is not sufficiently large, and neither can possible confounding factors account for the observed increase in risk.

It remains an established fact that ETS contains several human carcinogens as well as a number of experimental carcinogens and cancer promoters (see below) that are also present in mainstream (MS) and side-stream (SS) tobacco smoke. However, the mere presence of such agents hardly constitutes a meaningful prerequisite for classifying ETS as a human carcinogen – as was done by the US EPA. Fumes from frying and deep-frying likewise contain high levels of polyaromatic hydrocarbons (PAH), carcinogenic nitrosamines, and a number of tumour promoters (Gough et al. 1978; Fishbein, 1993). The grilling of bacon has been shown to give rise to air concentrations of 1 400 ng/m^3 the carcinogen N-nitrosopyrrolidine (Fishbein, 1993). This value may be compared to the total measured concentrations in smoky bars of the three most important tobacco-specific nitrosamines that ranged from 19 to 56 ng/m^3 (Brunnemann et al. 1992). Although it contains aflatoxins – some of the most potent carcinogens yet identified – we do not classify peanut-butter as a human carcinogen, nor do we so classify hamburgers, which contain high levels of PAH. The same goes for commercially grown mushrooms (*Agaricus bisporus*), that hold the potent experimental carcinogen agaritine, and for drinking water which contains variable amounts of arsenic and, in the case of chlorinated water, experimental carcinogens such as chloroform.

In view of what has been said above, the crucial issue from society's point of view is, clearly, whether or not the possible risk posed by exposure to ETS is sufficiently high to justify classifying it as a carcinogen and to regulate its creation, and, if so, at what cost. Although the decision to regulate is basically a political one, it should, nevertheless, be based on credible risk estimates. Taking into consideration the uncertainties involved, the Science Advisory Board of the US EPA advised against presenting definite numerical estimates (Stolwijk, 1993). Nevertheless, EPA (US EPA, 1992) claimed that exposure to ETS annually causes about 3 000 deaths from lung cancer, and OSHA has subsequently stated that its proposed new regulations on indoor air quality – mainly focusing on reducing exposure to ETS – would prevent between 5 583 and 35 502 deaths due to ETS (OSHA, 1994). In view of recent findings that are highly relevant to exposure misclassification bias and confounding by certain dietary and genetic factors, a reassessment of the basis for the observed statistical association between exposure to ETS and lung cancer in non-smokers seems warranted. Below, I propose to demonstrate that the apparent risks, as reported in major epidemiological studies, and used as a basis for quantitative risk estimation by US Government Agencies and others, are biologically implausible as well as grossly inflated, and that regulation of exposure to ETS on this ground should not be given priority by society in the prevention of human cancer. The present analysis revises and updates a previous communication on this subject (Nilsson, 1996).

The composition of environmental tobacco smoke

Tobacco smoke consists of a complex mixture of a large number of compounds, some of which may have significant adverse effects on human health above a certain exposure level. IARC (1985) mentions that about 3 800 different components have been detected in tobacco smoke. Tobacco-specific nitrosamines like 4-(nitrosomethylamino)-1-(3-pyridyl)-1-butanone (NNK), N′-nitrosonornicotine (NNN), N′-nitrosoanatabine (NAT), certain polyaromatic hydrocarbons (PAH), as well as benzene and formaldehyde constitute widely-cited examples of carcinogenic substances (US EPA, 1992). In addition to gasphase constituents, the undiluted cigarette smoke from one cigarette may deliver 15–40 mg of particulate matter. In Table 5.1, adapted from IARC (1985), some of the constituents that are most important from the toxicological perspective are presented.

Table 5.1. Concentrations of biologically active agents in non-filter cigarette mainstream smoke

Smoke constituent	*Amount/cigarette*
Total particulate matter	15–40 mg
Carbon monoxide	10–23 mg
Nicotine	1.0–2.3 mg
Acetaldehyde	0.5–1.2 mg
Acetone	100–250 1g
Methanol	90–180 1g
Nitrogen oxides	100–600 1g
Hydrogen cyanide	400–500 1g
Hydroquinone	110–300 1g
Catechol	100–360 1g
Benzene	20–50 1g
Acrolein	60–100 1g
Croton aldehyde	10–20 1g
Formaldehyde	70–100 1g
Pyridine	16–40 1g
3-Methylpyridine	20–36 1g
N′-Nitrosonornicotine	200–3000 ng
4-(Methylnitrosamino)-1-(pyridyl)-1-butanone (NNK)	80–770 ng
N′-Nitrosoanabasine	0–150 ng
N-Nitrosodiethanolamine	0–36 ng
N-Nitrosopyrrolidine	0–110 ng
N-Nitrosodimethylamine	2–20 ng
N-Nitrosopiperidine	0–9 ng
Hydrazine	32–43 ng
Urethane	20–38 ng
Vinyl chloride	1.3–16 ng
Benz[a]anthracene	20–70 ng
Benzo[b]fluoranthene	4–22 ng
Benzo[j]fluoranthene	6–21 ng
Benzo[k]fluoranthene	6–12 ng
Benzo[a]pyrene	20–40 ng
Indeno[1,2,3cd]pyrene	4–20 ng
2-Naphthylamine	1.7–22 ng
ortho-Toluidine	32–160 ng

However, with the exception of the compounds derived from alkaloids specific to *Nicotiana* and related species of the *Solanaceae* family, the chemical composition of tobacco smoke is very similar to smoke from the incomplete combustion of any other plant material. But for the exceptions just noted, chemical analysis of smoke from the burning of oak leaves would most certainly give a similar chemical 'fingerprint' with respect to its constituents. In spite of this theoretical similarity – however probable – there has, for obvious reasons, been little incentive to carry out studies that might confirm such a hypothesis with respect to leaves other than those from the tobacco plant (and perhaps from hemp, *Cannabis sativa*). Moreover, whereas a considerable effort has been devoted to the analysis of mainstream smoke (MS) and side-stream smoke (SS) from cigarettes, the composition of ETS is very little known. Obviously, this is a major shortcoming, especially when considering that the majority of the experimental investigations that have been designed to mimic the effects of ETS were based on MS or SS as surrogates.

It is true that a number of carcinogens present in MS and SS have also been detected in ETS, but it should also be noted that exhaled MS as well as SS undergoes complex processes of ageing and deposition that have been insufficiently investigated (NRC, 1986; Baker and Proctor, 1990; Gori, 1995). Nevertheless, certain important facts have been established. In one series of investigations, inhalation of MS caused more than 86 per cent of some hydrophilic and hydrophobic smoke constituents (except CO) to be retained (Dalhamn et al. 1986a, 1986b). Consequently, it can safely be assumed that there is a drastic reduction in most reactive components that are originally present in MS. In addition, an active smoker retains 90 per cent or more of respirable MS particles (Dalhamn et al. 1986b; Hiller, 1984), whereas a non-smoker exposed to ETS appears to retain a much smaller fraction of respirable particles. Circulation through ventilation systems etc. will further reduce the levels of some of the more reactive components. Furthermore, there is a coagulation of small particles to form larger aggregates (Hinds, 1978) which tend to be deposited on surfaces in an indoor environment (NRC, 1986) and also to some extent exhibit a different deposition pattern in the respiratory system.

Carcinogens in ETS; polyaromatic hydrocarbons and DNA-adducts

The presence of cancer initiators in ETS – exemplified by a polyaromatic hydrocarbon (PAH) such as benzo(a)pyrene (BaP) – has been used as especially incriminating evidence for ETS as the cause of lung cancer in exposed non-smokers. BaP has often been used as a dose surrogate to estimate exposure to products from the combustion of organic material (Swedish Cancer Committee, 1984b). BaP and other PAH are readily absorbed in the gut

(IARC, 1983), and it is, therefore, appropriate to mention that Frankfurter sausages grilled over an open log fire may contain up to 1 600 μg/kg of PAH, out of which some 80 μg/kg consists of BaP (Larsson et al. 1983). In smoked food products BaP levels up to 289 μg/kg have been reported (Torsteinson, 1969). Charcoal grilled hamburgers generally have lower levels, typically containing about 9 μg BaP per kg (van Maanen et al. 1994). As demonstrated below, regular consumption of such foods results in daily absorption of PAH far in excess of the likely uptake of PAH through inhalation of ETS.

Various estimates for the concentration of BaP in ETS-contaminated air have been published, generally ranging from 1 to 20 ng/m^3 (IARC, 1983). Higher estimates, based on monitoring grill kitchens (400 ng/m^3; Fishbein, 1993), restaurants with cooking fumes, or public places contaminated with automobile exhausts, etc., are of little value in this context since ETS is not the sole or even dominant source of BaP in such localities. However, if we base our assessment on a concentration of 10 ng/m^3 for ETS-contaminated air – which represents a high average value (US EPA, 1992, Fig. 3.3; Wilson and Chuang, 1989) – and assume 100 per cent absorption in the lungs, exposure of a 10-year-old child at such ETS levels for 10 hours would result in the uptake of 0.03–0.08 μg BaP (resting versus light activity), which compares favourably with intakes from the grilled meat products referred to above. Based on 80 per cent absorption in the gut, we obtain an uptake for BaP of 13 μg from a child's consumption of a 0.2 kg Frankfurter sausage grilled over an open fire (Larsson et al. 1983). For the same intake of charcoal grilled hamburgers bought at the butcher's, the intake would be 1.4 μg which is approximately equivalent to the dose that the child would receive from spending 2 hrs in a grill kitchen. Thus, the uptake of BaP by a child from eating these grilled meat products will be higher than from massive exposure to ETS for 10 hours by a factor of 18 to 433 (Figure 5.1). Or, to put it another way: in terms of BaP absorption, for a child, eating a charcoal grilled hamburger is roughly equivalent to being exposed to ETS for 10 hours a day every day for between two weeks and two months. Since the level of BaP can be used as a marker for total exposure to PAH, similar relations will hold also with respect to other PAH.

The formation of DNA-adducts by reaction with electrophilic agents – for example, benzo(a)pyrene diolepoxide formed from benzo(a)pyrene by metabolic reactions – appears to be a determinant for cancer initiation. The dose to DNA in target tissues, therefore, constitutes a relevant measure of dose (target dose) for quantification of carcinogenic risk. In some cases this dose may be derived indirectly from determination of haemoglobin adducts. For the past two decades the target dose concept, as originally elaborated in Sweden by Ehrenberg and co-workers, has been applied for risk assessments of a number of carcinogens (Ehrenberg et al. 1974, Osterman-Golkar et al. 1976; Wright et al. 1988; Segerbäck et al. 1994; Törnqvist et al. 1986, 1988; Törnqvist and Ehrenberg, 1990).

Thus, the determination of DNA-adducts may provide an even more adequate basis for comparison between exposure to ETS and other sources of

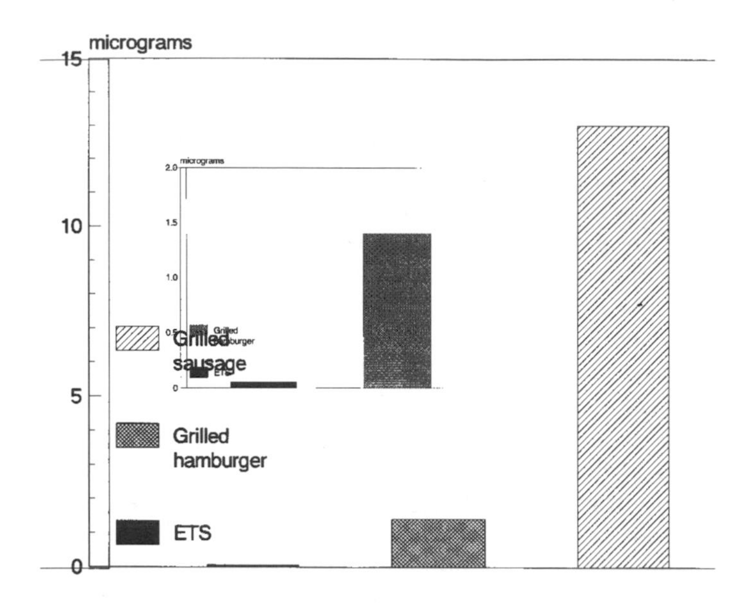

Figure 5.1 Child's intake of benzo[a]pyrene (BaP) from various sources

PAH. The formation of DNA-adducts in mutational hotspots of the p53 gene from exposure to benzo(a)pyrenediolepoxide has been cited as a 'direct link between a defined cigarette carcinogen and human cancer' (Denissenko et al. 1996), and these findings have been widely misquoted and misinterpreted. For intakes of, for example, benzo(a)pyrene, it should be emphasized that exposure by routes other than inhalation also gives the same DNA-adducts in various tissues, including lung (Qu and Stacey, 1996). Inasmuch as a PAH, such as benzo(a)pyrene, is formed by incomplete combustion of almost any type of organic matter, the results would be identical, irrespective of whether the source of benzo(a)pyrene is tobacco smoke, a coke oven furnace, smoke from the burning of wood, or ingestion of grilled hamburgers. Thus, increased DNA-adduct levels from exposure to PAH have been detected in active smokers, coke oven workers, foundry workers, fire-fighters, but also in consumers of grilled foods (Poirier and Weston, 1991; Strickland and Rothman, 1995). In 8 out of 24 subjects consuming grilled hamburgers with a mean BaP content of 9 μg/kg, and a content of pyrene of 27 μg/kg, an increase in the levels of DNA-adducts ascribed to benzo(a)pyrenediolepoxide could be detected by means of ^{32}P post-labelling (van Maanen et al. 1994). The possibility of detecting an increase in DNA-adducts is no doubt a function of fluctuating previous background adduct levels, but may also be influenced by variations in metabolic disposition. This no doubt explains the failure to detect an increase in adduct levels in some individuals (Kang et al. 1995). Given these limitations, an additional uptake even after massive exposure to ETS would hardly be detectable by DNA-adduct measurements using the super-sensitive ^{32}P-post-labelling assay. This conclusion is supported by the following experimental studies as well as by biomonitoring of exposed populations.

Lee et al. (1992, 1993) found no chromosomal aberrations in alveolar macrophages from rats after 14 or 90 days' exposure to high concentrations of aged and diluted SS (up to a concentration of 10 mg of total particulate matter per m^3). However, DNA-adducts were detected in lung, heart and larynx by means of the ^{32}P post-labelling assay, but only at the highest dose. Representative estimates of respirable suspended particles derived from ETS in homes are in the range 20–40 $\mu g/m^3$ (US EPA, 1992). Thus, the particulate level in SS required to produce measurable levels of DNA-adducts in Lee et al. (1993) were 250–500 times higher than those normally present in ETS.

In a large co-operative study sponsored by US EPA, and involving scientists from the National Health and Environmental Effects Research Laboratory, Research Triangle Park, NC working together with Czech institutions, various biomarkers for exposure to airborne carcinogens were determined in 51 non-smoking women residing in a polluted industrial district in the Czech republic. Whereas a significant increase in PAH-derived DNA-adducts and DNA damage (comet assay) in lymphocytes could be linked to exposure to ambient air pollutants, no significant effect from exposure to ETS could be detected (Binkova et al. 1996).

Should we, then, accept cancer risks from ETS just because arsenic in drinking water, hydrazines in mushrooms, and PAH in grilled Frankfurters also pose risks of similar, or even higher magnitudes? Before contemplating this not strictly scientific question, it may be appropriate first to see to what extent the risk estimates given for ETS are at all realistic.

The epidemiological evidence for ETS-induced lung cancer

Numerous studies have assessed the epidemiological evidence for a relationship between ETS and cancer. Apart from four cohort studies, three of which included a very limited number of cases, all the studies were of the case–control design. (In a cohort study, risk is expressed as relative risk, *RR*, whereas in a case–control study it is computed as an odds ratio, *OR**.) The extensive critical reviews analysing available epidemiological information have been presented by US EPA (1992), Lee (1992), CRS (1995), as well as by a European Working Group (EWG, 1996). The reader is referred to this material for further information on the individual studies. Although some quality aspects on selected investigations will be discussed later, this presentation will mainly focus on the reliability of the quantitative estimates provided in the literature.

Many of the published studies on ETS contain only a small number of cases and,

---

$$^{*}OR = \frac{\text{exposed cases/non-exposed cases}}{\text{exposed controls/non-exposed controls}}$$

consequently, the confidence intervals for the risk estimates will be broad, i.e. their ability to detect an increased risk is small. Thus, CRS (1995) points out that seven of eleven studies analysed by US EPA had only a 20 per cent chance of detecting a statistically significant risk of 50 per cent (i.e. $RR = 1.5$) using a 95 per cent confidence interval. It should, on the other hand, be emphasized that even if an association is weak, this does not rule out a cause–effect relationship. In the 30 studies examined by US EPA, an increased risk was found in 24, although statistical significance at the 95 per cent confidence level was only achieved in five. Of the other six studies, one actually found a statistically significant negative risk. Of the two largest studies so far conducted in the US that were relatively well designed, one found no statistically significant overall association (Brownson et al. 1992), whereas in the other the association attained statistical significance (Fontham et al. 1991, 1994). By combining the studies in a so called 'meta-analysis', US EPA obtained an overall relative risk of 1.19 for developing lung cancer in non-smoking women with a 90 per cent confidence interval of 1.04–1.35. Investigation of pooled results from US and Europe, where the relation between exposure to ETS in childhood and lung cancer was investigated, has not given *OR* that were significantly different from unity (Fontham et al. 1991, 1994; EWG, 1996).

Another method of analysing a collection of data of this kind, used by the European Working Group (EWG, 1996), is to use a funnel plot. This is basically a scatter plot where the natural logarithm of the relative risk (ln RR) or odds ratio (ln OR) is depicted as a function of the standard deviation of ln RR. If the collection of studies represent a random selection of all possible studies, a horizontally symmetrical funnel shaped cloud of points is obtained where the point of the funnel indicates the probable 'true' estimate.

This manner of presentation could reveal such things as publication or study size related bias but, as seen from Figure 5.2, this does not seem to be the case for the studies conducted in the US; the data form a scatter plot that widens symmetrically as precision decreases (indicating homogeneity). The situation is quite different for the Asian studies, where the data points in a similar plot deviate markedly from a funnel shape, suggesting that the collected data do not comprise a homogeneous population of studies (EWG, 1996). For this and other reasons (see below), the US studies seem to be the most suitable for a detailed analysis of risk estimates.

Risk–risk comparisons – what levels of risk are we dealing with?

Let us make the assumption that the epidemiologically based risk estimates for ETS that were cited by US EPA and by some other sources were true; how would this risk level compare with other common exposures of the general population to known human carcinogens such as active smoking?

US EPA attributes about 3 000 cases of lung cancer per year to ETS, which is around 2.5 per cent of the number of cases estimated by this agency to be

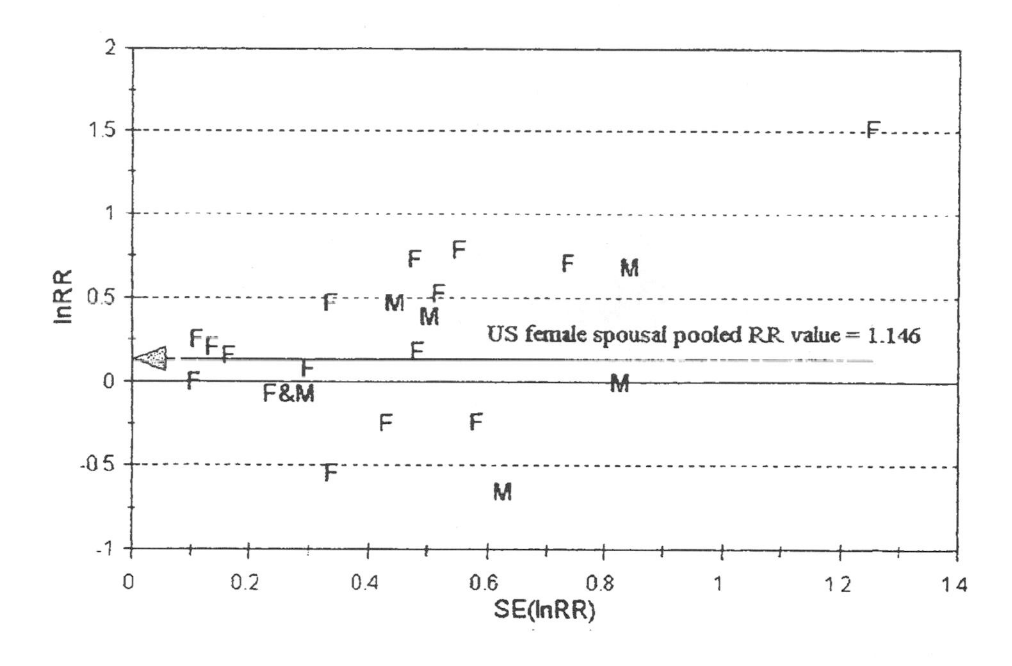

Figure 5.2 Funnel plot for lung cancer/ETS studies of spousal ETS exposure in the US. M = male study, F = female study. (From EWG, 1996)

caused by active smoking. In a population of 100 000 individuals with a normal age distribution representative of northern Europe or the US, we would normally expect a yearly total cancer incidence of around 450 cases. If we assume that roughly 50–60 per cent of this population are, or have recently been active smokers, then according to the estimates of the Swedish Cancer Committee (1984a) we could attribute some 68 (15 per cent) cases of tumours in lung and other target organs to the smoking of tobacco. However, approximately some 135 cases – or about 30 per cent of the total – would be due to diet, and 45 cases (10 per cent) to other lifestyle-related factors (Swedish Cancer Committee, 1984a). Let us then assume that at most 75 per cent of this population of 100 000 are exposed to ETS. According to the Swedish Cancer Committee (1984a), the background risk for lung cancer in non-smokers lies in the range 5–7 cases per 100 000 per year for both men and women. Somewhat higher incidences have been reported from the US. Using the average relative risk of 1.19 proposed by US EPA (1992), or the higher value of 1.34 used by OSHA (1994), for the 75 000 exposed, ETS would then at most cause some 1–2 additional cases of lung cancer per year (Figure 5.3). How does this cancer risk compare with risks associated with other common chemical carcinogens to which the general non-smoking population is exposed at low levels? Inorganic arsenic and commercially-grown mushrooms may prove instructive examples.

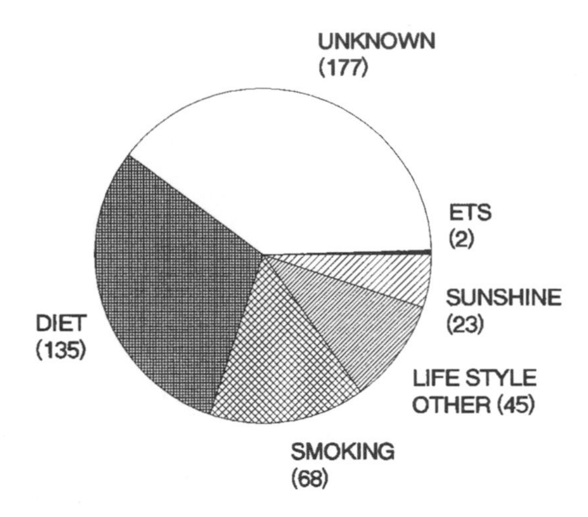

Figure 5.3 Estimated yearly incidence of cancer in a population of 100 000 individuals induced by various exogenous factors

Inorganic arsenic (As) is a proven human carcinogen, inducing skin and bladder cancer after ingestion, and lung cancer upon inhalation (WHO, 1981). As a result of its presence in bedrock, arsenic is present naturally in soils in varying quantities. Soils overlaying sulphide-ore deposits may contain several hundred ppm. Due to its ubiquitous presence in soil and rock, all living organisms, including man, are invariably exposed to various quantities of this toxic element. Not unexpectedly, drinking water frequently contains elevated levels even in the absence of any industrial pollution, and causes in some areas of Argentina, Chile, China, India, and Mexico endemic chronic arsenic poisoning, including skin cancer.

Whereas a survey of the drinking water in 18 000 communities in the US (McCabe et al. 1970) indicated that less than 1 per cent had arsenic levels exceeding 10 μg/l, levels of 70–1 700 μg/l have been found in wells in Oregon and Nevada (Goldblatt et al. 1963; Warner et al. 1994). In the vicinity of a copper smelter in the US, the mean inorganic arsenic concentration in tap water from homes was 90 μg/l (Morse et al. 1979). Further, based on the compliance monitoring data available through the US Federal Reporting Data System (FRDS), it has been estimated (US EPA, 1986), that more than 100 000 people in the US are receiving drinking water from public water supplies with arsenic levels above 50 μg/l, the current Maximum Contamination Level Goal (MCLG), of which approximately 84 per cent consume water from ground-water supplies.

Assuming that for the 100 000 Americans consuming drinking water above the US MCLG, that the average arsenic concentration in this water is 70 μg/l. Based on epidemiological data – that in comparison with those available for ETS are infinitely more convincing – US EPA has calculated a potency (slope) factor for skin cancer induction corresponding to 0.025 cases for an intake during one year of 1 mg per kg of body weight per day. Accepting a standard daily water intake of 2l/day for this population, the computed yearly incidence of skin cancer should be 5 cases (5 cases = (21 × 0.07 mg/l)/70 kg × 0.025 (mg/kg and day)$^{-1}$ × 100 000).

Marine fish commonly contain 2–20 mg/kg of arsenic (As) on a wet weight basis (Fowler, 1983), but much higher values have been reported (up to 50–100 mg/kg), especially for crustaceans (Munro, 1976). Although most of the As in seafood occurs as rapidly excreted (and apparently non-toxic) arsenobetaine, still about 1–5 per cent of the As in seafood is present as toxic, inorganic As. (Flanjak, 1982; Shinagawa et al. 1983; Norin and Vahter, 1984). In certain edible algae, like Hijiki (*Hizikia fusiforme*) or Arame (*Eisenia bicyclis*) that are common ingredients in the Japanese diet, a much higher proportion (about 50 per cent) of the total arsenic occurs as inorganic arsenate. Commercially available Hizikia contains As in the range 19–172 ppm (dry weight) with a mean of 112 ppm (Phillips 1994). Thus, a Japanese diet including on the average 5 g algae per day, will result in the ingestion of 280 μg of inorganic arsenic solely from this source. The contribution from other types of seafood will be less, but consuming 0.3 kg of this type of food (with an average total As content of 20 mg/kg, and where 2 per cent is present as inorganic As) three times a week, will add another 50 μg to the daily total. So, in a population of 100 000 on a typical Japanese seafood diet, using US EPA's potency estimate for skin cancer exposure to inorganic As, we should expect a yearly skin cancer incidence of 12 cases. (12 cases = (0.330 mg/70 kg) × 0.025 (mg/kg and day)$^{-1}$ × 100 000).

Added to this will be cancer of the bladder, for which no risk estimates are available. This risk is almost an order of magnitude higher than what was obtained for ETS above. Although some data suggest lower As intakes for the general Japanese population, the estimated risks is nevertheless considerably higher than those purportedly associated with ETS.

Whereas US EPA (1986) and a recent WHO (1993) expert group on drinking water quality have used linear extrapolation, it might be objected that the comparisons with arsenic may not be relevant, since in most cases skin cancer has a much better prognosis than for example, cancer of the lung, and further, that there probably exists a threshold for the induction of cancer by arsenic. Let us, therefore, take another example involving widespread exposure to a genotoxic carcinogen with strong initiator action in many target organs where US as well as EU regulatory agencies would recommend linear extrapolation to the low dose region.

The common commercially cultivated mushroom (*Agaricus bisporus*) contains a potent genotoxic carcinogenic hydrazine derivative, agaritine, that induces a high incidence of malignant tumours at multiple sites in the mouse, particularly in forestomach, liver, bone, and lungs (Toth and Erickson, 1986). Based on standard US EPA methodology, it can be estimated that the lifetime cancer risk from eating 50 g ordinary mushrooms twice a week corresponds to a yearly risk of about 3 in 100 000 (Nilsson et al. 1993). Again a risk is obtained that is significantly higher than that implied for ETS, even when using some of the higher risk estimates that have been cited in the

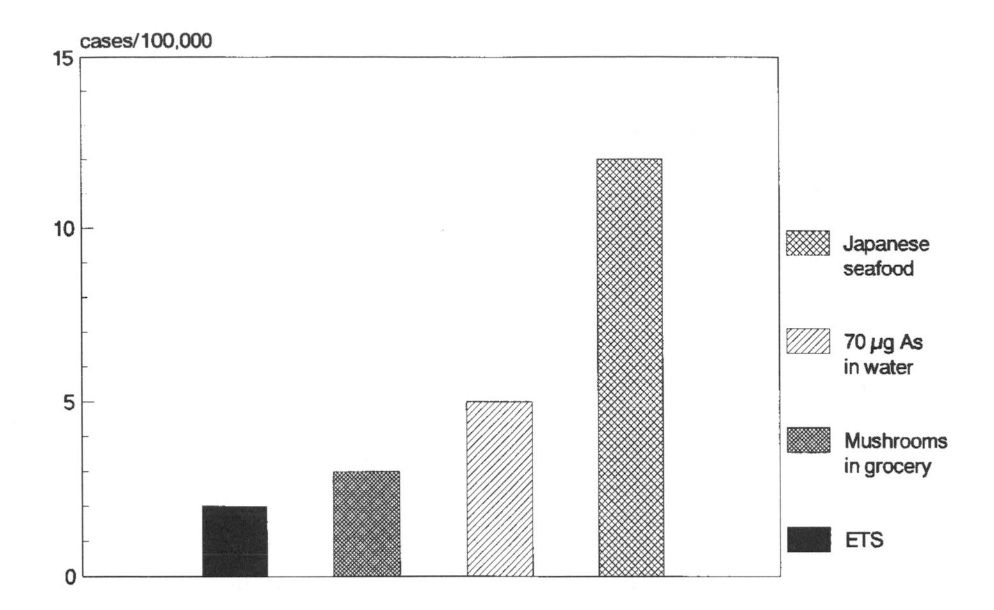

Figure 5.4 Estimated yearly cancer rate per 100 000 individuals due to ETS, consumption of commercially-grown mushrooms (50 g twice a week), arsenic in drinking water (701 g/l), and from arsenic from the regular consumption of a typical Japanese seafood diet

literature. Figure 5.4 provides a comparison of the estimated cancer risks with respect to exposure to ETS, ingestion of commercially grown mushrooms, as well as from inorganic arsenic in a typical Japanese seafood diet.

In other words, even when accepting the questionable risk estimates for ETS provided by US EPA and some epidemiologists, the cancer risk in terms of absolute numbers would be much lower than the cancer risk that more than 100 000 Americans are subjected to from mostly natural arsenic in their drinking water. It is also much lower than the computed cancer risk from eating 50 g of cultivated mushrooms twice a week, and about one order of magnitude lower than the risk from consuming Japanese seafood regularly.

Biological plausibility of current risk estimates

Epidemiological data

Epidemiologically derived risk estimates for ETS with respect to the highest exposure categories, pooled within countries, have given relative risks (*RR*), or odds ratios (*OR*) in the range of 1.38–3.11 (US EPA, 1992, pp. 5–38). The upper range of these estimates seems highly implausible considering that they would be higher than was, for example, found for women smoking 1–9 cigarettes per day in the American Cancer Society 25-state study, and for women smoking 1–14 cigarettes per day in the British Doctor Study

(USDHHS, 1982). Hirayama (1981a) reported that the association between exposure to ETS and lung cancer in women was particularly pronounced in agricultural homes, where an *RR* of 4.6 was obtained. This figure is higher than that reported for the active female smokers (*RR*=3.8). In the study reported from Greece by Trichopoulos et al. (1981), the risk found to be associated with passive smoking (*OR*=2.4) was about the same as for actively smoking women (*OR*=2.9).

Risk estimates based on data from active smokers

Cancer risks from ETS based on exposure assessment as related to tumour incidences in active smokers have provided more realistic estimates than those cited above. Because of the different chemical composition of MS and ETS, there are, admittedly, inherent limitations in the above comparisons of ETS-related risk estimates and data from active smokers, but they nevertheless provide an approximate frame of reference. NRC (1986, p. 289) extrapolated risk in a linear fashion from the relative levels of cotinine that had been measured in active smokers and ETS-exposed non-smokers, giving an expected relative risk for exposed non-smokers that would range from about 1.03 to 1.10. Using median values for cotinine levels from a study in 1537 subjects in UK, Lee (1987) arrived at a relative risk of 1.04 for women married to men who smoked.

CRS (1995) cites calculations presented by Hammond and co-workers according to which the amount of nicotine inhaled by a non-smoker in a relatively smoky restaurant for eight hours would correspond to smoking one-eighth of a cigarette (based on an average nicotine concentration in air of $18\mu g/m^3$). Giving a rather wide uncertainty range, and based on expected levels of benzo(a)pyrene and respirable particles, the Swedish Cancer Committee (1984b) estimated an exposure to ETS under extreme conditions (poor ventilation, 1 smoker in a room volume of $33m^3$, etc.) that corresponded to smoking about 0.5 cigarettes per day. This is in reasonable agreement with similar estimates based on determination of tobacco-specific N-nitrosamines in indoor air under extreme situations (Brunnemann et al. 1992). Based on *actual* measurements of several endpoints for exposure to ETS, and using personal monitors during 24-hour sampling periods, Phillips et al. (1994) could estimate that in a cohort of 255 non-smoking subjects exposed to ETS in UK, the median ETS exposure levels in terms of nicotine and ETS particulates equalled smoking about 0.01 cigarettes per day; for the most highly exposed subjects, this exposure corresponded at most to 0.2 cigarettes per day. Using the best-fitting, polynomial second-order equation as per Doll and Peto (1978) for cancer incidence as a function of cigarette consumption, this would translate into an increased risk of about 0.3 per cent for typical ETS exposure in the UK. In a recent, carefully designed study of exposure to ETS by personal monitoring conducted by the Corning Hazleton Laboratory in Great Britain in cooperation with the prestigious Karolinska Institute in

Stockholm (Phillips et al., 1996), the recorded exposures in smoking homes corresponded to smoking one two-hundredth to one three-hundredth of a cigarette per day. The extrapolated risks for ETS would be far too low to permit detection in any of the epidemiological studies so far conducted.

The significance of the promotive action of tobacco smoke

The Swedish Cancer Committee (1984a, 1984b) underlined that most of the carcinogenic action associated with tobacco smoke can be ascribed to unspecific promotive action, probably due to unspecific irritation that is reversible. Thus, 10 years after cessation of smoking, the risk has been reduced by about 75 per cent, and with increasing time the risk for ex-smokers approaches a value that is approximately double of that of a never-smoker, reflecting initiator-induced, non-repaired DNA damage remaining in lung tissue. This is in contrast to, for example, radiation-induced cancer where the integrated total dose is a major determinant of risk. For this reason, estimates based on exposure data and using linear extrapolation of cancer incidence in active smokers represent upper limits, and the true risks are likely to be considerably lower. Epidemiological data for active smokers are not adequate to establish the shape of the dose–response curve with a sufficient degree of accuracy for daily consumption of less than about 10 cigarettes per day. However, the shape of the dose–response curve is bound to be much less steep when the promotive action from tobacco smoke is greatly reduced, i.e. at normal exposures to ETS. Thus, the Swedish Cancer Committee (1984b) made the assumption that the slope factor at this low exposure range was about 1/3 of that observed after smoking about 10 cigarettes/day.

ETS is not *SS; relative carcinogenic potencies of MS, SS, and ETS*

If the risk increases reported in some epidemiological studies, and used by US EPA and others, are to be accepted as true, one would have to assume that ETS, in comparison with MS, is characterized by a carcinogenic potency that is almost an order of magnitude higher – an assumption that is highly implausible and which lacks concrete support. It has, nevertheless, been stated that SS is a more potent carcinogen than MS, incorrectly implying this to be true also for ETS (US EPA, 1992; OSHA, 1994). To support this notion it was claimed that, in comparison with MS, SS contains higher concentrations of potential carcinogens, such as certain nitrosamines. This line of reasoning is flawed from the following two points of view.

As to the higher concentrations in SS of certain initiators (e.g. nitrosamines) in comparison with MS, this could be less important owing to the dominating role of the promotive action of MS that is drastically reduced by elimination of reactive components during ageing, and in exhaled MS through inhalation. This is to some extent also supported by findings from *in vitro* tests, which seem to indicate that aged and diluted SS is less cytotoxic than fresh MS inhaled by the smoker (Sonnenfeld and Wilson, 1987). Other cited evidence to support a higher potency of SS in comparison with MS derive from a limited skin-painting study in mice by Wynder and Hoffman (1967). Whereas there was some disparity between MS and SS with respect to the induction of benign skin tumours in this investigation, no significant differences were evident for carcinomas. In addition, other reasons (e.g. lack of correlation of applied dose to skin with actually inhaled amounts), preclude any definite conclusions as to the relative potencies of MS and SS, and even less so for ETS.

The significance of the histological characteristics of lung tumours associated with ETS

Active smoking has mainly been associated with squamous cell and small (oat) cell carcinomas of the lung (Kreyberg Type I), whereas the connection between exposure to tobacco smoke and cell types like adenocarcinoma (Kreyberg Type II) has been found to be weak (Swedish Cancer Committee, 1984a). Thus, in the famous investigation by Doll and Hill (1964) of British physicians, no consistent increase was initially found for adenocarcinoma at all, although a small increase could later be detected (Doll and Peto, 1978). This is to be contrasted with a seven times increase in Kreyberg I tumours in the study of Hammond and Horn (1958) in active smokers.

In the Fontham et al. study (1991) on ETS, where the sample size was large and where considerable efforts were made to avoid misclassification for diagnosis and/or current exposure, an apparent increase in risk was only seen for adenocarcinoma, but not initially for squamous cell or small cell carcinomas. In the follow-up study (Fontham et al. 1994), an increase was also seen for primary carcinomas other than adenocarcinoma, but the *OR* was not statistically significant. The fraction of adenocarcinomas was almost 80 per cent, with 20 per cent belonging to other categories which is consistent with several of the previously published reports. However, in the Doll et al. (1957) study on active smokers the relation was reversed; in women only 20 per cent were adenocarcinomas and 80 per cent were of other histological cell types.

In some investigations an association was found for squamous cell carcinoma (Garfinkel et al. 1985; Pershagen et al. 1987; Svensson et al. 1989), and in Trichopoulos et al. (1981, 1983), cases of adenocarcinoma were deliberately excluded. However, the relevance of these findings is highly doubtful in view of the fact that there was insufficient or no verification of exposure in all of these studies; except for Garfinkel and co-workers (1985), the sample size was also limited, and the choice of controls less appropriate (see below).

Although Wynder and Goodman (1983) speculated as to the reasons for the above mentioned inconsistency in terms of different deposition patterns in the lungs for the components of MS and ETS, respectively, most authors do not seem to have realized its implications in a broader perspective. All established human lung carcinogens, including inorganic arsenic (Newman et al. 1976; Axelson et al. 1978), asbestos (Auerbach et al. 1984), chloromethyl methyl ether/bis-chloromethyl ether (IARC, 1974), coke-oven emissions (Kawai et al. 1967; Blot et al. 1983), chromium (VI) compounds (Baetjer, 1950; Bidstrup, 1967; Hueper 1966), certain nickel compounds (Sunderman, 1973; Sunderman et al. 1989), mustard gas (Yamada, 1963), alpha particles from radon daughters (Archer et al. 1973; Horacek et al. 1977; Saccomanno et al. 1971), as well as mainstream tobacco smoke, seem preferentially to induce tumours of the Kreyberg I type. *It is, therefore, biologically highly implausible, that in contrast to all other known human pulmonary carcinogens – including active smoking – ETS should induce mainly adenocarcinomas.* The objection could be raised that information on the agents described above have almost exclusively been derived from exposed male workers. In female active smokers the associated incidence of adenocarcinoma tends, on the other hand, to be higher than for males, but there is still a marked predominance of Kreyberg I type tumours. The relative increase in incidence of adenocarcinoma versus Kreyberg I type of tumours that has been noted in recent years in the US, probably reflects a decrease in the number of smokers (Devesa et al. 1991). It has been pointed out that the tobacco-specific nitrosamine, NNK, induces lung adenoma and lung carcinoma in rodents independent of the mode of application, and that an increase of the content of this carcinogen in some cigarette brands (from 120 to 160 ng/cigarette) could be responsible for the relative increase in adenocarcinoma in humans (Hoffmann et al. 1996). This argument is fallacious inasmuch as the histological type of tumour induced often varies with the species investigated. For example, bis(chloromethyl)ether mainly induces Kreyberg type I tumours in humans, lung adenomas in the mouse, and lung carcinomas in the Golden hamster (ATSDR, 1989).

The aetiology of adenocarcinoma in non-smoking women

The fact that adenocarcinoma in non-smoking women is much more prevalent than in non-smoking males, has been attributed by some authors to the operation of hormonal factors (Chaudhuri et al. 1982; Lubin and Blot, 1984; Gao et al. 1987). Others have advanced alternative explanations on the basis of the large differences for lung cancer prevalence in non-smoking women that exist between different ethnic groups. For instance, the incidence of lung cancer in non-smoking Chinese women is extraordinarily high, some 2–3 times higher than for US non-smoking women (IARC, 1990). In an investigation from Shanghai carried out jointly between Chinese institutions and the US National Cancer Institute (Gao et al. 1987), it was pointed out that for most cancers (primarily adenocarcinomas) smoking could only account for about one quarter of all newly diagnosed cases, and that factors other than exposure to tobacco smoke must be responsible for the rest. On the other hand, an association was found between pre-existing lung disease and hormonal factors, as well as exposure to cooking-oil vapours.

The elevated lung cancer rate in Chinese women is pronounced in those areas in China where smoky coal is used for cooking, for example in the Yunnan province (Mumford et al. 1987). In Juan Wei fuel burning occurs in shallow unvented pits which results in extremely high levels of indoor air pollution. Tobacco smoking here is common in males (40 per cent or more), but extremely rare in females (less than 0.1 per cent). However, in Shanghai (Gao et al. 1987; IARC, 1990), where smoky coal is not used in a similar manner to any significant extent, and where few women smoke, the incidence of lung cancer in women is, nevertheless, drastically elevated (age-standardized rate of about 20 per 100 000). A remarkable feature is that these high incidences – as well as the marked differences found between different ethnic groups of Chinese – also feature in the Chinese female populations in Singapore (Law et al. 1976; Coleman et al. 1993), Hong Kong (Hong Kong Cancer Registry, 1982; Coleman et al. 1993), as well as in the US (Fraumeni and Mason, 1974; Hinds et al. 1981; Green and Brophy, 1982). Furthermore, during the period 1970 – 1985, the cancer incidence rate among Singapore Chinese females has remained at a high and fairly stable level (Coleman et al. 1993). The age-standardized incidence rate per 100 000 females in Singapore was found (Law et al. 1976) to be highest in Cantonese (27.2), followed by Teochew (12.8) and Hokkien (12.4). Compared with, for example, Xuan Wei, Yunnan, the indoor pollution levels are low in Singapore as well as in Hong Kong, and the use of smoky coal for heating and cooking is virtually non-existent. The dietary habits of these ethnic groups are, on the other hand, mostly retained. Epidemiological data provide evidence that a high intake of vegetables exerts a marked protective effect with respect to lung cancer in Chinese Singapore women (MacLennan et al. 1977). In conclusion, there are strong indications that the high rate of lung cancer in these Chinese women is largely unrelated to air pollution, but due instead to dietary factors, possibly in combination with a genetic predisposition.

A number of investigations have focused on the relationship between a propensity to develop lung tumours and genetic polymorphism with respect to certain detoxifying enzymes, notably arylhydrocarbon hydroxylase (AHH; gene product of CYP1A1) and glutathione-S-transferase μ (GST-μ; gene product of GSTM1). AHH is involved in the conversion of PAH to procarcinogenic metabolic products, like diolepoxides, whereas GST-μ eliminates these reactive intermediates by conjugation with gluxtathione. About 50 per cent of Caucasians lack GST-μ, whereas AHH shows a trimodal distribution with various degrees of enzymatic activity. Although data have not been entirely consistent, a high activity of AHH, or lack of GST-μ has been associated with a somewhat higher susceptibility to lung cancer (d'Errico et al. 1996). However, a combination of lack of GST-μ and a mutant of AHH

(CYP1A1[m2/m2]) in the homozygous state, that is very rare in Caucasians, has been reported to be correlated with a markedly enhanced susceptibility (*OR* = 8–22) to smoking-related lung cancers of the Kreyberg I type in males (Kihara et al. 1995). Not only are AHH and GST involved in the activation and deactivation of tobacco derived constituents, but also in the metabolism of various other exogenous as well as endogenous substances (prostaglandins, steroids, hormones, fatty acids). Unfortunately, corresponding data for adenocarcinomas in women seem to be lacking, and the prevalence of CYP1A1[m2/m2] mutant in the Chinese population is not known. However, this type of genetic polymorphism could, nevertheless, contribute to the high background incidence of lung cancer seen in non-smoking Chinese women.

Bias and confounding

General

As an introduction it may be appropriate to cite some words of caution with respect to case–control studies from the authoritative textbook on basic epidemiology by Breslow and Day (1980) published by the WHO International Agency for Research on Cancer:

> The strength of association relates to causality. Relative risks of less than 2.0 may readily reflect some unperceived bias or confounding factor, those over 5.0 are unlikely to do so.

Since the epidemiological studies on ETS in general have given estimates of overall relative risk (*RR*), or odds ratios (*OR*) below 2, it is of primary importance to check for possible sources of bias and confounding. Potential sources of *bias* in an epidemiological study refer mainly to deficiencies in study design and data collection which result in differences with respect to inclusion of cases and controls, or regarding differences in quality of exposure information between the two groups to be compared. An example relevant to this presentation is the inclusion, among cases, of smokers misrepresenting themselves as non-smokers. Bias may also be caused by faulty diagnostic practice.

Confounding relates to causality, and in the case of ETS reflects the presence of other causes than ETS for lung cancer in exposed non-smokers. However, in order to affect the observed associations, the confounder must in some way also be linked to ETS exposure. An investigation of the association between dental disease and risk of coronary heart disease in 9760 subjects provides a good example of confounding (DeStefano et al. 1993). In this study a 25 per cent increased risk of coronary heart disease was found for patients with periodontitis as compared with those with minimal symptoms of this affliction. Obviously, periodontitis can hardly be incriminated as the cause of heart disease. However, it provides evidence of poor dental hygiene which, to a greater or lesser extent, reflects a less healthy lifestyle in general, and that is known to predispose people to coronary heart disease.

Diagnostic misclassification bias

In prospective as well as case–control studies, the use of proper diagnostic methods is essential, and experience also dictates that primary hospital records are often unreliable in this respect. Furthermore, unsatisfactory reporting of the diagnostic methods used, as well as lack of adequate verification of primary diagnosis, will raise doubts as to the quality of the investigation as a whole. Inadequate diagnostic routines, which, for example, lead to the inclusion of secondary tumours, may dilute an association between exposure to ETS and lung cancer, i.e. there will be an underestimation of risk. In a sample of non-smokers consisting of 49 diagnosed primary lung cancers from the US, histological verification showed that about 25 per cent had a primary cancer other than that of the lung (Auerbach et al. 1984). In 15 per cent of US death certificates as well as in the central Swedish cancer registry, cause of death is falsely listed as primary lung cancer.

Histologic (cytologic) verification is also important from another aspect. A large number of investigations have demonstrated that whereas active smoking is only associated with a slight increase in adenocarcinoma and large cell carcinoma of the lung (Kreyberg II), the increase in risk is much higher for other types of tumours like small cell carcinoma and squamous cell carcinoma (Kreyberg I). Thus, if ETS causes an increase in lung tumours in non-smoking individuals at all (and assuming that ETS has similar properties as MS), this increase should be small for adenocarcinomas – if it is detectable – but quite obvious for other tumours of the lung. In Fontham et al. (1991), reassessment of the primary records revealed that 20 per cent of lung cancers were originally misclassified according to histological type. As was done by Fontham et al. (1991), histological review should preferably be carried out by a pathologist specializing in pulmonary pathology.

The degree of misclassification owing to false diagnosis often varies greatly depending on the hospital in question, and provides an indicator of hospital competence. Thus, an extensive multiple regression analysis of cancer incidence in Sweden (Ehrenberg et al. 1985) found that marked geographical differences for lung cancer incidence were correlated – and could to a large part also be explained – with differences in 'diagnostic intensity'. In countries like Greece and China this effect would be expected to be even greater, but it is difficult to evaluate the extent and direction of the bias of such factors with respect to the investigations carried out in the various countries. Histological verification of primary diagnosis was lacking in many of the studies on ETS (see Table 5.2), but great efforts to validate diagnosis were undertaken in the more recent investigations from the US (Fontham et al. 1991, 1994; Brownson et al. 1992; Kabat et al. 1995) where this type of bias must be considered minimal.

Table 5.2. Deficiencies in 13 major epidemiological studies on ETS (study marked with code number) with respect to sample size, selection of controls, diagnosis, exposure verification, confounding owing to occupation, previous lung disease (family history of cancer), exposure to other indoor pollutants, socio-economic factors, intake of vegetables and fruit, intake of saturated fat, and intake of *selenium*.

Study No.														
Deficiencies with respect to	1	2	3	4	5	6	7	8	9	10	11	12	13	14
Sample size*		2			**5**		7	8	**9**		**11**		**13**	
Control selection	1	2			5	6	7		9	10	11	12	13	
Diagnosis		2				6			9	10		12	13	
Exposure verification	1	2		4		6	7				11	12		
Confounding														
Occupation				4					9			12		
Lung disease		2			5	6	7	8	9	10	11	12	13	
Exposure to other indoor pollutants	1	2	3		5	6	7	8	9	10	11	12	13	14
Socio/economic factors		2			5	6	7	8			11			
Fruit and vegetables		2		4	5		7	8		10	11	12	13	14
Saturated fat	1	2	3	4	5	6	7	8	9	10	11	12	13	14
Selenium	1	2	3	4	5	6	7	8	9	10	11	12	13	14

* Bold indicates a sample size of non-smoking of more than 50, but less than 100.

1=Brownson et al. (1992); 2=Correa et al. (1983); 3=Fontham et al. (1991, 1994); 4=Garfinkel et al. (1985); 5=Geng et al. (1987); 6=Hirayama (1981, 1984a, 1984b); 7=Humble et al. (1987); 8=Inoue and Hirayama (1988); 9=Kalandidi et al. (1990); 10=Lam et al. (1987); 11=Pershagen et al. (1987); 12=Stockwell et al. (1992); 13=Trichopoulos et al. (1981, 1983); 14=Kabat et al. (1995)

Exposure misclassification bias

Misclassification of smoking status represents perhaps the most serious source of potential error in the epidemiological studies that have been conducted with respect to ETS. Supported by a flawed methodological approach (Fontham et al. 1991, 1994), US EPA (1992) and OSHA (1994) have assumed misclassification rates of about 1 per cent, which to some undefined degree would be counterbalanced by underestimation of exposure to ETS outside the home. According to the computations of the National Research Council (NRC, 1986), assuming a 30 per cent increase in risk, an 8 per cent misclassification of smokers and ex-smokers would, theoretically, be sufficient to wipe out any significant association between ETS and lung cancer in non-smokers. The Science Policy Research Division of the Congressional Research Service (CRS, 1995) has amply illustrated this fact by statistical sensitivity analysis. At the highest level of exposures, 10 per cent misclassification of smokers as non-smokers will lower the crude *OR* in the Fontham study from 1.87 to 1.0, 'indicating no risk from ETS'. A 3 per cent non-differential misclassification rate would be sufficient to cause this risk level to lose statistical significance.

Misreporting of smoking status, which is influenced by a number of social and cultural factors is, obviously, especially problematic for ex-smokers, since there is no known biomarker of lifetime tobacco use. Past smoking history is important in this context, inasmuch as the increase in relative risk in ex-smokers has been reported to be consistently elevated in comparison with never-smokers. Although it rapidly decreases 10 years after cessation of smoking, the risk for an ex-smoker is still about 25 per cent of that he would face if he continued smoking. Considering that US EPA and OSHA assume a relative risk for ETS-induced lung cancer in the range 1.2–1.4, the inclusion of an appreciable number of ex-smokers among non-smoking cases would, obviously, introduce a considerable upward bias. The extremely low misclassification rates reported by Fontham et al. (1991, 1994) as well as by Riboli et al. (1990, 1995) have been challenged by a number of other investigators, and new data have given ample support to Lee's (1987) claim that the failure to control adequately for misclassification of ex-smokers and active smokers as never-smokers is, in principle, sufficient to provide an explanation for all of the purported association between ETS and lung cancer in non-smoking individuals.

US EPA (1992, p. B-10) has acknowledged the problem of inclusion of ex-smokers misrepresenting themselves as never-smokers, and estimated this misclassification to be 11.7 per cent of the true number of former smokers. However, in order to increase the total number of deaths attributable to ETS, the risk for these ex-smokers was based on 10+ years abstainers with an estimated 9 per cent excess risk of that computed for current smokers. This represents a gross underestimation. Since cotinine data cannot distinguish an active smoker that stopped 2–3 days ago from a true non-smoker, the risk for ex-smokers that misrepresent themselves as never smokers should rather be a weighted average from twice the background risk (very-long-time abstainers) all the way up to the same risk as for a current smoker.

Bias may also be introduced by the occasional misclassification of a non-smoker as smoker. As pointed out by US EPA (1992), *random exposure misclassification* dilutes a causal association. However, for low odds ratios the effect is relatively small, and cannot, for example, be compared with the effect by differential misclassification by inclusion of active smokers among purported non-smokers. Thus, according to Lee (1989), an *OR* of 1.38 is increased to 1.49 at 5 per cent random misclassification, and from 1.29 to 1.49 at 10 per cent random misclassification. Similar computations were made by CRS (1995, p. 38) giving even lower upwards adjustments: if 10 per cent of the cases or controls who state that their spouses smoke actually are not exposed to ETS, a measured *RR* of 1.87 would increase to 1.89; at 20 per cent this would increase to 1.9. In addition, and as demonstrated by Phillips et al. (1994), the subjective assessment of low or no exposure to ETS was in general found to be reliable.

Another type of exposure-related differential bias that is difficult to control for, is caused by the attitude of the investigator, i.e. *interviewer bias*. The finding in a supposedly non-smoking person of any type of lung tumour – including misdiagnosed secondary tumours – may trigger a more thorough search for an external cause, where special attention may be given to the possible exposure to ETS.

Recall bias and selection of controls

One important source of differential classification error with respect to exposure derives from the type of controls chosen. It is well-known that patients (cases) are more inclined than healthy controls to attach importance to exposure with respect to various external factors. Since it has been widely reported that tobacco smoke causes lung cancer, the cases (or next-of-kin) would, obviously, be more prone to highlight exposure to ETS than healthy controls. For this reason it is crucial that the matched controls should also have some serious illness which could possibly be associated with the exposure under investigation. However, many of the investigations on ETS have used healthy, population-based controls. Fontham et al. (1991, 1994) as well as Garfinkel et al. (1985) have chosen one set of controls consisting of colon cancer patients. However, it seems less likely that a patient with colon cancer would implicate tobacco smoke as a cause than is the case for a lung cancer patient, and even the Fontham group (1991) expressed some doubt as to the differences in recall bias which could exist between these two groups. In the case of Kalandidi et al. (1990) and Trichopoulos et al. (1981, 1983), hospital-based controls were chosen which were, obviously, totally inappropriate inasmuch as they were selected from orthopaedic wards. It is hardly likely that any patient will blame current or present exposures to ETS for his broken leg.

It should be noted, however, that the use of hospital-based controls is not entirely without problems. Patients in a hospital may not represent a random sample of the general population. However, in this context where recall bias with respect to smoking status would have such a profound effect, hospital-based controls are to be preferred. Early experience from case-control studies on the association between thrombosis and oral contraceptives has given ample support for this conclusion (Westerholm, B., personal communication).

An objection to the presence of differential classification owing to recall bias in the extensive studies on ETS conducted by the Fontham group would be that the results with colon-cancer controls reported in the first publication (Fontham et al. 1991), and the results using population-based controls, were very similar. However, these two control groups are *not* comparable, inasmuch as only 10 per cent of the information on smoking habits relied upon next-of-kin information for the colon cancer patients, compared with 34 per cent for lung cancer cases (in the follow-up study, information from next-of-kin respondents had increased to 36 per cent of the total number of cases). The quality of information on smoking status in the colon cancer controls was, in fact, almost comparable with that obtained for healthy controls.

The use of cotinine as biomarker of exposure

To avoid bias due to misclassification of active smokers as non-smokers, cotinine in saliva or urine has been used as a biological marker. Based on such data, Fontham et al. (1991, 1994) give estimates for such misclassification rates for cases in the range 0.6–0.8 per cent. When appraising this remarkably low proportion of untrue statements among the cancer cases, attention should be given to the 19 per cent of the living lung cancer cases and 20 per cent of the colon cancer cases in the first study who did *not* provide a urine sample for cotinine determination in connection with the interview. Obviously, failure to provide a urine sample was due either to lack of consent, or inability to provide urine at the time of interview. According to Fontham et al. (1994), about the same percentage of the controls (83 per cent) as of living patients (81 per cent) provided urine samples, and that the verification status from cotinine measurements of controls and cases were similar. This numerical consistency may very well have been coincidental. That 17 per cent of the controls did not provide such samples may not be surprising; these were healthy persons who had first been interviewed by telephone, and for whom delivery of a urine sample to the hospital would simply seem to be an extra nuisance. The underlying reason for about one fifth of the cases not to provide samples must, on the other hand, have been quite different. In some cases, the poor clinical condition of the patient may have been the cause, but it also seems likely that some of the patients were alerted as to the purpose of the urine sampling after the interview: if you are under treatment in a hospital, and have previously given your doctor false information on smoking, you do not want to be caught, especially if you can avoid this simply by not providing urine. After selecting individuals with elevated cotinine levels in urine from hospital centres in different countries, and correlating self-reported current smoking status with the analytical data, Riboli et al. (1990, 1995) cite similar low misclassifications rates. However, no analysis was made of the concordance between self-declared exposure to tobacco smoke in individuals with cotinine values below the defined cut-off value, nor were any data reported for the proportion of individuals who refused to provide urine samples.

There is no doubt that in the majority of cases cotinine determination adequately discriminates between active smokers and non-smokers (Lee, 1987). However, many epidemiologists, NRC (1986), US EPA (1992), as well as OSHA (1994), have all made the cardinal mistake of assuming that determination of cotinine would be sufficiently accurate for use in *this* context, i.e. assuming that misclassification in the range 5–10 per cent does not occur. An inadequate relation between exposure and cotinine levels was found by Coultas et al. (1988), and strikingly poor correlations between ETS, nicotine exposure, and saliva cotinine levels could also be confirmed in a study using personal samplers (Phillips et al. 1994). Furthermore, the half-life of cotinine is short (18–20 h; Willers, 1994) and, consequently, its measurement in (for example) urine can only confirm the smoking status of a person during the last two days prior to sampling. Secondly, cotinine is, obviously, without value for identifying ex-smokers and deceased cases, where interviews remain the only possible alternative. Finally, significant inter-individual differences in the rate of metabolism of nicotine, that may occasionally have profound effects on recorded cotinine levels, have almost universally been disregarded.

The consequences of genetically based variability in cotinine metabolism

Epidemiologists using cotinine as a biological marker seem to have overlooked other conversion products such as *trans*-3'-hydroxycotinine and its glucuronide (Neurath and Pein, 1987; Cholerton et al. 1994), that obviously also must be taken into consideration. This is especially so since cross-reactions with anti-cotinine antiserum may introduce analytical complications (Schepers and Walk, 1988). More important, though, are some recent findings on individual variations and genetic determinants for nicotine metabolism that have given a new perspective on the use of cotinine data in determining smoking status. A most surprising finding in Jarvis and co-workers (1987) cited above was that 9 (4 per cent) self-reported *active* smokers had plasma cotinine levels *below* the established cut-off limit. Similar findings were reported also by Lee (1987). Recent information on inter-individual variations in the metabolism of nicotine provides an explanation for this apparent anomaly. Neurath and Pein (1987) conducted a controlled study of 9 subjects to investigate individual variations in metabolite patterns. Although each of the subjects smoked the same number of cigarettes per day (19) for 6 days, the plasma nicotine concentration varied from 7 to 44 ng/ml (6-fold), the cotinine from 41–344 (8-fold), and the *trans*-3'-hydroxycotinine from 24–160 (7-fold). Although the number and intensity of 'puffs' may vary, this could hardly constitute the main reason for these large inter-individual differences, which are more likely to be due to genetic factors.

A study of genetic polymorphism with respect to the metabolism of nicotine by Cholerton et al. (1994) has given ample evidence that genetic predisposition can, indeed, provide an explanation for the extreme variability seen in Neurath and Pein (1987), and can also explain why some active smokers do not score positive for cotinine. To avoid confounding by individual smoking variables, Cholerton et al. (1994) gave 2 mg nicotine orally by capsule to 124 healthy, unrelated, non-smoking volunteers (80 of which were female). In percentage of the administered dose, the following levels in urine were found: for nicotine, 0.04–12.6, for cotinine, 1.8–28.4, for *trans*-3'-hyxdroxycotinine, 0–71.8, and for *trans*-3'-hydroxycotinine glucuronide, 0–25.4. *Thus, in spite of the fact that each subject had received exactly the same dose of nicotine, there was a 16-fold difference between the highest and the lowest value for urinary excretion of cotinine during a 24 hour period.* For the other metabolites, even larger variations were found, in as much as some subjects did not hydroxylate cotinine at all to produce *trans*-3'-hydroxycotinine, or the associated glucuronide.

The cytochrome P-450 component (the gene product of CYP) of the mixed function oxygenase system of liver is responsible for the conversion of nicotine to cotinine, and it is comprised of a complex of different isoenzymes with varying substrate specificities. Out of these, debrisoquine 4-hydroxylase (the gene product of CYP2D6) is involved in the metabolism of a wide range of nitrogen-containing compounds, including debrisoquine. In the Cholerton et al. study, 118 of the volunteers were tested with polymerase chain reaction (PCR) and restriction fragment length polymorphism (RFLP) methods for the presence of the CYP2D6 wild-type allele as well as mutant alleles which give rise to CYP2D6 polymorphism. Five subjects were homozygous for CYP2D6 mutations, and they were found not only to be poor metabolizers of debrisoquine, but also extremely poor metabolizers of nicotine. Although not all subjects who were homozygous for the CYP2D6 mutations were poor metabolizers (conversion by alternative pathways do exist), the fact remains that at least in 5 per cent of the tested individuals active smoking would not have been detected by urinary cotinine determination. This is in close agreement with the findings of Jarvis et al. (1987) cited above. What can be concluded from these data is that the use of cotinine as a marker may – even when sampling has been conducted in a correct manner – result in the inclusion of active smokers at a level that would have a significant impact on computed misclassification rates.

Haemoglobin adducts as indicators of exposure

Whilst it has not been possible to detect elevations in DNA-adducts from PAH due to exposure to ETS, haemoglobin adducts have proved to be more sensitive in this respect. Thus, high misclassification rates as well as evidence of the inadequacy of cotinine as a marker for exposure have been provided by such measurements. In a group consisting of 61 purported non-smokers, 4 subjects (7 per cent) were excluded as presumptive active smokers based on cotinine measurements (Maclure et al. 1989). Of the remaining subjects, one had 4-aminobiphenyl (ABP) adduct levels in haemoglobin consistent with a consumption of 15–29 cigarettes, and 6 exhibited levels corresponding to smoking 1–14 cigarettes or more per day. Out of these 7 subjects, 4 reported no or low exposures to ETS. Although it cannot be ruled out that other sources of exposure, not related to smoking, could have contributed to the measured adduct levels (e.g. work in rubber industry), in most cases the elevated adduct levels were most probably due to a misclassification that could not be detected by cotinine measurements. Bartsch et al. (1990) also found increased levels of ABP adducts in subjects who reported no exposure to ETS.

The individual variability in the capacity to activate carcinogens in tobacco smoke has been strikingly demonstrated by the measurements in smokers of haemoglobin adducts derived from the tobacco specific nitrosamines 4-(metylnitrosamino)-1-(3-pyridyl)-1-butanone (NNK) and N′-nitrosonornikotin (NNN). Whereas there was a good correlation between such adducts and the dose of NNN or NNK in an inbred strain of rats, only about 20 per cent of 100 smokers had haemoglobin adduct levels higher than those in controls (Hecht et al. 1993).

Recall bias: reliability of self-reported smoking status

In the epidemiological studies on ETS, information on smoking status has, with a few notable exceptions, been derived solely by interview. The unreliability of information on smoking status obtained in this manner has been well documented, especially for previous smoking habits. Among the eight purportedly non-smoking male cases reported by Humble et al. (1987), four later admitted that they were actually smokers (US EPA, 1992); according to a report by Auerbach et al. (1984), upon re-interview of women who were listed as non-smokers in hospital records, 40 per cent admitted that they had actually smoked. NRC (1986, p. 235) cites one study where 4.9 per cent of the subjects said that they had never smoked as much as one cigarette a day in 1982, when in fact they had previously smoked and reported so in interviews. In an extensive study from the UK, reported by Lee (1987), that included 1537 subjects, 10 per cent of whom in 1985 claimed to be never-smokers had, five years earlier, admitted being active smokers or ex-smokers. Even higher frequencies (17 per cent) of inconsistent reporting were found in the WHO MONICA (Multinational Monitoring of Determinants and Trends in Cardiovascular Disease) study in Germany (Wynder, 1993). Lee and Forey (1995, 1996) were able to confirm that very high proportions of smokers later claimed never to have smoked: 11–18 per cent of men and 5–7 per cent of women.

Pershagen et al. (1987) checked for consistency of reported smoking at different times (but no data were presented), and claimed results to 'indicate that misclassification of non-smokers was a minor problem'. That study was quite small and, in addition, it seems rather obvious that smokers who previously denied smoking would be inclined to do so again at a second interview, as amply demonstrated by an investigation by Heller et al. (1993). In this study of 1 343 subjects who reported never having smoked in 1987–88, smoking status was checked in 1984–85 and 1987–88, both by interview as well as by cotinine determination. Twenty-one subjects that were clearly regular smokers, as judged by cotinine sampling, denied smoking at both time points, whereas 1.8 would have been expected if denial of smoking were random (Lee and Forey, 1995, 1996). Thus, there is a clear tendency towards consistent false statements about current smoking habits, and the same would also certainly be expected to be true for past smoking habits, although this cannot be measured directly.

In the Fontham et al. (1994) study, 2 cases (0.6 per cent) and 25 controls (2.3 per cent) had cotinine concentrations 100 ng/mg creatinine or higher and were removed from the study, indicating a higher misclassification rate among controls. However, a much greater similarity between cases and controls was found in the cotinine concentration range 55–99 mg/kg, levels that can hardly be ascribed to ETS exposure, and these subjects should, therefore, also have been excluded. Here 9 cases (2.5 per cent) and 29 controls (2.7 per cent) were identified, i.e. the rates were virtually identical. Fontham et al. (1994) noted the discordance between these misclassification rates and those found for healthy subjects, and cites an EPA source giving a 1 per cent misclassification rate for female lung cancer cases, and 5.7 per cent for healthy individuals.

Several efforts to estimate objectively the reliability of self-reported current smoking status by measuring cotinine in saliva have cast doubt on the validity of the low values reported by Fontham et al. (1991, 1994) and by Riboli et al. (1991, 1995) at IARC. Evidence based on individual differences in nicotine metabolism as well as adduct measurements have already been mentioned, and in the extensive study by Jarvis and co-workers (1987) on verification of smoking status in 215 outpatients attending cardiology and peripheral clinics, 21 self-reported non-smokers (10 per cent) had plasma cotinine levels similar to those of smokers, providing evidence for a much higher systematic underestimation of unreported active smokers in this population. Even higher estimates of misclassification were obtained in two more recent investigations. In a thoroughly-conducted study from the UK on a well-documented group of subjects, out of 327 self-reported non-smoking volunteers, 53 (16 per cent) were identified as likely smokers (Phillips et al. 1994). Similarly, in an investigation from the US where cotinine in urine was measured in a total of 282 purported non-smokers to be enlisted in phase one human clinical trials for new pharmaceutical preparations, 45 (16 per cent) scored at cotinine levels implying that they were, in fact, active smokers (Apseloff et al. 1994). Recent data obtained by Jenkins (1994) of the Oak Ridge National Laboratory indicate a range from 2.5 to 4.6 per cent depending on the classification criteria used for current smokers based on cotinine determination.

A German study (Heller et al., 1993) revealed a significant difference in misclassification rates of current smokers as non-smokers for different age groups. Whereas this rate was 2.9 per cent in subjects aged 25–44, it was three times higher (9.2 per cent) in subjects aged 45–64 years, i.e. within the age bracket when lung cancer will appear. As expected, the proportion of current occasional smokers that reported themselves as non-smokers was very high (26–27 per cent), a finding that has been supported by several other investigations. Also, the proportion of ex-smokers detected as smokers was greater than the fraction of never-smokers detected as smokers.

There are obviously several different reasons for providing inaccurate statements of smoking habits, and there seems to exist certain differences between cases and normal controls with respect to misclassification rates that cannot be wholly ascribed to methodological shortcomings. According to Lee and Forey (1995, 1996) the studies by Apseloff et al. and by Phillips et al. (1994) included paid participants who may have concealed their smoking for financial reasons. In the investigation by Apseloff et al. (screening for phase one clinical trials) such motives may have been involved, but it seems hard to believe that this would also apply to the last-mentioned study. To avoid the possibility of influencing their behaviour, the selected subjects were told that the purpose of the study was to assess indoor air quality.

The differences in misclassification rates based on cotinine determinations between cases and controls brings into focus *possible lifestyle changes – including dietary and smoking habits – induced by the detection of disease* that may introduce a substantial differential bias. A change of dietary habits, limiting the usefulness of food-intake questionnaires in probing the past dietary habits of cancer patients, is to be expected, but the impact on tendencies to misrepresent smoking status is more complex. For some individuals the search for an external cause of the disease may decrease such tendency, but for others the opposite may be true. Before discovery of the disease, the would-be cases will, obviously, be as prone as the rest of the general population to misreport smoking status, and in many cases diagnosis of disease will not restore a person's truthfulness, especially if he is residing in a country where he may have to consider previous statements made with respect to insurance policies with differentiated premiums (CRS 1995). Actually, there is evidence that misclassification rates are much higher among patients who have been advised to quit smoking than among individuals from the general population (Velicer et al. 1992). In comparison with smoking habits that may have existed for a prolonged time period before diagnosis, non-smoking as revealed by cotinine determination after confirmation of disease is obviously less relevant with respect to accumulated cancer risk. While not admitting to have recently been an active smoker, some cases will stop smoking after detection of the disease, a change of habit that will go undetected. Finally, the unexpectedly high proportion of cancer cases that were unwilling to provide urine samples for cotinine determination in the reports from the Fontham group (1991, 1994) strengthens a suspicion that some of these patients were active smokers but wanted, for various reasons, to conceal this fact. Fontham et al. (1994) acknowledged these problems but failed to modify their estimates of misclassification rates accordingly. In the opinion of this author, it would appear more appropriate to use the misclassification rates for 'healthy' controls from the general population rather than the 'apparent' values from cases when adjusting relative risks, or odds ratios.

Some of the highest reported lung cancer risks in non-smoking women derive from Asia and Greece. While US EPA claims (1992, pp. 5–25), that 'the biases are low in East Asia, or in any traditional society as Greece, where female smoking prevalence is low and the female smoker risk is low', the opposite is actually true. In societies where women's smoking is less socially accepted, women who smoke tend to be less candid about such habits. In an investigation by Matsakura et al. (1984) on non-smokers, the urinary cotinine levels were about one-seventh the levels in average Japanese smokers. Since urinary nicotine levels in active smokers typically lie in the range 1 000–3 000 mg/ml, this implies a nicotine level of 140–430 mg/ml in non-smokers. This is far above the cut-off points employed to distinguish smokers from non-smokers in any of the epidemiological studies that have been conducted. Even when taking into account the smaller Japanese rooms, that give higher levels of ETS, this can only account for a part of the elevation in non-smokers. Undoubtedly, a substantial number of the self-reported non-smokers in this study had, in fact, been smokers. Further, Akiba et al. (1986) reported that among 187 men who reported not smoking during the period 1964–1968, 96 (51 per cent) reported in 1982 that they had in fact smoked.

Lee (1995) has recently published data that give ample support for a high misclassification rate in Japanese woman. 400 married Japanese women were selected for the study. Out of 106 with a cotinine/creatinine ratio (*CCR*) indicating current smoking (>100 ng/mg), 22 reported having never smoked. These misclassified smokers had a median *CCR* (1 408 ng/mg) similar to the 78 self-reported current smokers (1 483 ng/mg). Furthermore, among 264 confirmed non-smokers (with *CCR* <100 mg/ng), *CCR* was lower, although not significantly so, if the husband smoked (12 ng/mg) than if the husband did not smoke (18 ng/mg). The response rate was low (33 per cent), and demonstrates the difficulties associated with obtaining this type of information in Japanese society. Assuming, nevertheless, that all of the subjects who refused to participate would provide true statements – which is highly improbable – this would still leave about 7 per cent misclassified subjects using the cut-off level of 100 ng/mg, and 10 per cent when using the cut-off level of 50 ng/mg.

The data cited above seem to indicate that high misclassification rates in Japan will make any conclusions concerning the association between ETS and lung cancer in non-smoking women extremely doubtful, if not invalid. Undoubtedly, similar situations exist in other 'traditional societies' like China, and to a lesser extent also in Greece. Using cotinine determination in saliva, Ewers et al. (1995) compared self-reported smoking with results from 783 immigrant men and 620 women from Cambodia, Laos, and Vietnam residing in Ohio. The misclassification rate was found to be 6 per cent for males, but 62 per cent of the females who were found to be current smokers stated that they did not smoke.

The use of surrogate sources for information

A related problem is the use of surrogate subjects as the only source of exposure information. As lung cancer is so often fatal, differential misclassification between cases and controls has often occurred with respect to vital status. That is, surviving relatives of the deceased 'case' are interviewed about the case's smoking habits. As a consequence, there is a higher percentage of surrogate responders among cases than among controls. Garfinkel et al. (1985) took great care to verify diagnosis and histological data, but the authors heavily relied on proxy respondents (88 per cent) to obtain information on the smoking habits of the cases. An increase in *OR* was here seen for Kreyberg I as well as Kreyberg II tumours. However, any significant association between ETS exposure and lung cancer disappeared when, instead of proxy respondents, the case, or her husband, were used as the source of exposure information. Together with the fact that a marked increase was seen for squamous cell carcinoma, this strongly indicates that significant misclassification of deceased smokers had occurred.

Although the proportion of proxy respondents for lung cancer cases was more than one third of the total (about 36 per cent), Fontham et al. (1994) characterized this fraction as 'small'. With respect to possible smoking misclassification bias, and especially for *past* smoking habits, this does not accurately assess the situation. It is likely that a considerable number of the 241 proxy sources of information – especially in those cases where the next-of-kin was not the spouse – had very little knowledge of the previous smoking habits of the deceased patients. It is not true that the direction of the bias introduced by information on smoking status from proxy respondents cannot be predicted (US EPA, 1982). If, for example, a deceased mother was a non-smoker for many years before her death, common sense dictates that it is far more likely that the children will describe her as a life-long non-smoker than as an ex-smoker.

In the thorough epidemiological investigation recently conducted by the American Health Foundation, that included 41 male and 69 female never-smoking cases, i.e. a total of 110 cases, no increase in risk was found to be associated with ETS exposure (Kabat et al. 1995). The study is much smaller than those of Fontham et al. and Brownson et al. but larger than, for example, those of Kalandidi, Shimizu, Trichopoulos, Pershagen, and Correa. Although the statistical power of resolution was limited, the quality of the study was in many respects superior to most previously conducted studies. Only newly diagnosed patients were included and they were interviewed directly in the collaborating hospital at the time of *initial* diagnosis, i.e. before pathological confirmation was obtained. Thus, misclassification due to the use of proxy respondents was completely avoided. Although cotinine was not used to confirm present smoking status, a discriminating questionnaire was used to characterize each exposure source. The answers would then be checked for internal consistency. Reproducibility was checked by re-interview of 48 subjects. A previous history of lung disease (chronic bronchitis, asthma, emphysema, tuberculosis) was certified. Apart from being a confounder, absence of such ailments in *all* cases and less than 4 per cent among controls could be used to some degree as an instrument of verification, at least of the absence of present (past) moderate to heavy smoking. Instead of only using 'living with a smoker' as basis for classification, much more meaningful qualifiers such as 'spouse smokes in bedroom', were included. Given the limited power of resolution, the authors in particular looked for consistency between different measures of exposure and for evidence of an elevated risk in subgroups that most probably were highly exposed, for example, women married to men smoking in the bedroom, but 'no such associations were found'.

Exposure to ETS outside the home

Exposure to ETS outside the home undoubtedly does occur to an appreciable extent, but Phillips et al. (1994) estimated this source to be less important than exposure at home, a conclusion that has been given support by a large international study by Riboli et al. at IARC (1990), and several other investigators (Morris, 1995). US EPA (1992) claims on the other hand, that 'Background ETS exposure accounts for 2 200 (72 per cent) and spousal smoking for 860 (28 per cent) of the total fatalities due to ETS' (pp. 6–21). Furthermore, most epidemiological investigations assume a significant downward bias introduced by underestimation of exposure to ETS outside the home, for example, at the workplace and during travel, that compensates for upward bias due to misclassification.

Due to the extreme dilution of ETS in comparison with MS, it is hard to imagine that such exposures can noticeably counterbalance significant misclassification of active smokers as non-smokers. Also, it seems highly remarkable that out of some 20 studies reporting outside the home only 2 (Kabat and Wynder, 1984; Fontham et al. 1994) have provided data that give some indication of a positive association, an outcome (10 per cent) that can be attributed to chance. However, even in the case of the two last mentioned studies, the evidence is far from convincing. In its report to US Congress, CRS (1995) noted that:

> OSHA based its risk assessment on a workplace estimate by Fontham et al. which indicated an increased risk and chose not to use the remaining estimates which found no overall association between workplace exposure and lung cancer. Moreover, it assumed that workplace exposure is comparable to residential exposure, though studies that measured cotinine levels in non-smokers suggest that residential and other non-workplace ETS exposure may be more important than workplace exposure.

Kabat and Wynder (1984) include very few cases, and a small, but statistically not significant positive association between exposure at the workplace and lung cancer in non-smokers was found for males but not for females. In the Fontham et al. study (1994), a significantly increased risk with respect to *years of ETS-exposure* was found for a pooled subset of 385 cases, where most of the women were exposed at home as well as at work. When, on the other hand, exposure to ETS at home was expressed in the same manner – i.e. in terms of duration – no statistically significant increase in risk was seen even for the group with the longest exposures. The aggregated group with combined exposures was then compared with controls who were not exposed either at work or at home. One can well imagine the differences in lifestyles between females belonging to these two different categories. Furthermore, the fact that the *OR* for women who were solely exposed at work cannot be deduced is a matter for concern. The initial crude *OR* was only 1.12, but when the data were submitted to multiple regression analysis involving 10 (ten) potential confounders – an exercise that would daunt any expert in mathematical statistics – this odds ratio was increased to 1.39. Citing Butler, the European Working Group on Environmental Tobacco Smoke and Lung Cancer (EWG, 1996) pointed out that:

> Small spurious associations can be generated in such multiple logistic regression analyses if individual confounders are not correctly parameterized. . .Thus, there is no assurance that the adjusted estimate of 1.4 for workplace ETS is not a statistical artefact of their adjustment process.

Another anomaly that tends to underscore a lack of internal consistency in these data is the fact that an even larger *OR* (1.5) was found for ETS exposure during social settings. It is very difficult to believe that exposure to ETS associated with occasional visits to bars etc. over the course of 1–15 years (*OR*=1.45) would have a larger impact on cancer incidence than exposure to tobacco smoke at home during 10–14 hours per day for more than 30 years. In line with the analysis presented below, the last mentioned inconsistency may, on the other hand, very well be explained on the basis of confounding caused by lifestyle-related factors.

Confounding

General aspects

Assuming a similar distribution of histological types of lung tumours for ETS exposure as for active smoking, most of the increase in Kreyberg I tumours, as well as a part of the increase in adenocarcinomas, could be ascribed to misclassification. However, to account for the discrepancy between risk derived by extrapolation from data on active smokers and the epidemiological evidence for the impact of ETS, the role of confounding must be addressed. If factors other than exposure to ETS do explain part of the observed increased risks for lung cancer in exposed non-smokers – i.e. act as confounders – they must in some way be related to ETS exposure. Among the various factors taken into consideration, occupation, previous lung disease, and exposure to other air pollutants may have been serious sources of confounding in some of the smaller studies with few non-smoking cases, and where such factors were not controlled for. It should be noted, for example, that about half of the adenocarcinoma cases in non-smokers in the Wu et al. (1985) study could be 'explained' by previous exposure to coal smoke and lung disease in childhood. Especially in some of the Asian studies, a background of lung disease and exposure to smoke from coal and cooking may also have played a role. Furthermore, it has been known for a long time that disposition to lung cancer is influenced by hereditary disposition (see literature cited by Roots et al. 1992), a fact that was not considered to any significant extent in most of the studies on ETS.

The important role of socio-economic factors

Socio-economic factors of the type that are usually recorded (income, education, occupation) are causally indirectly related to lung cancer, because low social class is often associated with lifestyle factors, such as inappropriate diet with low vitamin intake and high fat intake, a background of lung disease, hereditary disposition, etc., all of which can be more directly linked to lung cancer. A higher prevalence of diseases of various kinds has been associated with low socio-economic status (SES) (Schmitz, 1973; Swedish

Cancer Committee, 1984a; Ehrenberg et al. 1985); some researchers have even associated SES with increased incidence of childhood cancer (Poole and Trichopolous, 1991). Thus, unequal inclusion among cases and controls of small groups of disease-prone individuals from low socio-economic backgrounds may give rise to a considerable differential bias.

In many studies on ETS no matching for socio-economic variables was carried out. However, even when considered, this was done in a superficial manner. In the Fontham et al. (1991, 1994) studies, selection of a major part of the population-based controls was carried out by random digit dialling. As pointed out by Poole and Trichopoulos (1991), persons of very low socio-economic status are difficult to identify, contact, and recruit as controls by this method, but such cases do appear for treatment in the type of hospitals from which the ETS cases were recruited by the Fontham group. Upon closer inspection of their data (Fontham et al. 1994), some interesting differences become obvious: cases had lower education and lower incomes than the controls. Although *averages* were not that disparate, differences become significant for the low-income brackets. There were about 50 per cent more cases than controls with household incomes below $8 000. In addition, around 60 per cent more cases than controls did not even have high school education. For some reason, Fontham et al. (1994) did not provide the number of ETS-exposed individuals in each of these categories. However, there is reason to believe that if the 216 cases (out of 653) and 266 controls (out of 1253) who did not even have a high school education, or if the 103 cases and 144 controls with family incomes below $8 000 were excluded from the study, the difference in total cancer risk between those that were exposed to ETS and those who were not, would be much lower. From the information on income that we are given, it can also be concluded that the cases (or their spouses) belonging to this sub-set (16 per cent) lacked regular employment. The absence of such disease-prone cohorts from investigations on ETS exposure at the workplace may offer an explanation as to why, in spite of considerable efforts, no significant association between exposure to ETS at the workplace and lung cancer has been found.

In Kabat et al. (1995), referred to above, where no association between lung cancer and exposure to ETS could be detected, there were no significant differences in socio-economic status between cases and controls. More importantly, there was also no evidence of the inclusion of a significant proportion of disease-prone cases from the lowest strata of society. 93 per cent of the male cases, and 64 per cent of the female cases were characterized as professional/skilled, and only 5 per cent of the male cases and 7 per cent of the female cases were 'unskilled'. This contrasts markedly with the characteristics of the Fontham cases.

Dietary factors: general aspects

If we wish to obtain reliable information on dietary habits, a superficial questionnaire is clearly not adequate (IARC, 1990). Note also that for the cohort studied by Alavanja et al. (1993), where a striking association between lung cancer and intake of saturated fat was recorded (see below), initially no effect of diet could be detected (Brownson et al. 1992). In the follow-up, however, an elaborate questionnaire was used in the in-home interviews for all cases over a period of 4 years prior to the first diagnosis of lung cancer (a period during which preclinical symptoms of disease should not have affected dietary practices). As mentioned above, diagnosis of cancer may often precipitate drastic changes in dietary habits as prescribed by the physician. It would seem that in the studies reported by Fontham et al. (1994), who found some effect of diet, this information had been collected and analysed in a less than optimal fashion, and that a more adequate approach would most probably have greatly enhanced the knowledge of the extent of confounding by dietary factors in this multi-centre study.

Intake of fruits and vegetables

In a co-operative study between medical institutions in Hong Kong and Karolinska Institute, Stockholm (Koo et al. 1988), the authors note that, '[n]on-smoking families had healthier lifestyles than wives with smoking husbands', thereby bringing into focus what may be the true nature of this type of confounding. Koo et al. (1988) observe, in accordance with the report by Sidney et al. (1989), that dietary carotene intake was lower in non-smokers exposed to passive smoke at home than in non-smokers who were not exposed. Similar findings were presented by LeMarchand et al. (1991).

Despite the fact that data have not always been consistent – mainly because of various methodological inadequacies – there is convincing evidence demonstrating that intake of fruits and vegetables (in particular, carotenoids, which contain vitamin A, and selenium) protects against lung cancer (see literature cited in Fraser et al. 1991, Stähelin et al. 1991, Candelora et al. 1992, and van den Brandt et al. 1993). Of particular interest was the case–control study including 124 cases of histologically confirmed lung cancer in lifetime non-smoking women by Candelora et al. (1992). Data from this investigation indicated that high vegetable intake and intake of carotene in particular had a strong protective effect. In comparison with the lowest quartile of consumption, individuals in the highest quartile of vegetable consumption had the greatest decrease in risk with an *OR* of 0.2 (CI = 0.1–0.5).

In those few ETS studies on lung cancer where dietary data were collected and used for analysis, with the exception of Hirayama's cohort study, all were designed as case–control studies. As such they suffered from the disadvantage that dietary information of reasonable reliability will be limited to the time period shortly before or after diagnosis. In view of the very long latency periods for induction of lung cancer, and given the likely changes in lifestyle following detection of the disease, such information is obviously of little utility. Therefore, it is not surprising that some investigators (Correa et al. 1983; Shimizu et al. 1988) have failed to detect any effect from diet.

Although Hirayama (1981a) stated in his first publication, that he detected no impact from diet in his cohort study, he found a marked protective effect from green–yellow vegetables in his later analysis (Hirayama, 1984a). Both in the Hirayama (1984a) cohort study, as well as in the investigation by Wu et al. (1985), a high intake of vegetables rich in ß-carotene had an apparent protective effect. In the last-mentioned study a RR of 1.5 was associated with a low intake of ß-carotene. Kalandidi et al. (1990) also reported that a high consumption of fruits was inversely related to risk for lung cancer, and Svensson et al. (1989) found a significant protective effect for high intakes of carrots.

Lifestyle and diet: intake of saturated fat

Low intake of fruit and vegetables may in itself constitute a risk factor. However, in Western societies a lack of such dietary constituents is often coupled to an increased intake of animal (saturated) fat (Wynder, 1993; Becker, 1995). Smokers generally have a higher fat consumption and a lower intake of fruits and vegetables. This means that, on average, if you lived with a smoker, the fat consumption of that household would be different from that if you lived in a family that had no smoker in the household (Wynder, 1993). There is long-standing and well-supported evidence, including five case–control studies and one cohort study, which link an increased risk of lung cancer with increased fat intake (see literature cited in Alavanja et al. 1993). Furthermore, experimental studies indicate that dietary fat can act as a promoter after treatment with an initiator (Birt, 1986; Birt and Pour, 1983; Beems and van Beek, 1984).

In an extensive case–control study comprising 429 non-smoking women with lung cancer and 1 021 controls, including an in-depth analysis of the possible role of various dietary factors (Alavanja et al. 1993), passive smoking did *not* affect risk estimates. On the other hand daily intake of fat ($p = 0.02$), saturated fat ($p = 0.0004$), oleic acid ($p = 0.07$), proportion of calories from fat ($p = 0.02$) were univariately associated (trend, $p < 0.10$) with an increasing risk of lung cancer, while the proportion of calories from carbohydrates ($p = 0.09$) was associated with decreased risk. Whereas in previous investigations, cholesterol and total fat consumption appeared to be relatively weak risk factors, this study revealed that they did not have an effect independent of saturated fat. Previously observed modest decreases in risk associated with the consumption of vegetables and fruits could also be confirmed.

In contrast to the studies on ETS, it is extremely unlikely that the observed dose-related association found by Alavanja et al. (1993) in their investigation is due to bias and confounding. The relative risk among non-smoking women observed with increased saturated fat consumption *was more than six-fold greater* for the highest quintile than for the lowest quintile. The effect of fat intake was more pronounced for adenocarcinoma than for other cell types. *For adenocarcinoma there was a 11-fold elevation in risk* in the highest versus

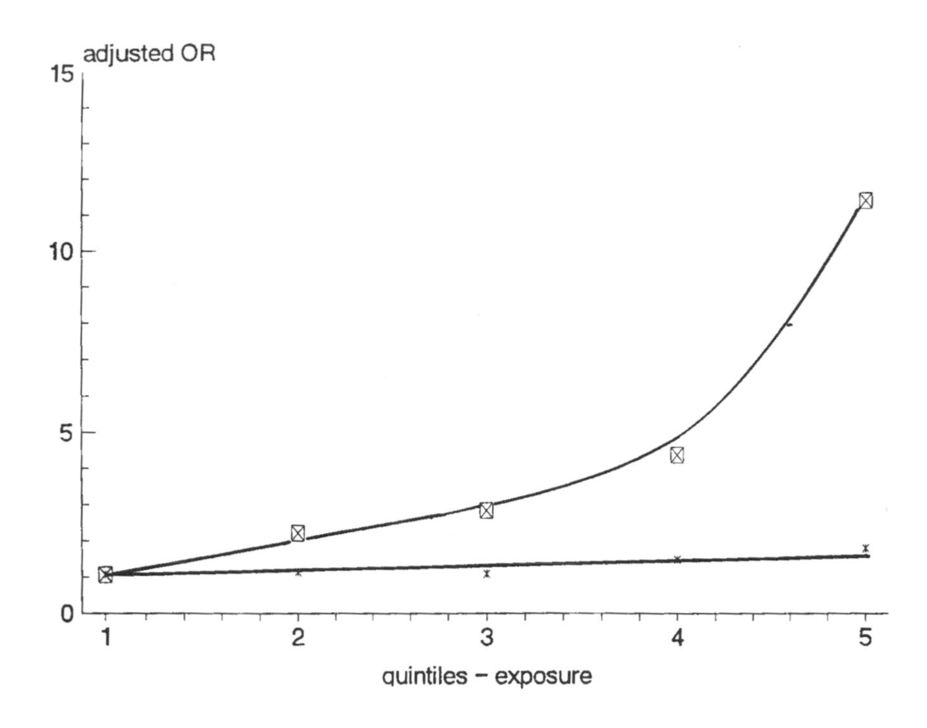

Figure 5.5 Risk of lung cancer associated with ETS (Fontham et al. 1994) and with intake of saturated fat (adenocarcinoma; Alavanja et al. 1993) as a function of exposure

lowest quintiles of saturated fat consumption. In Figure 5.5 the adjusted odds ratios for adenocarcinomas associated with ETS from the latest Fontham et al. (1994) paper is depicted on the same scale together with those from ingestion of unsaturated fat (Alavanja et al. 1993) as a function of increasing exposure. Using this scale of reference, the alleged effect of ETS is hardly visible in comparison with the effect of saturated fat. The graph also clearly shows that the ETS data for *RR* (*OR*) would have little meaning, unless cases and controls have been carefully analysed for intake of saturated fat. Neither saturated fat intake, nor selenium intake, was controlled for in any of the epidemiological studies on ETS.

Even though the 'true' cause for a disease is linked to lifestyle-related factors, correction of confounders like social status or diet often reduces the crude *OR*, but does not eliminate the association. One reason for a limited influence by such corrections is that the measured variables are inaccurate descriptors of the underlying cause. Lee (1989) has provided the following example which is highly relevant in this context. A disease in children is induced by an agent that is inaccurately measured by social class. Misclassification amounts to 20 per cent in both directions (compare with Wynder's (1993) figure of a bias of some 20–30 per cent with respect to caloric intake). Maternal smoking does not affect the disease, but is correlated with the agent. Crude *OR* for association with mother's smoking is 1.31, which is only reduced to 1.19 when the *OR* is standardized with respect to social class. In other words, because the social class as descriptor inaccurately mirrors the causative factor, an apparent association with the mother's smoking is retained even after adjustment for social class (see Tzonou et al. 1986).

In the follow-up by Fontham et al. (1994), high intake of fruits and vegetables and supplemental vitamins was found to be associated with decreased risk for lung cancer, while dietary cholesterol was associated with increased risk. Nevertheless, adjusting for such effects, the association between ETS exposure and an increase in lung cancer in non-smoking women remained. However, the crucial question with respect to the results obtained by these authors, who for some reason did not mention Alavanja et al. (1993), is *how* information on diet was collected and analysed. According to Wynder (1993) of the American Health Foundation, investigations on food intake indicate that the average individual underreports caloric intake by 20 per cent, and for fat intake a downward bias of some 30 per cent is not unusual. In some studies cholesterol in serum was measured (Hole et al. 1989). It should be noted, however, that due to the operation of feedback mechanisms regulating cholesterol uptake and biosynthesis, the levels of this lipid in plasma is often not a reliable index of dietary intake (Brown and Goldstein, 1984), even when sampling is performed in a correct manner.

One objection could possibly be raised with respect to the claim that intake of saturated fat is responsible for a large proportion of the increased incidence of lung cancer seen in ETS exposed non-smoking women. That is: it has been claimed that a high fat diet is also associated with colorectal cancer. If it is true that ETS exposure is linked to a less healthy lifestyle in lung cancer cases (implying more exposure to ETS and a higher intake of saturated fat) relative to healthy controls, a higher proportion of ETS-exposed individuals would then also be expected for the colon cancer group. This did not seem to be the case (Fontham et al. 1991). However, apart from the fact that nothing is known about the actual fat intake in these two groups, a critical review of the literature also reveals that no simple relationship exists between high fat intake and colorectal cancer. This is especially so since the risk of colorectal cancer seems also to be influenced by fibre intake (Dales et al. 1979). It is true, that some case–control studies from Puerto Rico, Canada, the UK and the US have demonstrated an increased risk of colorectal cancer in subjects with a high intake of fat (IARC, 1990). On the other hand, in two case–control studies conducted in France (Macquart-Moulin et al. 1986) and Australia (Potter and McMichael, 1986) no such association was found and in one study in Belgium (Tuyns et al. 1988) a reduced risk was observed in association with high intake of polyunsaturated fat. Although high meat consumption is not equivalent to high fat intake, they are closely associated. Three prospective cohort studies failed to demonstrate an association between high intake of meat and risk of colorectal cancer. In Hirayama's large study, a decreased risk was found in individuals who reported frequent intake of meat (Hirayama, 1981b), and this was also the case for a study by Stemmermann et al. (1984).

In a cohort study of Seventh-Day Adventists, Phillips and Snowdon (1985) did not find any association between meat consumption and risk of colorectal cancer. This cohort is unique, in so far as a large proportion are mostly lacto-vegetarians, providing a wide range of fat intakes. Further, it represents a well-defined group with respect to lifestyle, where for example, confounding by alcohol abuse and tobacco consumption is greatly reduced. For colon cancer there is, finally, a strong hereditary component that in some cases can be associated with the familial adenomatous polyposis gene. Hereditary non-polyposis colorectal cancer, which may be responsible for about 15 per cent of all colon cancer patients (Rumsby and Davies, 1995), is even more important. To be able to assess the effect of diet on colon cancer, these groups must first be properly categorized, which does not seem to have been carried out in any of the epidemiological studies on diet and colon cancer, including the Fontham et al. studies (1991, 1994).

Quantifying exposure

Dose–response relationships

In the ETS studies used for risk assessment by US EPA and OSHA, the husband's smoking has, in general, been used as a surrogate measure for intensity of exposure, and a purported dose–response relationship has been claimed to be one of the strongest arguments for a causal association between ETS exposure and lung cancer. In the Fontham et al. studies (1991, 1994), a significant increase in risk was only associated with heavy smoking by the husband, not with light smoking, and as for other studies, the dose–response relationships presented are not convincing. This would, in fact, not be expected also when accepting the observed increases in cancer risk attributed to ETS as true. In real life, actual exposure will be determined by a number of factors other than the spouse's consumption of tobacco such as number of hours exposed to husband's smoking at home, exposure to other smokers, distance (location) to smoke source, ventilation, etc.

The European Working Group (1996) as well as CRS (1995) have emphasized that the use of statistical tests by which a positive trend was found in some studies is not equivalent to the establishment of a dose–response relationship. CRS points out that (p. 29) 'If there was a threshold effect, then a trend test which included the zero exposure level might show a trend even if an analysis which included only exposures above zero did not show such a trend. In other words a sharp rise at some exposure level above zero could incorrectly be interpreted as a dose–response trend overall exposure levels'. Below, are presented alternatives to the existence of a dose threshold that are related to lifestyle factors in homes where the spouse's cigarette consumption is about two packs per day or more. Finally, unlike the overall results, tests for such trends did not include adjustment for smoker misclassification, whereby the significance for any dose-related trends would have been considerably lower.

A criticism of published epidemiological studies of ETS is that virtually all failed to include a direct measurement of exposure. Coultas et al. (1988) demonstrated the great variability in indoor air quality within homes and among homes. A linear regression model that included variables for the number of smokers in the home, the number of hours of smoking, and the season explained only a small part of the variability of the respirable particle and nicotine concentrations. Furthermore, the number of smokers, number of hours of exposure, the season, and age group could only explain 8 per cent of the variability of the urinary cotinine levels.

In occupational hygiene it is axiomatic that exposure must be verified in an objective manner, preferably by the use of, for example, personal air sampling. In Phillips et al. (1994) cited above, direct measurements of exposure were, in general, in agreement with purported zero or low exposures to ETS. However, for exposures judged by the subjects to be moderate or high, there was a poor correlation with the objective measurements. For 29 per cent of subjects with a smoking partner, exposure to ETS particles was less than the mean exposure level of subjects with a non-smoking partner. The authors concluded that spousal smoking status is definitely not a reliable means to assess the level of ETS exposure. These results are in agreement with a previous investigation, by Coghlin et al. (1989) who made an attempt to validate questionnaires on ETS exposure. The main conclusion was that the response to simple questions that have been used most frequently in epidemiological studies, such as whether a subject lived with a smoker or the number of hours the subject was exposed, were poorly correlated with the measured exposures.

Measurement of adducts in haemoglobin derived from tobacco smoke extends verification of exposure to up to about four months (the lifetime of the human erythrocyte). Attempts in this direction have been made using haemoglobin adducts from ABP (Maclure et al. 1989; Bartsch et al. 1990) and ethylene oxide (Törnqvist et al. 1986). The main problem with ABP seems to be the convex dose–response curve resulting in a lack of linearity between the degree of exposure and adduct level (Bartsch et al. 1990). In the case of ethylene oxide, although adequate to detect active smoking, endogenous background levels as well as other sources of exposure could make measurement of exposure to ETS difficult. Work on finding alternative, more adequate, adduct markers is in progress at Stockholm University.

The phenomenological approach

The rationale behind this method to determine a potential association between lung cancer and ETS in non-smokers has been advocated by Repace and Lowrey (1990). It is based on the fact that certain religious groups, exemplified by Seventh-Day Adventists (SDAs), abstain from smoking and, consequently, misclassification of smoker status as well as exposure to ETS should be very low among this population. In line with this concept a comparison was made between a demographically and educationally matched group of average never-smoking citizens (from the American Cancer Society Study 1960–1971) with a corresponding group of SDAs whereby a lower risk for lung cancer was reported for the SDAs in comparison with the non-SDAs, a difference that was attributed to ETS. For the following reasons this conclusion is not warranted. The comparison referred to by Repace and Lowrey is based on very few cases (25 SDA deaths and 84 non-SDA deaths) and the strength of the association is, therefore, very imprecise (Phillips, et al. 1980). On the other hand, the existence of such a statistical correlation is probably true, but is for several reasons likely to be unrelated to ETS exposure. The rates of misclassification between the two groups are clearly not comparable. The strong religious convictions of the SDAs would most probably result in a more truthful representation of smoking habits for this group. Further, marked differences exist between the two groups with respect to a number of other lifestyle-associated factors, like diet (Phillips et al. 1980; Fraser et al. 1991). This is clearly borne out by the fact that SDAs (both males and females) are significantly healthier in most respects (Table 4, Phillips et al. 1980). Compared with non-SDAs, both male and female SDAs have a significantly lower death rate for all causes of death; they are less prone to develop cerebrovascular disease and less prone to get fatal colon rectal cancer (certainly not smoking related). What is particularly suggestive of a diet-related aetiology is that in addition to a lower rate in colorectal cancer, SDA non-smoking females have a significantly lower risk of developing breast cancer than non-SDA non-smokers. One factor that was not consistent with this hypothesis was that the reduction in coronary heart disease (CHD) was only visible for males, but not for females. CHD is, on the other hand, not only affected by diet, but also by other factors, like regular exercise. Whereas data related to this aspect (e.g. obesity) were available for men, unfortunately, such information was not presented for the SDA women.

Pooling of risk estimates

Quality control of data

Out of a total of 30 studies, US EPA (1992) claimed that 24 showed an association between ETS and lung cancer. Using the Agency's somewhat unusual criteria, this association was considered statistically significant in 9 of those 24 studies. However, to attain statistical significance, US EPA (1992) as well as OSHA (1994) utilized one-tailed statistical analysis at $p<0.1$ (corresponding to 90 per cent confidence), instead at the usual $p<0.05$ (corresponding to 95 per cent confidence). Furthermore, although strict rules apply with respect to comparability of study design and quality when aggregating data in a meta-analysis, this operation was performed – against advice from the US EPA Science Advisory Board (Stolwijk, 1993) – with investigations that differed widely in many crucial respects.

Various definitions of the terms 'smoker', 'ex-smoker', and 'non-smoker' were sometimes used in the different studies, and the efforts to verify smoking status varied from none at all to measurement of cotinine in urine. In the studies by Humble et al. (1987) and Shimizu et al. (1988), non-smokers may actually have had a history of smoking. Trichopoulos et al. (1981, 1983), as well as Kalandidi et al. (1990) classified ex-smokers as non-smokers. Not unexpectedly, in most of the studies reporting the highest apparent increases in risk, little or no effort had been made to validate exposure objectively (for example, Correa et al. 1983; Hirayama, 1981a; Inoue and Hirayama, 1988; Kalandidi et al. 1990; Trichopoulos et al. 1981, 1983; Lam et al. 1987; Geng et al. 1988; Hole et al. 1989).

Likewise, a dependence on a high percentage of surrogate responders for exposure information would also weaken, or invalidate the conclusions, especially when ascertaining ex-smoking status. Very high proportions of 'proxy' respondents were used to derive exposure information for example, in the studies of Akiba et al. (1986), Brownson et al. (1992), Humble et al. (1987), Garfinkel et al. (1985), Pershagen et al. (1987), and Stockwell et al. (1992).

For some studies, common sense would dictate that they be excluded from the start. The case–control study by Akiba et al. (1986) was not characterized as positive but, according to OSHA (1994), it purportedly provided an 'equivocal positive trend' for an association between cancer and exposure to ETS. However, the investigated cohort included survivors from the atomic bomb explosions over Hiroshima and Nagasaki, and was part of a larger study designed primarily to investigate modulation of the effects of ionizing radiation on cancer incidence. Apart from a number of other shortcomings, for example, lack of adequate exposure verification, it is obvious that this population would be less suitable for a study of ETS. Still, US EPA did include the study in the tier 2 – tier 1 having the highest degree of 'utility' for the Agency's assessment. Tier 2 also contained Shimizu et al. (1988), which provided a most remarkable result. Here, no association was observed between the risk of lung cancer and the husband's smoking, or exposure to ETS at work. However, a strong association was found for smoking by the mother as well as by fathers-in-law.

In Table 5.2 (page 116) an overview is made of 13 of the most important studies listed by OSHA (1994) as providing positive evidence for an association between lung cancer and ETS in non-smoking women. The table includes deficiencies (study marked with code number) with respect to sample size, selection of controls, diagnosis, exposure verification, confounding due to occupation, previous lung disease (family history of cancer), exposure to other indoor pollutants, socio-economic factors and intake of vegetables and fruit, intake of saturated fat, as well as intake of selenium. The descriptor exposure verification covers likely misclassification of smokers as non-smokers as well as likely bias due to a high number of proxy respondents. In some of the publications it was noted that one or more of the previously mentioned confounding factors had, in fact, been recorded. However, if there was no specific explanation of how these data were used, and what the conclusions were, the study is, nevertheless, penalized with respect to the factor in question.

US EPA (1982), OSHA (1994), and others have claimed that a statistically significant association between ETS and an increase in lung cancer has consistently been found for populations in Asia, Europe, as well as in the US. This is not correct. Although major problems remain unresolved (the degree of smoker and exposure misclassification, confounding by socio-economic factors, differences in diet, etc.), only the Fontham et al. (1991, 1994) multi-centre studies and the combined studies of Brownson (using a uniform design) seem to include a sufficient sample size and to be of adequate study quality to detect increases in risks in the range of 30–50 per cent. These investigations were conducted in the US.

Aggregation of data – meta-analysis

As mentioned before, strict rules apply to the combination of individual studies in a 'meta-analysis', and this method can only be used when the studies are unbiased as a group and are homogenous. This is not the case for the various studies on ETS, where data from five different geographical regions (USA, China, Japan, Hong Kong, and Europe) have been combined, and where the factors causing bias and confounding are expected to differ greatly. Combining studies in this manner would narrow the confidence interval, and may thereby demonstrate statistical significance for low risk estimates. However, if ten studies are combined that more or less share the same sources of bias, ten times bias will still be bias.

Before aggregating risk estimates from different sources, US EPA made an individual analysis of the quality of each study to be included using certain defined criteria. However, not only did the system employed for quality ranking into four groups feature – according to the Agency's wording – 'a high degree of subjectivity', but it was also inadequate for its purpose. For a study to be at all considered as supportive of either a causal relationship, or of a risk estimate, not all variables can be expected to have equal importance, but certain basic quality criteria must, clearly, always be met. Considering that a very low misclassification rate would suffice to obliterate any statistical association, for a study to be considered at all it is of paramount importance that information on smoking status and exposure to ETS must be considered as reliable. This was not the case; although penalty scores were given for example, for inclusion of smokers, questionable exposure criteria, and the use of more than 10 per cent proxy respondents, unverified smoking status was not penalized; the studies by Fontham et al. (1991) and Pershagen et al.

(1989) were given bonus scores in this respect. Even if a study was penalized for including a high number of former smokers, using questionable exposure criteria, or containing a high proportion of proxy respondents (sometimes a combination of all of these), it could, according to the Agency's scoring system, theoretically still qualify for ranking in the highest tier of quality.

The limitations of epidemiological methodology

Many of the general problems associated with the epidemiological studies cited in this chapter have recently been discussed in the report *Epidemiology Faces Its Limits* that was recently published in the prestigious journal *Science* (Taubes, 1995). Upon being interviewed, Ken Rothman, editor of the journal *Epidemiology*, frankly admitted that 'We're pushing the edge of what can be done with epidemiology'. Epidemiologists conceded that their studies are so plagued with biases, uncertainties, and methodological weaknesses that they may be inherently incapable of accurately discerning weak associations with a purported increase in risk of a magnitude similar to that reported for ETS. For this reason, any attempt to assess risk due to exposure to ETS should be based on a dosimetric approach that exploits data from active smokers as an upper boundary for plausible estimates.

Although it is biologically plausible that ETS has a contributory role in the induction of lung cancer in non-smoking individuals, most of the increase in risk reported in the epidemiological studies linking ETS with lung cancer can be adequately explained by misclassification of smoking status, disproportionate inclusion among cases and controls of groups of disease-prone individuals from low socio-economic background, as well as by certain confounding factors related to lifestyle and diet. Using data from active smokers and actual measurements of ETS exposure by means of personal monitoring, the extrapolated risk levels obtained are compatible with a possible increase in the relative rate of lung cancer that is about one order of magnitude lower than that indicated by the epidemiological studies used to support regulatory action in the US.

Other potentially adverse health effects

The analysis of other potentially deleterious health effects that have been associated with ETS lies outside the scope of this review. It is well established that ETS may induce severe irritation of the respiratory tract in sensitive individuals, as well as exacerbation of symptoms in asthmatics. It is also possible that passive smoking may decrease resistance to respiratory infections in children. However, for effects on the cardiovascular system or on lung function the evidence is even less well substantiated than the association between lung cancer and ETS (Armitage, 1993; Trédaniel et al. 1994; Gori, 1995). Further, to the knowledge of this author, studies on significant adverse effects in experimental animals have only been observed at exposures to mainstream or side-stream smoke at concentrations that are not relevant to the human exposure situation.

Non-scientific considerations

Rather than objectively assessing the possible strengths and shortcomings of the evidence linking ETS to lung cancer, some physicians have chosen to refute critical arguments by maligning well-known and respected scientists who have expressed their doubts on the role of ETS as the hired guns of the tobacco industry (Smith and Phillips, 1996). It should also be a matter of great concern for the scientific community that, when assessing the potential effects of ETS, certain government institutions seem to exclude data which do not support certain preconceived opinions (Gori, 1994, 1995; LeVois and Layard, 1994). No doubt this phenomenon to some degree reflects the general predicament for experts employed in government agencies (and industry) where the pursuit of scientific objectivity is secondary to other goals (Nilsson et al. 1993). One reason for the selective use of data on behalf of the US EPA (1992) can be found on pages 5–14 in its extensive supporting document, where the 'utility' 'for assessing questions related to ETS and lung cancer' is discussed:

> *Note.* Study utility does not mean study quality. Utility is evaluated with respect to research objectives of *this report*, while objectives of individual studies often differ.

Accordingly, Wu-Williams et al. (1990), although characterized by the US EPA reviewer as a 'large and basically well-executed study', is given the lowest possible 'utility' ranking score (4). This study showed no association between ETS and lung cancer (actually, there was a negative correlation), but provided information with respect to plausible alternative causes of lung cancer in non-smokers. Referring to the support document published by OSHA, that has been scathingly criticized for its bias, the Science Policy Research Division of the US Congressional Research Service (CRS, 1995) dryly remarked in its report to Congress that 'The agency may choose to make substantial revisions to the ETS risk assessment before releasing a final regulation'.

Led by a strongly held conviction that tobacco smoke is a major health risk, some scientists and physicians unfortunately seem to have shelved their efforts to analyse the possible effects of ETS in the rigorous and objective manner that scientific method requires. This may prove to be a short-sighted strategy. To those who believe that the end will justify the means in a crusade against tobacco, it should be pointed out that the controversy has yet other serious dimensions. Exposure to tobacco smoke is a major confounder in a large number of epidemiological investigations, and it is therefore extremely important to be able correctly to define the levels of risk involved. Another aspect is that some women, who have virtually no exposure to tobacco smoke (as well as to significant levels of various other types of potentially carcinogenic air pollutants), nevertheless develop lung cancer, mostly adenocarcinoma. As discussed above, dietary/hormonal factors and, possibly, hereditary disposition may be involved. Thus, the one-eyed preoccupation with ETS may prevent us from properly clarifying a truly multifactorial aetiology of lung cancer in non-smoking women.

References

Akiba, S. , Kato, H., and Blot, W. J. (1986). Passive smoking and lung cancer among Japanese women. *Cancer Res.* **46**, 4804–4807.

Alavanja, M. C. R., Brown, C. C., Swanson, C., and Brownson, R. C. (1993). Saturated fat intake and lung cancer risk among non-smoking women in Missouri. *J. Natl. Cancer Inst.* **85**, 1906–1916.

Apseloff, G., Ashton, H. M., Friedman, H., and Gerber, N. (1994). The importance of measuring cotinine levels to identify smokers in clinical trials. *Clin. Pharmacol. Ther.* **56**, 460–462.

Archer, V. E., Wagoner, J. K., and Lundin, F. E. (1973). Uranium mining and cigarette smoking effects on man. *J. Occup. Med.* **15**, 204–211.

Armitage, A. K. (1993). Environmental tobacco smoke and coronary heart disease. *J. Smoking-Related Dis.* **4**, 27–36.

ATSDR (1989), Agency for Toxic Substances and Disease Registry, US Public Health Service. *Toxicological Profile for Bis(chloromethyl)ether*, Atlanta, GA.

Auerbach, O., Garfinkel, L., Parks, V. R., et al. (1984). Histologic type of lung cancer and asbestos exposure. *Cancer* **54**, 3017–3021.

Axelson, O., Dahlgren, E., Jansson, C. -D., and Rehnlund, S. O. (1978). Arsenic exposure and mortality: a case referent study from a Swedish copper smelter. *Br. J. Ind. Med.* **35**, 8–15.

Baetjer, A. M. (1950). Pulmonary carcinoma in chromate workers. I. A review of the literature and report of cases. *A. M. A. Arch. Hyg. Occup. Med.* **2**, 487–504.

Baker, R. R., and Proctor, C. J. (1990). The origins and properties of environmental tobacco smoke. *Environ. Int.* **16**, 231–245.

Bartsch, H., Caporaso, N., Coda, M., Kadlubar, F., Malaveille, C., Skipper, P., Talaska, G., Tannenbaum, S. R., and Vineis, P. (1990). Carcinogen haemoglobin adducts, urinary mutagenicity, and metabolic phenotype in active and passive cigarette smokers. *J. Natl. Cancer Inst.* **82**, 1826–1831.

Becker, W. (1995). Social and regional factors influencing food habits and nutritional intake. *Vår Föda* **47**, 4–12 (in Swedish).

Beems, R. B., and van Beek, L. (1984). Modifying effect of dietary fat on benzo[a]pyrene induced respiratory tract tumours in hamsters. *Carcinogenesis* **5**, 413–417.

Bidstrup, P. L. (1967). Bronchi and lungs – industrial factors. In *The Prevention of Cancer* (Raven, R. W. and Roe, F. J. C., eds.), Butterworths, London, pp. 193–203.

Binkova, B., Lewtas, J., Miskova, I., et al. (1996). Biomarker studies in northern Bohemia. *Env. Health Persp.* **104**, Suppl. 3, 591–597.

Birt, D. F. (1986). Introduction to the effects of neutral fats and fatty acids on carcinogenesis in experimental animals. *Progr. Clin. Biol. Res.* **222**, 249–253.

Birt, D. F., and Pour, P. M. (1983). Increased tumourigenesis induced by N-nitroso-

bis(2-oxopropyl)amine in Syrian golden hamsters fed high-fat diets. *J. Natl. Cancer Inst.* **70**, 1135–1138.
Blot, W. J., Brown, L. M., Pottern, L. M., Stone, B. J., and Fraumeni, J. F. (1983). Lung-cancer among long-term steel workers. *Am. J. Epidemiol.* **117**, 706–716.
Breslow, N. E. and Day, N. E. (1980). *Statistical methods in cancer research. Vol. 1. The analysis of case control studies.* IARC Sci. Publ. No. 32, Lyon, p. 36.
Brown, M. S. and Goldstein, J. L. (1984). How LDL receptors influence cholesterol and atherosclerosis. *Scientific American*, **Dec. 1984**, pp. 67–75.
Brownson, R. C., Alavanja, M. C. R., Hock, E. T., and Loy, T. S. (1992). Passive smoking and lung cancer in non-smoking women. *Am J. Publ. Health* **82**, 1525–1530.
Brunnemann, K. D., Cox, J. E., and Hoffmann, D. (1992). Analysis of tobacco-specific N-nitrosamines in indoor air. *Carcinogenesis* **13**, 2415–2418.
Candelora, E. C., Stockwell, H. G., Armstrong, A. W., and Pinkham, P. A. (1992). Dietary intake and risk of lung cancer in women who never smoked. *Nutrition and Cancer* **17**, 263–270.
Chaudhuri, P. K., Thomas, P. A., Walker, M. J., Briele, H. A. et al. (1982). Steroid receptors in human lung cancer cytosols. *Cancer Lett.* **16**, 327–332.
Cholerton, S., Arpanhal, A., McCracken, N., Boustead, C. et al. (1994). Poor metabolizers of nicotine and CYP2D6 polymorphism. *Lancet* January 1, 62–63.
Coghlin, J., Hammond, S. K. and Gann, P. (1989). Development of epidemiological tools for measuring environmental tobacco smoke exposure. *Am. J. Epidemiol.* **130**, 696–704.
Coleman, M. P., Esteve, J., Damiecki, P. et al. (1993). *Trends in Cancer Incidence and Mortality.* IARC Scientific Publications No. 121, Lyon. pp. 324–327.
Correa, P., Pickle, L. W., Fontham, E. et al. (1983) Passive smoking and lung cancer. *Lancet*, Sept. 10, 595–597.
Coultas, D. B., Howard, C. A., Peake, G. T. (1988). Discrepancies between self-reported and validated cigarette smoking in a community survey of New Mexico Hispanics. *Am. Rev. Respir. Dis.* **137**, 810–814.
CRS (1995). Redhead, C. S., and Rowberg, R. E. *Environmental Tobacco Smoke and Lung Cancer Risk.* Science Policy Research Division, Congressional Research Service. November 14, 1995.
Dales, L. O., Friedman, G. D., Ury, H. K., et al. (1979). A case–control study of relationships of diet and other traits to colorectal cancer in American blacks. *Am. J. Epidemiol.* **109**, 132–144.
Dalhamn, T., Edfors, M. -L., and Rylander, R. (1986a). Mouth absorption of various compounds in cigarette smoke. *Arch. Environm. Health* **16**, 831–835.
Dalhamn, T., Edfors, M. -L., and Rylander, R. (1986b). Mouth absorption of various compounds in cigarette smoke. Retention of cigarette smoke components in human lungs. *Arch. Environm. Health* **17**, 746–748.
Denissenko, M. F., Pao, A., Tang, M. -S., and Pfeifer, G. P. (1996). Preferential formation of benzo[a]pyrene adducts at lung cancer mutational hotspots in p53. *Science* **274**, 430–432.
d'Errico, A., Taioli, E., Chen, X., and Vineis, P. (1996). Genetic metabolic polymorphisms and the risk of cancer: a review of the literature. *Biomarkers* **1**, 149–173.
DeStefano, F., Anda, R. F., Kahn, H. S., Williamson, D. S., and Russel, C. M. (1993). Dental disease and risk of coronary heart disease and mortality. *Brit. Med. J.* **306**, 688–691.
Devesa, S. S., Shaw, G. L., and Blot, W. J. (1991). Changing patterns of lung cancer

incidence by histological type. *Cancer Epidemiol. Biomarkers Prev.* **1**, 29–34,
Doll, R., and Hill, A. B. (1964). Mortality in relation to smoking: Ten years' observations of British doctors. *Brit. J. Med.* **1**, 1399–1410.
Doll, R., and Peto, R. (1978). Cigarette smoking and bronchial carcinoma: Dose and time relationships among regular smokers and lifelong non-smokers. *J. Epidemiol. Commun. Health* **32**, 303–313.
Doll, R., Hill, A. B., and Kreyberg, L. (1957). The significance of cell type in relaxation to the aetiology of lung cancer. *Br. J. Cancer* **11**, 43–48.
Ehrenberg, L., Hiesche, K. D., Osterman-Golkar, S., and Wennberg, I. (1974). Evaluation of genetic risks of alkylating agents: Tissue doses in the mouse from air contaminated with ethylene oxide. *Mutat. Res.* **24**, 83–103.
Ehrenberg, L., von Bahr, B., and Ekman, G. (1985). Register analysis of measures of urbanization and cancer incidence in Sweden. *Env. Int.* **11**, 393–399.
EWG (1996). European Working Group on Environmental Tobacco Smoke and Lung Cancer (Benitez, J. B., Idle, J. R., Krokan, H. E. et al.). *Environmental Tobacco Smoke and Lung Cancer; an Evaluation of the Risk.* Trondheim, April 1996.
Flanjak, J. (1982). Inorganic and organic arsenic in some commercial east Australian Crustacea. *J. Sci. Food Agric.* **33**, 579.
Fishbein, L. (1993). Cooking, heating and air treatment pollutants in indoor environments. In *Environmental Carcinogens – Methods of Analysis and Exposure Measurements. Vol. 12,* Indoor Air (Seifert, B., et al. eds). IARC Sci. Publ. No. 109, WHO, Lyon, 1993, pp. 31–40.
Fontham, E. T., Correa, P., Wu-Williams, A. et al. (1991). Lung cancer in non-smoking women: A multicentre case–control study. *Cancer Epidemiol. Biomarkers and Prevention* **1**, 35–43.
Fontham, E. T., Correa, P., Wu-Williams, A., Reynolds, P., et al. (1994). Environmental tobacco smoke and lung cancer in non-smoking women: A multicentre study. *J. Am. Med. Assoc.* **271**, 1752–1759.
Fowler, B. A. (1983). Arsenical metabolism and toxicity to freshwater and marine species. In *Biological and Environmental Effects of Arsenic* (Fowler, B. A., ed.), Topics in Environmental Health, Vol. 6, Elsevier Amsterdam, 1983. pp. 155–170.
Fraser, G. E., Beeson, W. L., and Phillips, R. L. (1991). Diet and lung cancer in California Seventh-Day Adventists. *Am. J. Epidemiol.* **133**, 683–693.
Fraumeni, J. F., and Mason, T. J. (1974). Cancer mortality among Chinese Americans, 1950–1969. *J. Natl. Cancer Inst.* **52**, 659–665.
Gao, Y. -T., Blot, W. J., Zheng, W. et al. (1987). Lung cancer among Chinese women. *Int. J. Cancer* **40**, 604–609.
Garfinkel, L., Auerbach, O., and Joubert, L. (1985). Involuntary smoking and lung cancer: A case–control study. *J. Natl. Cancer Inst.* **75**, 463–469.
Geng, G. -Y., Liang, Z. H., Zhang, A. Y., and Wu, G. L. (1988). On the relationship between smoking and female lung cancer. In *Smoking and Health* (Aoki, M. et al. eds) Excerpta Medica, Elsevier, Amsterdam, 1988, pp. 483–486.
Goldblatt, E. L., Vandenburgh, A. S., and Marsland, R. A. ; The usual and widespread occurrence of arsenic in well waters of Lane county, Oregon, Oregon Department of Health, 1963.
Gori, G. B. (1994). Policy against science: The case of environmental tobacco smoke. Ann. Meeting Soc. Toxicol., Symposium on Environmental Tobacco Smoke, Dallas, TX.
Gori, G. B. (1995). Passive smoking and coronary heart syndromes. *Regul. Toxicol.*

Pharmacol. **21**, 281–295.
Gough, T. A., Webb, K. S., and Coleman, R. F. (1978). Estimate of the volatile nitrosamine content in UK food. *Nature* **272**, 161.
Green, J. P., and Brophy, P. (1982). Carcinoma of the lung in non-smoking Chinese women. *West. J. Med.* **136**, 291–294.
Hammond, E. C., and Horn, D. (1958). Smoking and death rates – Report on forty-four months of follow-up of 18,783 men. II. Death rates by cause. *J. Am. Med. Assoc.* **166**, 1294–1308.
Hecht, S. S., Carmella, S. G., Foiles, P. G. (1993), Tobacco specific nitrosamine adducts: studies in laboratory animals and humans. *Env. Health Perspect.* **99**, 57–63.
Heller, W. -D., Sonnewald, E., Gostomzyk, J. -G., et al. (1993). Validation of ETS exposure in a representative population in Southern Germany. In *Indoor Air '93. 6th Int. Conf. Indoor Air Quality and Climate*, Helsinki, pp. 361–365
Hiller, F. C. (1984). Deposition of side-stream smoke in the human respiratory tract. *Prev. Med.* **13**, 602–607.
Hinds, W. C. (1978). Size characteristics of cigarette smoke. *Am. Ind. Hyg. Assoc. J.* **39**, 48–54.
Hinds, M. W., Stemmerman, G. N., Yang, H. Y. et al. (1981). Differences in cancer risk from smoking among Japanese, Chinese, and Hawaiian women in Hawaii. *Int. J. Cancer* **27**, 297–302.
Hirayama, T. (1981a). Non-smoking wives of heavy smokers have a high risk of lung cancer: a study in Japan. *Brit. Med. J.* **282**, 183–185.
Hirayama, T. (1981b). A large-scale cohort study on the relationship between diet and selected cancers of the digestive organs. In *Gastrointestinal Cancer, Endogenous Factors* (Bruce, W. R., Correa, P., Lipkin, M., and Tannenbaum, S. R., eds), Banbury Report No. 7, Cold Spring Harbor, NY, CSH Press, pp. 409–429.
Hirayama, T. (1984a). Lung cancer in Japan: Effects of nutrition and passive smoking. In *Lung Cancer: Causes and Prevention* (Mizell, M., and Correa, P., eds), Verlag Chemie International, New York, pp. 175–195.
Hirayama, T. (1984b). Cancer mortality in non-smoking women with smoking husbands based on a large-scale cohort study in Japan. *Prev. Med.* **13**, 680–690.
Hoffmann, D., Melikian, A. A., and Wynder, E. L. (1996). Scientific challenges in environmental carcinogenesis. *Prev. Med.* **25**, 14–22.
Hole, D. J., Gillis, C. R., Chopra, C., and Hawthorne, V. M. (1989). Passive smoking and cardiorespiratory health in a general population in the west of Scotland. *Brit. Med. J.* **299**, 423–427.
Hong Kong Cancer Registry (1982): *Cancer Incidence in Hong Kong, 1982*. Hong Kong Government's Medical and Health Department Institute of Radiology and Oncology, Queen Elisabeth Hospital, Kowloon, Hong Kong.
Horacek, J., Placek, V., Sevic, J. (1977). Histological types of bronchogenic cancer in relation to different conditions of radiation exposure. *Cancer* **40**, 832–835.
Hueper, W. C. (1966). Occupational and environmental cancers of the respiratory system. Recent Results. *Cancer Res.* **3**, 1–214.
Humble, C. G., Samet, J. M., and Pathak, D. R. (1987). Marriage to a smoker and lung cancer risk. *Am. J. Publ. Health* **77**, 598–602.
IARC (1974). International Agency for Research on Cancer. *IARC Monographs on the Evaluation of the Carcinogenic Risks to Humans – Some aromatic amines, hydrazine and related substances, N-nitroso compounds and miscellaneous alkylating agents.* Vol. 4, WHO, Lyon, pp. 231–245.

IARC (1983). International Agency for Research on Cancer. *IARC Monographs on the Evaluation of the Carcinogenic Risks to Humans – Polynuclear Aromatic Compounds. Part 1. Chemical, Environmental and Experimental Data.* Vol. 32, WHO, Lyon.

IARC (1985). International Agency for Research on Cancer. *IARC Monographs on the Evaluation of the Carcinogenic Risks to Humans – Tobacco Smoking.* Vol. 37, WHO, Lyon.

IARC (1990), World Health Organization International Agency for Research on Cancer, *Cancer: Causes Occurrence and Control,* (Tomatis, L. et al. eds), Chapter 12, Diet, IARC Scientific Publ. No. 100, Lyon, pp. 201–228.

Inoue, R., and Hirayama, T. (1988). Passive smoking and lung cancer in women. In *Smoking and Health* (Aoki, M. et al. eds) Excerpta Medica, Elsevier, Amsterdam, pp. 283–285.

Jarvis, M. J., Tunstall-Pedoe, H., Feyerabend, C. et al. (1987). Comparison of tests used to distinguish smokers from non-smokers. *Am. J. Publ. Health* **77**, 1435–1438.

Jenkins, R. A. (1994). Addendum to comments on Proposed Rule-making Occupational Safety and Health Administration 29CFR parts 1910, 1915, 1926, and 1928. Indoor Air Quality; Proposed Rule, Oak Ridge National Laboratory, December 22, 1994, p. 30. cited in CRS (1995), p. 40.

Kabat, G. C., and Wynder, E. (1984) Lung cancer in non-smokers. *Cancer* **53**, 1214–1221.

Kabat, G. C., Stellman, S. D., and Wynder, E. (1995). Relation between exposure to environmental tobacco smoke and lung cancer in lifetime non-smokers. *Am. J. Epidemiol.* **142**, 141–147.

Kalandidi, A., Katsonuyanni, K., Voropoulou, N., Bastas, G., Saracci, R., and Trichopoulos , D. (1990). Passive smoking and diet in the aetiology of lung cancer among non-smokers. *Cancer Causes and Control* **1**, 15–21.

Kang, D. H., Rothman, N., Poirier, M. C., et al. (1995). Inter-individual differences in the concentration of 1-hydroxypyrene-glucuronide in urine and polycyclic aromatic hydrocarbon-DNA-adducts in peripheral white blood cells after charbroiled beef consumption. *Carcinogenesis* **16**, 1079–1085.

Kawai, M., Amamoto, H., and Harada, K. (1967). Epidemiologic study of occupational lung cancer. *Arch. Environm. Health* **14**, 859 864.

Kihara, M., Kihara, M., and Noda, K. (1995). Risk of smoking for squamous and small cell carcinomas of the lung modulated by combinations of CYP1A1 and GSTM1 gene polymorphism in a Japanese population. *Carcinogenesis* **16**, 2331–2336.

Koo, L. C., Ho, J. H. -C., and Rylander, R. (1988). Life-history correlates of environmental tobacco smoke: A study on non-smoking Hong Kong Chinese wives with smoking versus non-smoking husbands. *Soc. Sci. Med.* **26**, 751–760.

Lam, T. H., Kung, C. M., Wong, C. M. et al. (1987). Smoking, passive smoking and histological types in lung cancer in Hong Kong Chinese women. *Brit. J. Cancer* **56**, 673–678.

Larsson, B. K., Sahlberg, G. P., Erikssson, A. T. et al. (1983). Polycyclic aromatic hydrocarbons in grilled food. *J. Agric. Food Chem.* **31**, 867–873.

Law, C. H., Day, N. E., Shanmugaratnam, K. (1976). Incidence rates of specific histological types of lung cancer in Singapore Chinese dialect groups, and their aetioxlogical significance. *Int. J. Cancer* **17**, 304–309.

Lee, C. K., Brown, B. G., Reed, B. A., et al. (1992). Fourteen-day inhalation study in rats, using aged and diluted side-stream smoke from a reference cigarette. II. DNA-adducts and alveolar macrophage cytogenetics. *Fund. Appl Toxicol.* **19**, 141–146.

Lee, C. K., Brown, B. G., Reed, B. A., et al. (1993). Ninety-day inhalation study in rats, using aged and diluted side-stream smoke from a reference cigarette. II. DNA-

adducts and alveolar macrophage cytogenetics. *Fund. Appl Toxicol.* **22**, 393–401.
Lee, P. N. (1987). Passive smoking and lung cancer association: A result of bias? *Human Toxicol.* **6**, 517–524.
Lee, P. N. (1989). Problems in Interpreting Epidemiological Data. In *Assessment of Inhalation Hazards* (Mohr, U. ed.), Springer Verlag, Berlin, pp. 49–59.
Lee, P. N. (1992). *Environmental Tobacco Smoke and Mortality*, Karger, Basel.
Lee, P. N. (1995). 'Marriage to a smoker' may not be a valid marker of exposure in studies relating environmental tobacco smoke to risk of lung cancer in Japanese non-smoking women. *Int. Arch. Occup. Environm. Health* **67**, 287–294).
Lee, P. N., and Forey, B. A. (1995). Misclassification of smoking habits as determined by cotinine or by repeated self-report – A summary of evidence from 42 studies. *J. Smoking-Related Dis.* **6**, 109–129.
Lee, P. N., and Forey, B. A. (1996). Misclassification of smoking habits as a source of bias in the study of environmental tobacco smoke and lung cancer. *Statistics in Medicine* **15**, 581–605.
Lee, C. K., Brown, B. G., Reed, E. A. et al. (1993). Ninety-day inhalation study in rats, using aged and diluted side-stream smoke from a reference cigarette: DNA-adducts and alveolar macrophage cytogenetics. *Fund. Appl. Toxicol.* **20**, 393-401.
LeMarchand, LK., L., Wilkins, L. R., Hankin, J. H., and Haley, N. J. (1991). Dietary patterns of female non-smokers with and without exposure to environmental tobacco smoke. *Cancer Causes Control* **2**, 11–16.
LeVois, M. E., and Layard, M. W. (1994). Inconsistency between workplace and spousal studies of environmental tobacco smoke and lung cancer. *Regul. Toxicol Pharmacol.* **19**, 309–316.
Lubin, J., and Blot, W. J. (1984). Assessment of lung cancer risk factors by histologic category. *J. Natl. Cancer Inst.* **73**, 383–398,
MacLennan, R., Da Costa, J., Day, N. E., et al. (1977). Risk factors for lung cancer in Singapore Chinese, a population with high female incidence rates. *Int. J. Cancer* **21**, 854–860.
Maclure, M., Ben-Abraham Katz, R., Bryant, M. S. et al. (1989). Elevated blood levels of carcinogens in passive smokers. *Am. J. Publ. Health* **79**, 1381–1384.
Macquart-Moulin, G., Riboli, E., Cornee, J. et al. (1987). Colorectal polyps and diet: a case–control study in Marseilles. *Int. J. Cancer* **40**, 179–188.
Matsakura et al. (1984). Effects of environmental tobacco smoke on urinary cotinine excretion in non-smokers. *New Engl. J. Med.* **311**, 828–832.
McCabe, L. J., Symons, J. M. Lee, R. D., and Robeck, G. G. (1970) Survey of community water supply systems, *J. Am. Water Works Assoc.* **62**, 670–687.
Morris, P. D. (1995). Re: 'Inconsistency between workplace and spousal studies of environmental tobacco smoke and lung cancer. *Regul. Toxicol. Pharmacol.* **22**, 292–293.
Morse, D. L., Harrington, J. M., Houseworth, J., et al. (1979). arsenic exposure in multiple environmental media in children near a smelter. *Clin. Toxicol.* **14**, 389–399.
Mumford, J. L., He, X. Z., Chapman, R. S., Cao, S. R. et al. (1987). Lung cancer and indoor pollution in Xuan Wei, China. *Science* **235**, 217–220.
Munro, I. C. (1976). Naturally occurring toxicants in foods and their significance. *Clin. Toxicol.* **9**, 647–663.
Neurath, G. B. and Pein, F. G. (1987). Gas chromatographic determination of trans-3'-hydroxycotinine, major metabolite of nicotine in smokers. *J. Chromatogr.* **415**, 400–406.
Newman, J. A., Archer, V. E., Saccomanno, G. et al. (1976). Occupational

carcinogenesis. Histological types of bronchogenic carcinoma among members of copper mining and smelting communities. *Ann. N. Y. Acad. Sci.* **271**, 260–268.
Nilsson, R., Tasheva, M., Jaeger, B. (1993). Why different regulatory decisions when the scientific information base is similar? Part I – Human risk assessment. *Regul. Toxicol. Pharmacol.* **17**, 292–332.
Nilsson, R. (1996). Environmental tobacco smoke and lung cancer: a reappraisal. *Ecotoxixcol. Env. Safety* **34**, 2–17.
Norin, H., and Vahter, M. (1984). Organic arsenic compounds in fish, Report to the National Swedish Environment Protection Board, SNV PM 1892, 1984.
NRC (1986), National Research Council, *Environmental Tobacco Smoke.* Natl. Acad. Press, Washington, DC.
OSHA (1994). Occupational Safety and Health Administration, 29 CFR Parts 1910, 1915, 1926, and 1928; Indoor Air Quality; Proposed Rule. *Fed. Reg.* Vol. **59**, No. 65, April 5, pp. 15968–16039.
Osterman-Golkar, S., Ehrenberg, L., Segerbäck, D., and Hällström, I. (1976). Evaluation of genetic risks of alkylating agents: II Haemoglobin as a dose monitor. *Mutat. Res.* **34**, 1–10.
Pershagen, G., Hrubec, Z., and Svensson, C. (1987). Passive smoking and lung cancer in Swedish women. *Am J. Epidemiol.* **125**, 17–24.
Phillips, D. J. H. (1994). The chemical forms of arsenic in aquatic organisms and their interrelationships. In *Arsenic in the Environment. Part I, Cycling and Characterization* (Nriagu, J. O., ed.), pp. 263-288. John Wiley.
Phillips, K., Howard, D. A., Browne, D., and Lewsley, J. M. (1994). Assessment of personal exposures to environmental tobacco smoke in British non-smokers. *Environ. Int.* **20**, 693–712.
Phillips, K., Bentley, M. C., Havard, D. A., and Alvan, G. (1996). Assessment of air quality in Stockholm by personal monitoring of non-smokers for respirable suspended particles and environmental tobacco smoke. *Scand. J. Work Environm. Health.* **22**, Suppl. 1; 24 p.
Phillips, R. L., and Snowdon, D. A. (1985). Dietary relationships with fatal colorectal cancer among Seventh-Day Adventists. *J. Natl. Cancer Inst.* **74**, 307–317.
Phillips, R. L., Garfinkel, L., Kuzuma, J. W. et al. (1980). Mortality among California Seventh-Day Adventists for selected cancer sites. *J. Natl. Cancer Inst.* **65**, 1097–1107.
Poirier, M. C., and Weston, A. (1991). DNA adduct determination in humans. *Prog. Clin. Biol. Res.* **374**, 205–218.
Poole, C., and Trichopoulos , D. (1991). Extremely low-frequency electric and magnetic fields and cancer. *Cancer Causes and Control* **2**, 267–276.
Potter, J. D., and McMichael, A. J. (1986). Diet and cancer of the colon and rectum: a case–control study. *J. Natl. Cancer Inst.* **76**, 557–569.
Qu, S. -X., and Stacey, N. H. (1966). Formation and persistence of DNA-adducts in different target tissues of rats after multiple administration of benzo(a)pyrene. *Carcinogenesis* **17**, 53–59.
Repace, J. L., and Lowrey, A. H. (1985). Risk assessment methodologies for passive smoking-induced lung cancer. *Risk Analysis* **10**, 27–37.
Riboli, E., Preston-Martin, S., Saracci, R., et al. (1990). Exposure of non-smoking women to environmental tobacco smoke: a 10-country collaborative study. *Cancer Causes Control* **1**, 243–252.
Riboli, E., Haley, N. J., Trédaniel, J., Saracci, R., Preston-Martin, S., and Trichopoulos, D. (1995). Misclassification of smoking status among women in relation to exposure

to environmental tobacco smoke. *Eur. Respir. J.* **8**, 285–290.
Roots, I., Brockmöller, J., Drakoulis, N., and Loddenkemper, R. (1992). Mutant genes of cytochrome P-450IID6, glutathione S-transferase class Mu, and arylamine N-acetyltransferase in lung cancer patients. *Clin. Investigator* **70**, 307–319.
Rumsby, P., and Davies, M. (1995). Genetic events in the development of colon cancer. *Fd. Chem. Toxicol.* **33**, 328–330.
Saccomanno, G., Archer, V. E., Auerbach, O. et al. (1971). Histologic types of lung cancer among uranium miners. *Cancer* **27**, 515–523.
Schepers, G., and Walk, R. A. (1988). Cotinine determination by immunoassays may be influenced by other nicotine metabolites. *Arch. Toxicol.* **62**, 395–397.
Schmitz, W. (1973). Social factors in the occurrence of cancer. *Öff. Gesundh.-Wesen (Stuttgart)* **35**, 289–307.
Segerbäck, D., Osterman-Golkar, S., Molholt, B., and Nilsson, R. (1994) In vivo tissue dosimetry as a basis for cross-species extrapolation in cancer risk assessment of propylene oxide. *Regulatory Toxicol. Pharmacol.* **20** 1–14.
Shimizu, H., Morishita, M., Mizuno, K., et al. (1988). A case–control study of lung cancer in non-smoking women. *Tohoku J. Exp. Med.* **154**, 389–397.
Shinagawa, A., Shiomi, K., Yamanaka, H., and Kikuchi, T. (1983). Selective determination of inorganic arsenic (III), (V), and organic arsenic in marine organisms. *Bull. Japanese Soc. Sci. Fisheries*, **49**, 75.
Sidney, S., Caan, B., and Friedman, G. (1989). Dietary intake of carotene in non-smokers with and without passive smoking. *Am. J. Epidemiol.* **129**, 1305–1309.
Smith, G. D., and Phillips, A. N. (1996). Passive smoking and health: should we believe Philip Morris' 'experts'? *Brit. Med. J.* **313**, 929–933.
Sonnenfeld, G., and Wilson, D. M. (1987). The effect of smoke age and dilution on the cytotoxicity of side-stream (passive) smoke. *Toxicol. Lett.* **35**, 89–94.
Stemmermann, G. N., Nomura, A. M. Y., and Heilbrun, L. K. (1984). Dietary fat and the risk of colorectal cancer. *Cancer Res.* **44**, 4633–4637.
Stähelin, H. B., Gey, K. F., Eichholzer, M. et al. (1991). Plasma antioxidant vitamins and subsequent cancer mortality in the 12-year follow-up of the prospective Basel study. *Am. J. Epidemiol.* **133**, 766–775.
Stockwell, H. G., Goldman, A. L., Lyman, G H. et al. (1992). Environmental tobacco smoke and lung cancer in non-smoking women. *J. Natl. Cancer Inst.* **84**, 1417–1422.
Stolwijk, J. (1993). EPA Science Advisory Board Experience. 1993 Toxicology Forum Summer Meeting, Aspen, CO. Toxicology Forum, Washington, DC
Strickland, P. T., and Rothman, N. R. (1995). Molecular dosimetry of polycyclic aromatic hydrocarbons. In *Biomarkers of Environmental Exposure* (Haley, N. J., and Hoffmann, D., eds), CRC Press, Boca Raton.
Sunderman, F. W. (1973). The current status of nickel carcinogenesis. *Ann. Clin. Lab. Sci.* **3**, 156–180.
Sunderman, F. W., Morgan, L. G., Andersen, A., Ashley, D., and Forouhar, F. A. (1989). Histopathology of sinonasal and lung cancers in nickel refinery workers. *Ann. Clin. Lab. Sci.* **19**, 44–50.
Svensson, C., Pershagen, G., and Klominek, J. (1989) Smoking and passive smoking in relation to lung cancer in women. *Acta Oncol.* **28**, 623–629.
Swedish Cancer Committee (1984a). Cancer – Causes and Prevention, Report to the Ministry of Social Affairs from the Cancer Committee. SOU 1984:67, Stockholm, p. 238. (English translation: Taylor and Francis, London, 1992)
Swedish Cancer Committee (1984b). Cancer – Causes and Prevention, Report to the Ministry of Social Affairs from the Cancer Committee. Ehrenberg, L. Estimation of

Cancer risk due to passive smoking. SOU 1984:67, Appendix DsS 1984:5, Stockholm.
Taubes, G. (1995). Epidemiology faces its limits. *Science* **269**, 164–169.
Törnqvist, M., Osterman-Golkar, S., Kautiainen, A. et al. (1986). Tissue doses of ethylene oxide in cigarette smokers determined from adduct levels in haemoglobin. *Carcinogenesis* **7**, 1519–1521.
Törnqvist, M., Kautiainen, A., Gatz, R. N., and Ehrenberg, L. (1988). Haemoglobin adducts in animals exposed to gasoline and diesel exhausts. 1. Alkenes. *J. Appl. Toxicol.* **8**, 159–170.
Törnqvist, M., and Ehrenberg, L. (1990). Approaches to risk assessment of automotive engine exhausts. In *Complex Mixtures and Cancer Risk* (Vainio, H., Sorsa, M. and McMichael, A. J., eds). IARC Sci. Publ. 104, 277–287.
Torsteinson, T. (1969). Polycyclic hydrocarbons in commercially and home-smoked food in Iceland. *Cancer Lett.* **23**, 455–457.
Toth, B., and Erickson, J. (1986). Cancer induction in mice by feeding of the uncooked cultivated mushroom of commerce *Agaricus bisporus, Cancer Res.* **46**, 4007–4011.
Trédaniel, J., Boffetta, P., Saracci, R., and Hirsch, A. (1994). Exposure to environmental tobacco smoke and adult non-neoplastic diseases. *Eur. Respir. J.* **7**, 173–185.
Trichopoulos , D., Kalandidi, A., Sparros, L., and MacMahon, B. (1981). Lung cancer and passive smoking. *Int. J. Cancer* **27**, 1–4.
Trichopoulos , D. Kalandidi, A., and Sparros, L. (1983). Lung cancer and passive smoking. *Lancet* Sept. 17, 677–678.
Tuyns, A. J., Kaaks, R., and Haelterman, M. (1988). Colorectal cancer and the consumption of foods: a case–control study in Belgium. Nutr. *Cancer* **11**, 189–204.
Tzonou, A., Kaldor, K., Smith, P. G. et al. (1986). Misclassification in case–control studies with two dichotomous risk factors. *Rev. Epidemiol. Santé Publique* **34**, 10–17.
US EPA (1986). Risk Assessment Forum; Special Report on Ingested Inorganic Arsenic and Certain Human Health Effects, Washington, DC
US EPA (1992). Respiratory Health Effects of Passive Smoking: Lung Cancer and Other Disorders. EPA/600/6–90/006F, Washington, DC, 1992.
USDHHS (1982). US Department of Health and Human Services. The health consequences of smoking: Cancer. A report of the Surgeon General. USDHHS Public Health Service, Washington, DC
USDHHS (1986). US Department of Health and Human Services. The health consequences of involuntary smoking A report of the Surgeon General. USDHHS Public Health Service, Washington, DC DHHS Publ. No. (PHS) 87–8398.
van den Brandt, P. A., Goldbohm, R. A., van't Veer, P. et al. (1993). A prospective cohort study on selenium status and the risk of lung cancer. *Cancer Res.* **53**, 4860–4865.
van Maanen, J. M. S., Moonen, E. J. C., Maas, L. M. et al. (1994). Formation of aromatic DNA-adducts in white blood cells in relation to urinary excretion of 1-hydroxypyrene during consumption of grilled meat. *Carcinogenesis* **15**, 2263–2268.
Velicer, W. F., Prochaska, J. O., Rossi, J. S., and Snow, M. G. (1992). Assessing outcome in smoking cessation studies. *Psychological Bull.* **111**, 23–41.
Warner, M. L., Moore, L. E., Smith, M. T. et al. (1994). Increased micronuclei in exfoliated bladder cells of individuals who chronically ingest arsenic-contaminated water in Nevada. *Cancer Epidemiol. biomarkers & Prevention* **3**, 583–590.
Wewers, M. A., Dhatt, R. K., Moeshberger, M. L. et al. (1995). Misclassification of smoking status among Southeast Asian adult immigrants. *Am. J. Respir. Crit. Med. Care*, **152**, 1917–1921.)
WHO (1981): International Programme on Chemical Safety; Environmental Health

Criteria No. 18, Arsenic, WHO, Geneva.
WHO (1993). *Guidelines for Drinking-Water Quality*, 2nd edn, Vol. 1, Recommendations. WHO, Geneva, p. 41.
Willers, S. (1994). Environmental tobacco smoke: Cotinine in urine as a biomarker and some effects. Doctoral thesis, University of Lund, Sweden.
Wilson, N. K., Chuang, J. C. (1989). Indoor levels of PAH and related compounds in an eight-home pilot study. *Proc. Eleventh Int. Symp. Polynuclear Aromatic Hydrocarbons*, Gaithersburg, MD, Sept. 1987, Gordon & Breach, New York.
Wright, A. S., Bradshaw, T. K., and Watson, W. P. (1988). Prospective detection and assessment of genotoxic hazards: A critical appreciation of the contribution of L. Ehrenberg. In *Methods for Detecting DNA Damaging Agents in Humans: Applications in Cancer Epidemiology and Prevention* (Bartsch, H., Hemminki, K. and O'Neill, I. K., eds), IARC Sci. Publ. 89, 237–248.
Wu, A. H., Henderson, B. E., Pike, M. C., and Yu, M. C. (1985). Smoking and other risk factors for lung cancer in women. *J. Natl. Cancer Inst.* **74**, 747–751.
Wu-Williams, A. H., Dai, X. D., Blot, W. et al. (1990). Lung cancer among women in north-east China. *Brit. J. Cancer* **62**, 982–987.
Wynder, E. L., and Hoffmann, D. (1967). *Tobacco and Tobacco Smoke: Studies in Experimental Carcinogenesis*. Academic Press. New York.
Wynder, E. L., and Goodman, M. T. (1983). Smoking and lung cancer: Some unresolved issues. *Epidemiol. Revs*. 5, 177–207.
Wynder, E. L. (1993). Environmental tobacco smoke. 1993 Toxicology Forum Summer Meeting, Aspen, CO. Toxicology Forum, Washington, DC
Yamada, A. (1963). On the late injuries following occupational inhalation of mustard gas, with special reference to carcinoma of the respiratory tract. *Acta Pathol. Jap.* **13**, 131–155.

6 Beneficial ionizing radiation

Zbigniew Jaworowski

Summary

Contrary to the prevailing wisdom, human exposure to ionizing radiation – from whatever source, be it nuclear fallout or x-ray diagnosis – exhibits a threshold effect: below a certain exposure level, impacts are no longer detrimental. Moreover, significant evidence supports the hypothesis that low doses of radiation are actually beneficial. This is known as the hormetic effect.

This conventional wisdom on thresholds for radiation and hormesis owes itself largely to the International Commission on Radiological Protection (ICRP) which has, since 1959, dictated that regulations be based on an assumption that the impact of low dose radiation may be derived by linear extrapolation from high dose radiation. The absurdity of this assumption can be seen by considering the impact of exposing individuals to 10 000 millisieverts (mSv) of radiation. Such a dose would be likely to cause death in a matter of hours. Extrapolation would imply that 1 mSv would result in the death of one out of every 10 000 people exposed. This is ridiculous: 1 mSv might cause one DNA lesion in one cell in the human body, but natural, spontaneous DNA lesions occur at a rate of about 70 million per cell per year. If we can survive this relatively high rate of spontaneous DNA lesions, there must be mechanisms that are capable of combating the minuscule insult caused by 1 mSv of radiation.

Dramatic radiation hormesis was observed in 1943, during the early days of the Manhattan Project, when animals were exposed to uranium dust at levels that were expected to be fatal. Instead, the animals lived longer, appeared healthier and had more offspring than the non-contaminated controls. Despite the rejection of threshold and hormetic effects by international and national bodies, research into radiation hormesis has continued apace, with over 1000 academic papers on the subject published.

Until very recently, reports by national and international organizations, including the United Nations Scientific Committee on the Effects of Atomic Radiation (UNSCEAR) and the American National Research Council Committee on the Biological Effects of Ionizing Radiation (BEIR), consistently failed to acknowledge the hormetic effect. However, Appendix 2 of the 1994 UNSCEAR report discusses numerous studies which indicate that ionizing radiation has a hormetic effect at low doses.

Of particular interest are the findings of several studies which strongly indicate that radon levels in homes are inversely correlated with incidence of lung cancer. If there was a linear relationship between dose of radiation and effect, the opposite would be expected. This suggests that the US EPA policy on limiting radiation in homes is actually against the public interest.

Introduction

Since Wilhelm von Roentgen announced his discovery of invisible x-rays or ionizing radiation nearly 112 years ago and, Antoine Henri Becquerel reported on mysterious and penetrating 'uranium rays', the social status of ionizing radiation has oscillated between enthusiastic acceptance and emphatic rejection. Following the publication in Nature, in January 1896, of an x-ray photograph of the hand of Dr Roentgen's wife and the announcement by Pierre Curie, in 1902, that the new radiation destroyed sick cells in animals – the first step in the development of 'Curie-therapy' – ionizing radiation became widely accepted as an efficient tool for the diagnosis and cure of many ailments. 100 years ago W. Shrader, at the University of Missouri, found that guinea pigs infected with diphtheria bacteria and then exposed to ionizing radiation survived, whereas non-irradiated controls died within 24 hours (Luckey, 1996). Soon it was recognized that radiation is beneficial not only in high doses applied for treatment of malignant neoplasms, but also in small doses used in diagnostics, balneology etc. However, Becquerel himself remained somewhat sceptical and once, according to Eve Curie (daughter of Marie Sklodowska-Curie) stated, 'I love radium, but the stuff snarls at me'. This was after he had suffered a skin burn on his chest in 1900, caused by a radium vial which he carried in the vest pocket. The first paper on adverse acute effects (dermatitis, erythema, deep skin lesions, epilation) of the new radiation was published in the year after Roentgen's discovery, and contained a description of the first victim, a German engineer (Marcuse, 1886). During the first five years of its use – before simple measures of precaution were adopted – 170 cases of radiation injury were recorded (UNSCEAR, 1958). The first students and users voluntarily exposed themselves to high radiation doses, which resulted in the deaths of 406 persons, whose names (including Marie Sklodowska-Curie) are recorded in a 'Book in Honour of Radiologists of All Nations', published in Berlin in 1992. This early experience showed the need for protection against high doses of radiation.

In the 1920s the concept of 'tolerance dose' was introduced, defined as a fraction of the dose causing reddening of skin. This fraction corresponded initially to an annual dose of 700 mSv (in modern units); in 1936 this was reduced to 350 mSv, and in 1941 it was reduced further to 70 mSv. The concept of tolerance dose was based on observations of exposed persons, mainly roentgenologists, showing that a person can be continuously irradiated up to a certain level without apparent adverse effects. For the next three decades this concept, which was effectively a statement of threshold, served as a basis for radiation protection standards (Kathren, 1996). Until the end of the Second World War, ionizing radiation and radionuclides were generally regarded as a blessing for humankind. It was after the war and the development of nuclear weapons, that this opinion changed to radiophobia – an irrational fear that any level of radiation is dangerous. The concern about great doses of ionizing radiation, such as could be encountered at a nuclear battleground, in a beam of cyclotron radiation, or received by the 28 fatal victims of the burning Chernobyl reactor, is obviously justified. However, the fear of small doses of radiation, such as those absorbed as a result of fallout from Chernobyl by inhabitants of Central or Western Europe, is about as justified as a fear that a temperature of 20°C may be hazardous because at 200°C one can easily get third degree burns. Radiophobia, perhaps the most widespread and influential superstition of the second half of the twentieth century, has as its main driving force the hypothesis that there is a no-threshold, linear relationship between radiation and its effects on the living organism.

The linear hypothesis

The linear hypothesis assumes that harmful post-irradiation effects, such as neoplasms and genetic diseases, appear not only after high doses but also after extremely low doses of radiation, and only that the frequency of the effects is proportional to the dose (Figure 6.1). The hypothesis extrapolates information about the impact of very high doses, absorbed over a short period of time (i.e. at high dose rate) – and about which there are epidemiological data with a high degree of statistical confidence attached – to an unknown region of extremely low dose rates. According to this hypothesis, the dose/effect relationship is a straight line, so that even a dose close to zero has some detrimental effect. Thus, the linear hypothesis assumes that there is no dose threshold or limit below which the effects observed at high doses cease to appear. In addition, the 'effects' in this hypothesis are assumed to be only the detrimental ones, such as decrease in life duration, occurrence of cancers, and genetic damage. It also implies that no new effects occur at low doses. One of the consequences of this hypothesis was the development of the sievert unit, which is a dose-independent quasi risk factor (Waligórski, 1997); another is the concept of collective dose, which enables aggregation, to frightening imagined levels, of trivial individual doses received by whole populations of particular regions or even the whole world. The hypothesis justifies adding the doses absorbed by a person at various times and in various parts of the body, without taking into account duration of periods between particular doses; it also allows aggregation of the doses received by any number of persons in current and future generations. For example, applying the linear hypothesis it would, in theory, be possible to sum the risk faced by the whole population of the Earth for eternity, and so forecast 'exactly' the health detriment in any period until the end of time. Of course, such an exercise would obviously require omniscience, or else it would be based upon daring assumptions unfounded in reality.

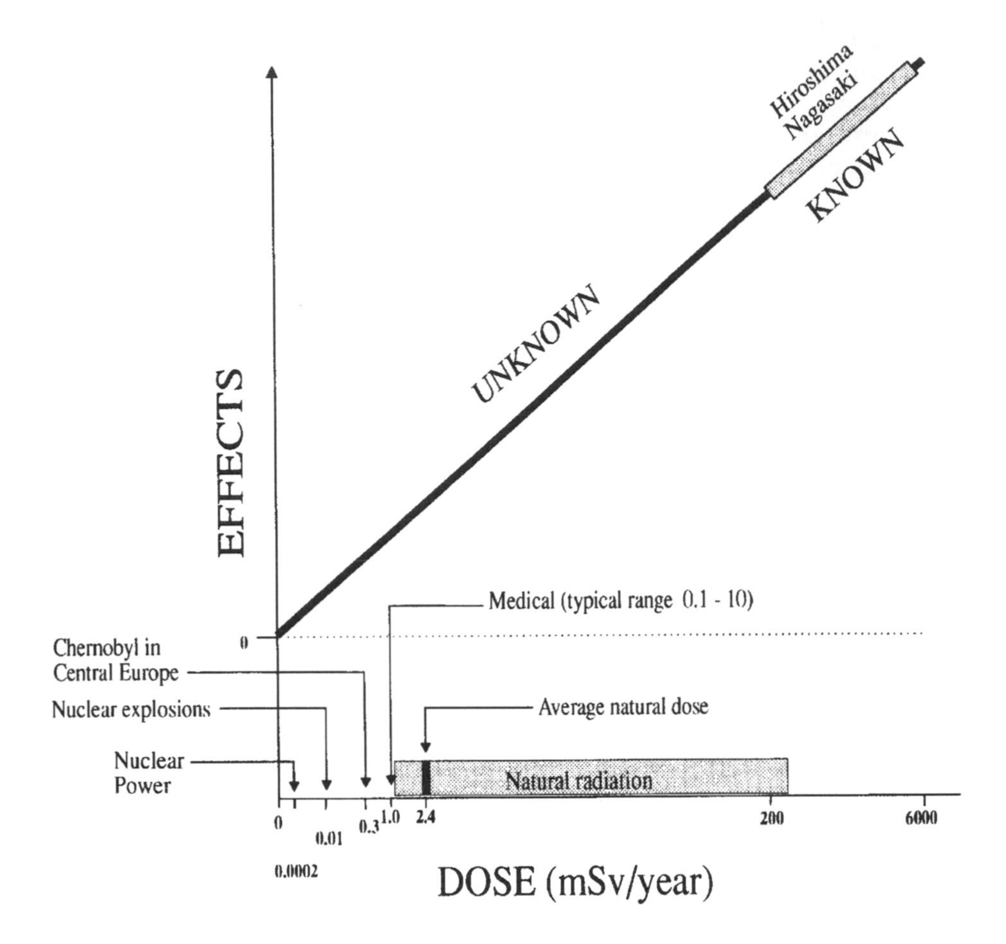

Figure 6.1. Linear hypothesis extrapolates epidemiological data on survivors from Hiroshima and Nagasaki, irradiated with high doses of ionizing radiation at extremely high dose rate, to near zero doses and very low dose rates. Average annual doses from Chernobyl fallout in Central Europe (first year), current doses from past nuclear weapon tests, nuclear power, medical irradiation, and natural radiation, after UNSCEAR (1988). Maximal range of natural radiation dose after Sohrabi (1993).

The linear hypothesis was accepted in 1959 by the International Commission on Radiological Protection (ICRP) as the philosophical basis for radiological protection (ICRP, 1959). This decision was based on the first report of the then recently established United Nations Scientific Committee on the Effects of Atomic Radiation (UNSCEAR, 1958). Since its very beginning, in 1955, UNSCEAR gained rather a privileged political position, reporting directly to the General Assembly, and over the years became the most distinguished international scientific body on matters of ionizing radiation. A large part of its first report was dedicated to a discussion of linearity and of the threshold dose for adverse radiation effects. About forty years ago, UNSCEAR's stand on this subject was formed as a result of an in-depth debate that was influenced heavily by the political atmosphere and issues of the time. For example, the Soviet delegation submitted a proposal (supported by Czechoslovakia and Egypt, but rejected by the ten remaining delegations, with two abstaining) in which a no-threshold assumption was used as a basis for recommending the immediate cessation of test explosions of nuclear weapons.

In 1958 UNSCEAR stated that contamination of the environment by nuclear explosions would increase radiation levels all over the world, posing new and unknown hazards for the present and future generations. These hazards supposedly could not be controlled and 'even the smallest amounts of radiation are liable to cause deleterious genetic, and perhaps also somatic, effects'. This sentence had a spectacular career over the next few decades – repeated in a plethora of publications out of the context of the UNSCEAR report and taken as an article of faith by the public. But in the final conclusions the UNSCEAR made two important points:

(1) 'Linearity has been assumed primarily for purposes of simplicity'; and
(2) 'There may or may not be a threshold dose. Two possibilities of threshold and no-threshold have been retained because of the very great differences they engender'.

Throughout the whole report until its final conclusions, the original UNSCEAR view on linearity and threshold remained ambivalent. For example, UNSCEAR (1958) accepted as a threshold for leukaemia a dose of 4000 mSv (page 42). However, at the same time the Committee accepted the risk factor for leukaemia of 0.52 per cent per 1000 mSv, assuming no threshold and linearity (page 115). The Committee quite openly presented this difficulty, showing in one table (page 42) its consequences: continuation of nuclear weapon tests in the atmosphere was estimated to cause up to 60 000 leukemia cases world-wide if no threshold is assumed, and zero leukemia cases if a threshold of 4000 mSv exists.

In the ICRP document of 1959 no such controversy appears. The problem was simplified and it was arbitrarily assumed that there is no threshold for harmful effects of ionizing radiation. The epistemological problems related to the lack of any possibility of finding these effects at very low doses were ignored. These problems were later taken into account by Weinberg (1972) and Walinder (1987; 1995), who discussed the fundamental limits to our knowledge in physics and biology. In his analysis of questions which science can ask and yet cannot answer, Weinberg (1972) used as an example the well known fact that to determine experimentally at a 95 per cent confidence level that 1.5 mSv will increase the mutation rate by 0.5 per cent, as predicted by the linearity assumption, one would need 8 000 000 000 mice; at a 60 per cent confidence level the number is 195 000 000. Thus, the question about the effects predicted for 1 mSv is unanswerable by science; it transcends science. Weinberg proposed a term 'trans-scientific' for such questions. Laymen, politicians, civic leaders etc., often look to scientists to provide scientific answers when scientists can offer only trans-scientific answers. In the field of ionizing radiation, trans-science has often, during the past few decades, been cloaked in the mantle of respected science.

Over the years, the working assumption of ICRP (1959) came to be regarded as a scientifically documented fact by mass-media, public opinion, regulatory bodies, and even many scientists. The no-threshold principle, however, belongs to the realm of administration and is not a scientific principle. Also, in later years UNSCEAR stated that a no-threshold, linear dose/effect relationship is only 'a cautious assumption, the validity of which has not yet been established. . . . It cannot be assumed that the values of (risk factor) which can be derived from observations at high doses and dose rates also apply to small dose increments at low dose rates' (UNSCEAR, 1977). Professor W.V. Mayneord, one of the most notable persons in radiation protection and a former head of the United Kingdom delegation to UNSCEAR, stated: 'I have always felt that the argument that because at higher values of dose an observed effect is proportional to dose, at very low doses there is necessarily some 'effect' of dose, however small, is nonsense' (Mayneord, 1964).

Estimates of risk factors for radiation carcinogenesis based on the linearity principle were commonly based upon the results of epidemiological studies of about 90 000 survivors of nuclear attacks in Japan. These studies indicated that cancers are induced by single radiation doses, delivered in 10^{-8} second, and thousands of times higher than the world-wide average annual dose from natural background radiation (2.4 mSv). Whether any cancers are induced by background radiation or ordinary exposure of the general population to man-made radiation, however, was not indicated by the Hiroshima and Nagasaki studies.

The ICRP assumption on linearity was not very realistic. It was generally accepted, however, because it simplified regulatory work. The original purpose was to regulate the exposure to radiation of a relatively small group of occupationally exposed persons, and it did not involve exceedingly high costs to society. In the 1970s, however, ICRP extended the no-threshold principle to exposure of the general population to man-made radiation, and in the 1980s it extended the principle to limiting the exposure to natural sources of radiation. The dose limit for the public was set at 50 mSv over a lifetime (ICRP, 1984a). This value is less than one-third of the global average lifetime dose from background radiation (168 mSv) and many tens or hundreds of times lower than the lifetime dose in many regions of the world.

Limiting exposure below the levels of natural radiation, at which millions of people have lived since time immemorial, is a logical consequence of the administrative assumption from 1959; if each dose is detrimental, then one should also attempt to decrease the risk of background radiation from Mother Nature herself (as from man-made radiation) even at such trivial doses as 1 mSv per year. Yet such reasoning was less than palatable to many scientists associated with radiation protection. This was not only because of the epistemological problem of trespassing beyond the limits of knowledge and because of growing evidence of the beneficial effects of low doses of radiation, but also because of the many absurd practical, as well as morally repugnant, consequences (such as the forced migration of whole populations away from their native soil).

The limit of 1 mSv per year, recommended by ICRP, was supposed to protect the general population against the appearance of so called stochastic effects (read: neoplastic and genetic diseases), caused by radiation damage of DNA. 1 mSv may induce about 1 DNA lesion in 1 cell of the human body, whereas the number of natural, spontaneous DNA lesions (due to thermal and oxidative insult) is about 70 million per cell per year (Billen, 1990). This comparison suggests that it is absurd to propose to defend anybody against the DNA insult of 1 mSv.

The probability of tumour formation for which several changes must occur in an irradiated cell was calculated by Walinder (1995), taking into account results of modern oncological studies. If one assumes that the three changes (continuous growth stimulation, immortalization and changes on the cell surface) must first have taken place in order to induce a complete tumour transformation, the probability of cancer occuring after a dose of 1 mSv is 5×10^{11}. This means that five cancer deaths may be expected per 100 billion people after each of them were irradiated with a dose of one mSv - a far cry from the ICRP estimate of five cancer deaths per 100 000 per mSv, based on the assumption that a single event genome could transfer a cell into malignant phenotype.

The absurd linearity principle was brought to light by the Chernobyl accident in 1986. On the basis of this principle, and following the ICRP recommendations, about 400 000 persons were resettled in Belarus, Ukraine and Russia. These people are actual 'linearity victims'. Resettlement caused unspeakable suffering, including the mass occurrence of psychosomatic diseases and a reduction in the average life span, owing to chronic stress and a decrease in the standard of living. The material losses were estimated to be tens of billions of dollars. In the impoverished Belarus alone the costs of resettlement, rents and compensation paid to these 'linearity victims' will, by 2015, reach US$91.4 billion (Rolevich et al., 1996). The intervention level for evacuation was first set as a 70-year lifetime radiation dose of 350 mSv, or about twice the world-wide average natural background dose. In 1991, five years after the relocation of the first 135 000 persons, the Supreme Soviet lowered this level to 70 mSv (Ilyin, 1995). This was in agreement with ICRP (1984b) recommendations for relocation when the expected last year dose may surpass 50 mSv, what in the Chernobyl dose corresponds to about a 150 mSv lifetime dose. However, no increase in cancer death rate has ever been observed after irradiation with such small doses received over 70-years lifetime, nor has any such increase been observed after natural doses tens and even hundreds of times higher (Sohrabi, 1990).

Given actual figures of natural radiation, one might ask why governments of various countries do not relocate populations living in areas where lifetime exposure to natural radiation exceeds 50 mSv, or 350 mSv? For example, why isn't everyone evacuated from Norway, where the average lifetime dose is 365 mSv (Henriksen and Saxebøl, 1988) and in some districts 1500 mSv (Baarli, 1995)? Should not regions of India with average lifetime dose of 2000 mSv (Sunta, 1990) and Iran with more than 3000 mSv (maximum 17 000 mSv) (Sohrabi, 1990) be depopulated?

Using the linearity principle to calculate precise numbers of imaginary cancer deaths due to Chernobyl fallout at a global or regional scale is behaviour which Dr Lauriston Taylor, the former president of the US National Council on Radiological Protection and Measurements, defined as 'deeply immoral uses of our scientific heritage' (Taylor, 1980). More recently Professor Gunnar Walinder, an eminent Swedish radiobiologist, stated that: 'The hypothetical nature of this calculational method makes it completely unscientific and I consider it to be more or less criminal to specify figures of this kind, bearing in mind the damage and the anxiety they can provoke and how confusing they are ...' (Walinder, 1995).

Radiation hormesis

Beneficial and protective effects of low doses of ionizing radiation were known long ago. In algae such effects were found by Atkinson (1898), soon after the discovery of Roentgen radiation. He noticed an increased growth rate in bluegreen algae exposed to x-rays. This was confirmed 82 years later (Conter et al., 1980).

As already stated, in 1943, during the early days of the Manhattan Project, it was found that the animals exposed to inhalation of uranium dust at levels that were expected to be fatal actually lived longer, appeared healthier, and had more offspring than the non-contaminated control animals. For years these results were treated as an anomaly but later studies produced similar results (Brucer, 1989). The first UNSCEAR Report presented the results of the experiments showing longer survival times of mice and guinea pigs exposed to small doses of gamma radiation and fast neutrons. These results were interpreted by UNSCEAR as indicating the existence of a threshold, but the hormetic effect was not noticed (UNSCEAR, 1958).

Since the 1960s, such effects, which are in direct contradiction to linear hypothesis, have been ignored in radiation protection practice, while research on stimulating and adaptive effects of radiation, the radiation hormesis (from the Greek word *hormein*, to excite) has continued over several decades, and the number of published papers on this subject has surpassed 1 000.

Radiation hormesis goes beyond the notion that radiation has no deleterious effects at small doses: at small doses new stimulatory effects occur that are not observed at high doses, and these new effects may be beneficial to the organism (Figure 6.2). In 1994 UNSCEAR published a review of 405 papers on stimulating and adaptive effects of low doses of radiation. The reader of this document can see how far we are from a full understanding of hormetic

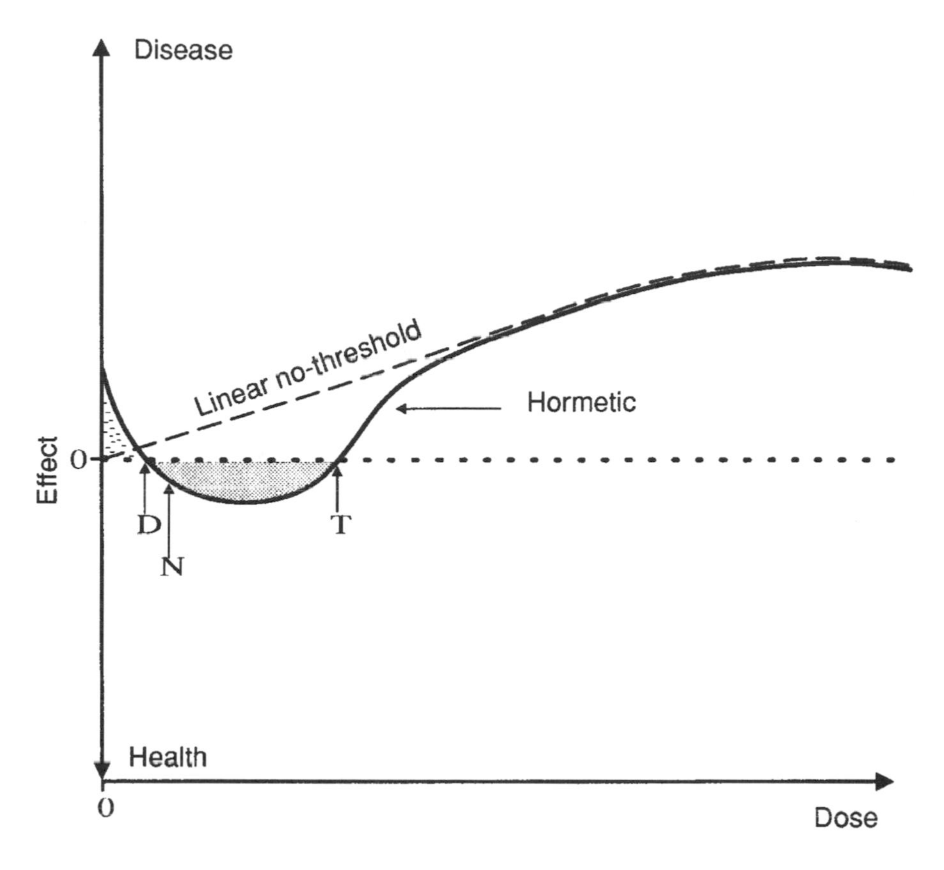

Figure 6.2 Generalized biological response to chemical and physical agents. Deficiency symptoms are caused by a deficit of an agent (dose less than D). Small doses (between D and T) are vital for good health (shaded area). Doses higher than T cause toxic or other harmful effects. N is the average global natural radiation dose. Dotted and solid lines represent linear no-threshold and hormetic dose-effect relationship, respectively.

processes and their mechanisms. At the top of the list of these mechanisms are stimulation of DNA repair, protein synthesis, gene activation, production of stress proteins, detoxification of free radicals, activation of membrane receptors and the release of growth factors, stimulation of immune system and proliferation of splenocytes. The hormetic effects were found at biochemical, cellular, and organic levels, in cell cultures, bacteria, plants, and animals. In mammals, radiation hormesis enhances defence reactions against neoplastic and infectious diseases, increases longevity, and improves fertility. The UNSCEAR (1994) report presented ample documentation of these effects and for the first time recognized their existence. This started a real ferment in the radiation protection rank and file, who for decades witnessed a seemingly endless process of decreasing radiation limits set by ICRP and implemented by national regulatory bodies (see Goldman, 1987; Mossman et al., 1996; Académie, 1995; Becker, 1997).

Experimental evidence

Of special interest among the research reviewed in UNSCEAR (1994), is a group of French studies started in the 1960s. They indicate that protozoa and bacteria exposed to artificially *lowered* levels of natural radiation demonstrate deficiency symptoms expressed as dramatically decreased proliferation (Figure 6.3). This indicates that ionizing radiation may be essential for life. Indeed, this might be expected. Living organisms developed under constant exposure to ionizing radiation, which 3.5 billion years ago was about 3 times higher than now (Jaworowski, 1997). Near natural nuclear reactors, such as that found at Oklo in Gabon, the radiation dose rate was probably about 47 Gy per hour. It was estimated that about 100 million such natural reactors existed on Earth in past geological epochs (Draganic et al., 1993). Early organisms had to protect themselves against the adverse effects of radiation to survive and they probably

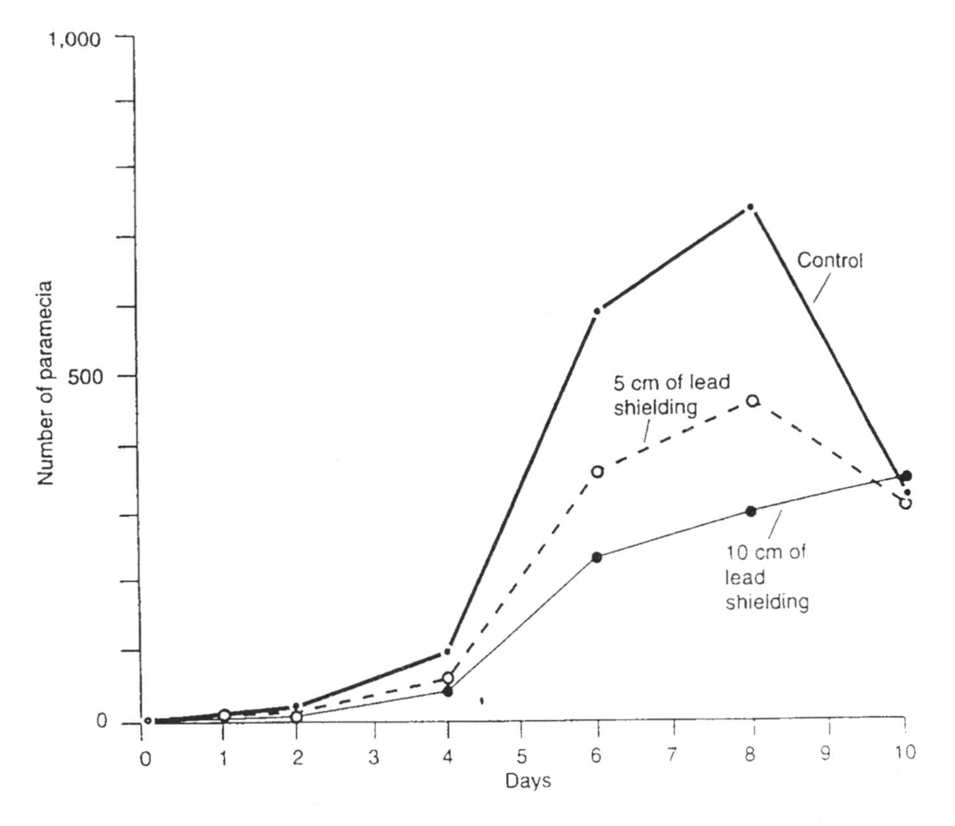

Figure 6.3 Effect of shielding on proliferation of Paramecium tetraurelia cultured in identical chambers which were shielded from background radiation with either 5 or 10 cm of lead. The nonshielded control animals were exposed to the normal natural gamma radiation rate of 1.75 mGy per year, and animals shielded 10 cm of lead were exposed to 0.3 mGy per year. On the eighth day, the proliferation of paramecia in the chambers shielded with 5 and 10 cm of lead was 60 and 40 per cent, respectively, of proliferation of the nonshielded control animals. After Planel et al. (1987).

began to use the ionizing radiation to their advantage. Similarly, some 500 million years after the ascent of cellular organisms, algae equipped with chlorophyll learned how to benefit from high dose ultraviolet radiation in the process of photosynthesis, which became the basis for almost all life.

The lack of sense organs for ionizing radiation in higher organisms is probably an effect of defence mechanisms superfluously covering the whole range of natural radiation levels. The range of natural doses: from <1 to 376 mSc/year (UNSCEAR, 1993; Sohrabi, 1990) is greater than, for example, the range of normal exposure to thermal energy, covering about 50°C. Increasing the water temperature in a bath by a factor of 1.27 from about 293 K to 373 K (i.e. from 20°C to 100°C) may cause death. But a single lethal dose of ionizing radiation, delivered in about 1 hour, which for man is 3000 to 5000 mSv, is more than 10 million times higher than the average natural radiation dose received in that time. The fractionated lethal dose is even higher. The defence mechanisms developed during the course of evolution, provide to the living organisms with an enormous safety margin for natural levels of radiation, and for man-made radiation from peace-time sources under control.

UNSCEAR (1994) reviewed numerous studies on mammals. For example, in an experiment with mice the incidence of leukemia, cancers, and sarcomas was lower in animals irradiated with cesium-137 gamma radiation doses of 2.5 to 20 mSv than it was in non-irradiated controls. The number of all malignant neoplasms in animals exposed to a single dose of 10 mSv was more than 30 per cent lower than in non-irradiated controls (Table 6.1). In several experiments, small initial radiation doses have been shown to improve the survival of animals subsequently irradiated with large, near lethal doses. In other experiments, an increased life span was found in animals irradiated with doses between 250 and 3000 mSv.

Table 6.1. Incidence of neoplasms (%) after a single exposure of the three-month-old mice to ^{137}Cs gamma radiation. Bold numbers show hormetic effect. (In parentheses number of leukemias expected according to linearity assumption). Adapted from Maisin et al. (1988).

Dose, mGy	*Leukaemia*	*Carcinomas and sarcomas*	*All neoplasms*
0	20.93	16.3	33.19
250	**18.18** (21.92)	14.04	28.51
500	**15.48** (22.91)	**8.79**	**23.01**
1000	**15.04** (24.89)	**8.94**	**21.95**
2000	**13.82** (28.85)	14.29	26.27
4000	26.57 (36.76)	16.08	39.16
6000	44.68 (44.68)	14.36	55.32

Table 6.2. Incidence of neoplasms (%) in RFM mice irradiated with ^{137}Cs gamma rays and neutrons. Bold numbers show hormetic effect. After Ullrich et al. (1976).

	Gamma-rays (450 mGy/min.; 15,256 animals)						
Dose (mGy)	0	100	250	500	1000	1500	3000
Reticulum cell sarcoma	42.3	**40.9**	**31.4**	**28.4**	**25.0**	**24.6**	**23.2**
Lung adenoma	30.2	**23.9**	**20.2**	**22.0**	**22.5**	**25.0**	37.1
Ovarian tumour	2.4	**2.0**	6.4	35.5	35.1	42.4	47.8
Mammary tumour	2.8	2.8	**1.5**	4.0	2.1	3.4	2.2
	Neutrons (50 mGy/min.; 767 animals)						
Dose (mGy)	0	48	96	192	470		
Lung adenoma	23.9	**10.1**	23.8	30.1	45.8		
Endocrynic tumour	7.0	5.2	10.7	50.4	54.2		

A clear hormetic effect was found by Ullrich et al. (1976) from Oak Ridge National Laboratory, in a study on mice irradiated with various doses of cesium-137 gamma radiation (dose rate 450 mGy/min.) and neutron radiation from a nuclear reactor (dose rate 50 mGy/min.) (Table 6.2). After gamma irradiation, in the case of reticulum cell sarcoma and lung adenoma, hormetic effect was observed at a wide range of doses between 100 and 1500 mGy, and at 2.0 and 1.5 mGy for ovarian and mammary tumours. After a 48 mGy neutron dose, a 58 per cent decrease in lung adenoma incidence was observed and a 26 per cent decrease in endocrine tumours. The authors commented: '...decreased incidence was observed in this experiment at 50 mGy. Using these data, linearity may be rejected ($P<0.05$) over the 0 to 470 mGy dose range'. This important paper was cited in UNSCEAR (1993) 18 years after publication, but both this comment and the hormetic effect were ignored.

Epidemiological evidence

Perhaps most interesting are the results of studies on human populations. Such studies were carried out for the population of Japanese nuclear attack survivors, inhabitants of regions with high natural radiation, inhabitants of houses with high concentration of radon-222, persons irradiated for medical reasons, nuclear industry workers, and inhabitants of a highly contaminated region in East Ural.

The obvious hormetic effects were sometimes not noticed by the authors themselves (see for example Table 6.3). One of the factors influencing interpretation of epidemiological data is a publication bias (Dickersin, 1990). For example, if a 5-year study on adverse effects of mammography in a group of 1000 women found no effect, then the results would not be published, since reviewers and editors would claim that one would not expect an effect

Table 6.3. Death from breast cancer among 31 710 Canadian women, exposed to radiation at fluoroscopic examinations. Authors concluded that 'The data were most consistent with linear dose response relation' and did not comment on the hormetic effect evident at 100–190 dose range. Adapted from Miller et al. (1989).

Dose (mGy)	*Standardized rate per 10^3 person-years*
0–90	578.6
100–190	**421.8**
200–290	560.7
300–390	650.7
400–490	610.0
700–990	1 362
1000–2900	1 382
3000–5990	2 334
6000–10 000	8 000
>10 000	20 620

at such a dose, sample size and length of observation. In the same situation, data showing a strong, although random, increase of cancer risk would certainly be published (Modan, 1993).

The linearity paradigm stands behind a selective treatment of epidemiological data in publications of respected scientific committees. For example, the BEIR V report of the American National Research Council Committee on the Biological Effects of Ionizing Radiation (BEIR, 1990) presents the number of leukemia cases among 3224 participants in the 1957 nuclear test 'SMOKY'. These persons were irradiated with an average dose of 6 mSv and only 1 per cent with a dose higher than 50 mSv. BEIR V reported that, until 1977, 9 leukemia cases were registered among the participants, whereas 3.5 cases were expected in a normal population group of the same size. BEIR V commented that this was a 'significant increase . . . in leukemia incidence and mortality'. However the report also stated that among participants of other tests studied by Robinette et al. (1985) no significant excess of leukemia appeared. But BEIR V did not mention an important observation from the Robinette et al. (1985) study which showed a hormetic effect. Participants of the 'OPERATION GREENHOUSE' test series received an average dose of 13 mSv, with 3 per cent receiving doses higher than 50 mSv. Among the 3%, only one case of leukemia was registered, whereas in the normal population of the same size 4.43 cases were expected. The UNSCEAR (1993) report shows the same bias: 'SMOKY' data are discussed but not data from 'OPERATION GREENHOUSE', showing a hormetic effect.

BEIR V discussed results of Holm's et al. (1988) epidemiological study of 35,074 Swedish patients diagnosed with radioactive iodine-131. These received an average dose on their thyroid of 500 mSv. BEIR V noticed that among them 50 thyroid cancers were observed, and 39.37 were expected. Thus the ratio of observed to expected cancers was 1.27. But BEIR V did not inform the readers that there were two groups in the Holm et al. (1988) study –: patients suspected for thyroid cancer and patients diagnosed for other ailments. The ratio of 1.27 is an average for both groups. In the group not suspected of having thyroid cancers this ratio was 0.62, i.e. in this group 38 per cent fewer thyroid cancers were observed than in a normal population. Appendix A of UNSCEAR (1994), 'Epidemiological studies of radiation carcinogenesis', treated the Holm et al. (1988) data in the same way. However, in the 'revolutionary' Annex B of UNSCEAR (1994), 'Adaptive Responses to Radiation in Cells and Organisms', the hormetic effect found in Holm et al. (1988) was correctly described. Similar results were reported by Hall et al. (1996) who studied 34 104 patients diagnosed with iodine-131. In patients not suspected of having thyroid tumours, the incidence of thyroid cancers was 25 per cent lower than in a normal population.

The effects of multiple chest X-ray fluoroscopy was studied recently by Howe (1995) in more than 64,000 Canadian tuberculosis patients in relation to the lung tissue dose. The lung cancer mortality between 1950 and 1987 of patients receiving fractionated doses of between 10 and 990 mSv, was 10 to 12 per cent lower than among the general Canadian population (Table 6.4). This evident hormetic effect was not discussed by the author.

The UNSCEAR (1994) report states that among nuclear attack survivors from Hiroshima and Nagasaki who received doses lower than 200 mSv, there was no increase in the normal number of total cancer deaths. In fact, mortality caused by leukemia was lower in this population at doses below 100 mSv than among the non-irradiated inhabitants of these Japanese cities, which was statistically not significant (Figure 6.4). Japanese epidemiological data indicate that among the women who survived nuclear attack and were irradiated with small doses, general mortality in older age groups was about 40 per cent lower than among the non-irradiated women. Also among the women irradiated with doses higher than 10 mSv, general mortality was lower than among those irradiated with doses lower than 10 mSv (Table 6.5).

Table 6.4. Lung cancer mortality in the Canadian fluoroscopy study (1950-1987). Bold numbers show hormetic effect. After Howe (1995).

Lung dose (mSv)	*Observed*	*Expected*	*Standardized mortality ratio*
<10	723	707.33	1.02
10-490	**180**	**200.85**	**0.90**
500-990	**92**	**104.24**	**0.88**
1000-1990	114	110.77	1.03
2000-2990	41	33.58	1.22
>3000	28	24.25	1.15

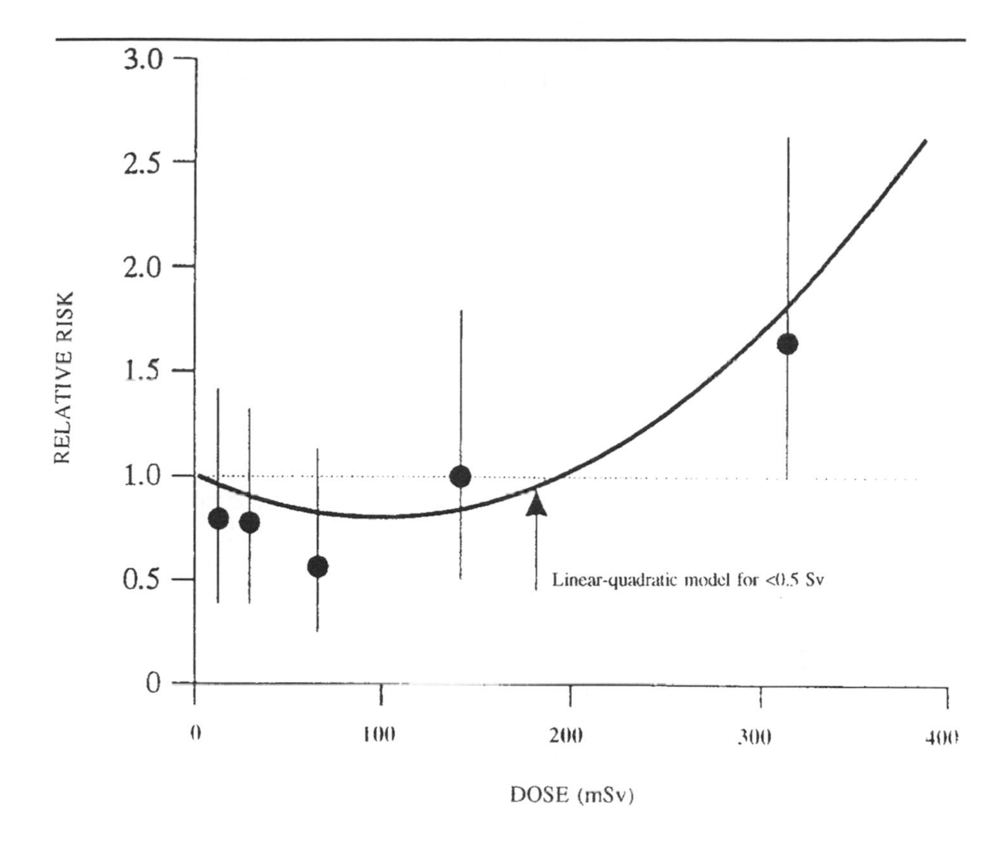

Figure 6.4. Mortality risk from leukemia in Hiroshima and Nagasaki. Dotted line shows normal risk in non-irradiated inhabitants of these cities. After UNSCEAR (1994).

Table 6.5. Annual mortality rates due to all causes for female survivors of nuclear attack in Nagasaki (per 100 000) in 1970–1976. After Kondo (1993).

Age	*Unexposed*	*Exposed*	
		<5mSv	*>10mSv*
30-39	103	87	78
40-49	223	224	218
50-59	510	569	428
60-69	1516	1303	833
70-79	5305	4161	3242
>80	19634	12626	13158

The Hiroshima and Nagasaki data indicate that a single irradiation with doses between 400 and 600 mSv did not cause detrimental effects in the next generation. Unexpectedly, rather positive effects appeared. Among the infants of parents who survived the nuclear attack the mortality was 4 per cent lower than in children of non-irradiated parents, there was 23 per cent less aneuploidy, 29 per cent fewer chromosomal aberrations, and 30 per cent fewer mutations in blood proteins (Table 6.6). Similarly unexpected results were obtained in one of the best studies in human genetics carried out in Hungary before and after the Chernobyl accident. Several serious congenital anomalies occurred with lower frequency after the Chernobyl accident than before the accident (Table 6.7).

The first large-scale radio-epidemiological study at low doses was carried out for the white population in the United States by Frigerio and Stowe (1976). In each of the 50 states, the population was divided into three categories, according to average natural radiation dose (most of radon dose not included). In the states with an average annual radiation dose of 2.1 mSv (population 5 735 000) the mortality rate (per 100 000 persons) caused by all malignancies was 126.3, in states with 1.7 mSv dose (population 16 897 000) 132.2 and the highest mortality rate of 146.8 was found in states with the lowest dose of 1.18 mSv (population 59 683 000). The authors concluded that their results showed a consistent and continuous decrement in malignant mortality with increasing background radiation. Similar results were obtained for mortality from congenital malformations. This study was not cited in UNSCEAR or BEIR documents.

A similar study was carried out more recently in China. Between 1970 and 1986, 74 000 people in Yagjiang county, which has a high level of natural background radiation (5.5 mSv per year) were compared with 77 000 people in two adjacent low-background counties (Enping and Taishan, 2.1 mSv per year). In the high-background Yangjiang county, the inhabitants receive a 70-year lifetime dose of 385 mSv, which is higher than the intervention level for evacuation adopted for Chernobyl, and 5.5 times higher than the dose limit recommended for population by ICRP and now implemented in the regulations of several countries as well as the European Union.

Table 6.6. Frequency of genetic effects of radiation (%) in children of atomic attack survivors in Hiroshima and Nagasaki (parents exposed to dose of 400– 600 mSv).

Effect	*Unexposed*	*Exposed*	*References*
Death of live-born children 1946-1958	7.35	7.08	Yoshimoto et al., 1991
Chromosomal aberrations	0.31	0.22	Awa et al., 1989
Aneuploidy	0.30	0.23	Awa et al., 1989
Mutations in blood proteins	6.4×10^{-6}	4.5×10^{-6}	Neel et al., 1988

Table 6.7. Number of birth defects and anomalies (per 10 000) in Hungary following the Chernobyl accident. After Czeizel, 1989. Average Chernobyl radiation dose in various parts of Hungary ranged from about 60 to >500μ per caput (Feher, 1988).

Birth defect	1980-85	1.5.1986 30.4.1987	1.5.1987 30.4.1988
Number of live births	807,939	126,708	125,514
Down's syndrome	8.44	7.27	6.77
Retinoblastoma	0.32	0.16	0.08
Wilms tumour	0.80	0.47	0.56

Should the Chinese government follow the Soviet example and evacuate Yangjiang county? The epidemiological data show that there is no reason to do so. In an age group of 10 to 79 years the general (non-leukemia) cancer mortality was 14.6 per cent lower in the high-background county than in the low-background counties. The leukemia mortality among men was 15 per cent lower and among women 60 per cent lower in Yangjiang. No difference in the frequency of various genetical diseases was noticed between the counties (Wei et al., 1990). Similar results were also observed in Japan (Iwasaki et al., 1990) and France (Tirmarche et al., 1988).

Most of the epidemiological studies of a relationship between radon levels in homes and lung cancer also seem to be in disagreement with the no-threshold principle, and some may suggest a hormetic effect. Of 19 studies reviewed in UNSCEAR (1994) 8 showed an increase in lung cancer mortality or incidence in populations exposed to higher-than-average radon levels, and 11 studies showed lower mortality or incidence or no association of lung cancer and radon levels.

In the United States, a study that covered 89 per cent of the population found that the people living in houses with radon air concentrations that were higher than the average level had a lower mortality from lung cancer (Cohen, 1993).

In China, a meticulous study measured the radon level for one year in the houses of several hundred women with lung cancers and in houses of a similar number of healthy women. The results demonstrated at a 95 per cent confidence level that women who lived in high-level radon houses (more than 350 Bq/m^3) had a 30 per cent lower lung cancer risk than those living in low-level radon houses (4 to 70 Bq/m^3).

This result contradicts the no-threshold principle estimate, according to which the lung cancer risk in the high-radon houses should be 80 per cent higher than the normal risk (Blot et al., 1990).

Similarly, in one region of Japan with an average indoor radon level of 35 Bq/m^3, the lung cancer incidence was 51 per cent of that in a low-level radon region (11 Bq/m^3), and the mortality caused by all types of cancer was 37 per cent lower (Mifune et al., 1992). Similar results showing a lack of positive correlation between lung cancer and indoor radon levels, were reported from Canada, Sweden, Denmark, Finland, France, and Great Britain (see UNSCEAR, 1994 for references).

Despite the evidence from these studies, the US Environmental Protection Agency (EPA) has recommended remedial action when indoor radon concentrations reach 150 Bq/m^3. The EPA considers that remedial action at any level down to 70 Bq/m^3 would be cost effective, claiming that the cost of reducing the level from 150 to 70 Bq/m^3 is approximately $2 million per life hypothetically saved (Schiager, 1992).

From among several studies of people occupationally exposed to low radiation doses discussed in UNSCEAR (1994), here data are presented on mortality in Canadian, American and British nuclear industry. Gribbin et al. (1992) studied 13 491 employees of the Atomic Energy of Canada Limited. Among these persons 5504 were not exposed to gamma radiation. The mean radiation dose of exposed persons was 49 mSv for men and 5.5 mSv for women. As shown in Table 6.8 the mortality due to all leukemias in the exposed group was only 32 per cent of that in the general Canadian population. The observed mortality among employees of AECL from all cancers and from all non-cancer diseases was also less than expected.

Workers from American nuclear shipyards were divided into three study groups: 28 542 workers receiving an average working lifetime dose higher than 5 mSv; 10 462 workers with doses lower than 5 mSv; and 33 352 non-irradiated workers. Mortality from all causes in both irradiated groups was 28 per cent lower than in the non-irradiated group, and the mortality due to leukemia in the group irradiated with doses higher than 5 mSv, was 58 per cent lower (Matanowski, 1991).

Similar results were obtained in a study of 95 000 British workers with an average individual dose of 34 mSv. The mortality rate for all (non-violent) causes of death was 16 per cent lower than in the average British population and mortality from malignant neoplasms (in 23 organs) was 14 per cent lower (Kendall et al., 1992).

Some recently declassified data showing hormetic effects in humans have come from the former Soviet Union. In September 1957, inhabitants of 22

Table 6.8. Standardized mortality ratios (SMR) for leukaemia in 9997 male employees of Atomic Energy of Canada Limited. After Gribbin et al., (1992).

Cause of death	*Non-exposed*		*Exposed*	
	SMR	*Observed/ expected*	*SMR*	*Observed/ expected*
Lymphocytic leukaemia	2.25	2/0.85	-	0/2.40
Myeloid leukaemia	1.47	2/1.36	0.57	2/3.50
Other leukaemia	0.72	1/1.40	0.28	1/3.59
All leukemia	1.37	5/3.64	0.32	3/9.50

villages in the Eastern Urals were irradiated with high radiation doses of up to 1 500 mSv, the result of a radioactivity release from a thermal explosion in a Soviet military nuclear facility. About 10 000 people were evacuated and their cancer mortality was studied during the next 30 years.

From this group, 7852 of the persons studied were divided into three exposure groups: those who received average doses of 496 mSv, 120 mSv, and 40 mSv. Tumour-related mortality in the 496 mSv group was 28 per cent lower than in the non-irradiated control population from the same region; in the 129 mSv group it was 39 per cent lower, and in the 40 mSv group it was 27 per cent lower. In the first two groups, the difference from the controls was statistically significant (Kostyuchenko and Krestinina, 1994).

Each human life hypothetically saved by implementing the US Nuclear Regulatory Commission's regulations costs about $2.5 billion (Cohen, 1992). Such spending is morally questionable. Studies of radiation hormesis suggest that such expenditures are probably futile and may actually have an adverse effect on the health of the population.

The recognition by UNSCEAR, the most distinguished international scientific body on the subject of ionizing radiation, of the possibility that low doses of radiation may result in changes in cells and organisms, which reflect an ability to adapt to the effects of radiation, may inspire the authorities to switch to a more realistic approach to problems of estimating and managing the risks of ionizing radiation.

References

Académie des Sciences (1995). Problemes liés aux effets des faibles doses de radiations ionisantes. Rapport Académie des Sciences, Paris, 29 mai, 1995.

Atkinson, G. F. (1898). Report upon some preliminary experiments with Roentgen rays on plants. *Science*, **7**, 7.

Awa, A. A., Honda, T. Neriishi, S. (1989). Cytogenic study of the offsprings of atomic bomb survivors, Hiroshima and Nagasaki. Radiation Effects Research Foundation, Hiroshima, Report No. RERF TR 21–88, pp. 1–20.

Baarli, J. (1995). Private information.

BEIR (1990). *Health Effects of Exposure to Low Levels of Ionizing Radiation, BEIR V.* Committee on the Biological Effects of Ionizing Radiations, Board on Radiation Effects Research, Commission on Life Sciences, National Research Council. National Academy Press, Washington DC.

Becker, K. (1997). *Schwelle oder Nicht-Schwelle, ist das hier die Frage?* Strahlenschutzpraxis, in press.

Billen, D. A. (1990) Spontaneous DNA damage and its significance for the 'negligible' dose controversy in radiation protection. *Radiation Research*, **124**, 242-245.

Blot, W. J., Xu, Z.-Y., Boice Jr et al. 1990). Indoor radon and lung cancer in China. *J. Natl. Cancer Inst.*, **82**(12), 1025–1030.

Brucer, M. (1989). Letter to the Editor of Time magazine, quoted in *Access to Energy*, **16**(7), March.

Cohen, B. L. (1992). Perspectives on the cost effectiveness of life saving. In J. H. Lehr *Rational Readings on Environmental Concerns*, Van Nostrand Reinhold, New York, pp. 461–473.
Cohen, B. L. (1993). Relationship between exposure to radon and various types of cancer. *Health Physics*, **65**(5), 529–531.
Conter A., Dupouy D., and Planel, H. (1980). Demonstration of a biological effect of natural ionizing radiation. *Int. J. Rad. Biol.*, **43**, 421–432.
Czeizel, A., (1989). Hungarian surveillance of germinal mutations. *Hum. Genet.*, **82**, 359–366.
Dickersin, K. (1990). The existence of publication bias and risk factors for its occurrence. *JAMA* **263**, 1385–1389.
Draganic, I. G., Draganic, Z. D. and Adloff, J. P. (1993). *Radiation and Radioactivity on Earth and Beyond.* CRC Press, Boca Raton.
Feher, I. (1988). Experience in Hungary of the radiological consequences of the Chernobyl accident. In Z. Jaworowski (ed.) Chernobyl Accident: Regional and Global Impacts. *Special Issue of Environ. Internat.*, **14**(2), 113–135.
Frigerio N. A. and Stowe R. S. (1976). Carcinogenic and genetic hazard from background radiation. Biological and Environmental Effects of Low-level Radiation. IAEA, Vienna, vol. II, pp. 385–393.
Goldman, M., Catlin, R., and Anspaugh, L. (1987). Health and environmental consequences of the Chernobyl Nuclear Power Plant accident. US Department of Energy, Washington, DC Report No. DOE/RR–0232.
Gribbin, M. A., Howe, G. R., Weeks, J. L. (1992). A study of the mortality of AECL employees. V. The second analysis: Mortality during the period 1950–1985. Atomic Energy of Canada Limited, Report No. AECL–10615.
Henriksen, T., Saxebøl, G. (1988). Fallout and radiation doses in Norway after the Chernobyl accident. In Z. Jaworowski Chernobyl Accident: Regional and Global Impacts *Environment International, Special Issue* **14**(2), 157–163.
Hall, P., Mattsson, A. and Boice, Jr. J. D. (1996). Thyroid cancer after diagnostic administration of iodine–131. *Radiation research*, **145**, 86–92.
Holm, L. E., Wiklund, K. E., Lundell,G. E. et al. 1988). Thyroid cancer after diagnostic doses of iodine–131: a retrospective cohort study. *J. Natl. Cancer Inst.*, **80**, 1132–1138.
Howe, G. R. (1995). Lung cancer mortality between 1950 and 1987 after exposure to fractionated moderate-dose-rate ionizing radiation in the Canadian Fluoroscopy Cohort Study and a comparison with lung cancer mortality in the Atomic Bomb Survivors Study. *Radiation Research*, **142**, 295–304.
ICRP (1959). *Recommendations of the International Commission on Radiological Protection.* ICRP Publication No. 1, Pergamon Press, London.
ICRP (1984a). *Principles for Limiting Exposure of the Public to Natural Sources of Radiation.* Statement from the 1983 Washington Meeting of the ICRP. ICRP Publication No. 39, Annals of the ICRP, 14(1): i–vii.
ICRP (1984b). *Protection of the public in the Event of Major Radiation Accidents.* Principles for Planning. ICRP Publication 40. Pergamon Press, Oxford.
Ilyin, L. A. (1995). *Chernobyl: Myth and Reality*. Megapolis, Moscow.
Iwasaki T., Minowa, M., Hashimoto, S., Hayashi, N. et al. (1993). Cancer mortality rates by geographic distribution and levels of natural radiation dose in Japan. In: M. Sohrabi, J. U. Ahmed and S. A. Durrani (eds) *High Levels of Natural Radiation*. MAEA, Vienna, 503–514.
Jaworowski, Z. (1997). Radiation 3. 5 billion years ago. *21st Century Sci. Technol.*, **10**(1):4.

Kathren, R. L. (1996). Pathway to a paradigm: The linear non-threshold dose–response model in historical context: the American Academy of Health Physics 1995 Radiology Centennial Harman Oration. *Health Physics*, **70**(5), 621–635.
Kendall, B. M., Muirhead, C. R., MacGibbon B. H. et al., (1992). First analysis of the national registry for radiation workers: occupational exposure to ionizing radiation and mortality. National Radiological Protection Board, Chilton, UK. Report No. NRPB-R251.
Kondo, S. (1993), *Health Effects of Low-level Radiation*. Kinki University Press, Osaka, Japan.
Kostyuchenko, V. A., Krestinina, L. Yu., (1994). Long-term irradiation effects in the population evacuated from the East-Urals radioactive trace area. *The Sci. Total. Environ.*, **142**, 119–125.
Luckey, T. D. (1990). *Radiation Hormesis*, CRC Press, Boca Raton, Florida.
Luckey, T. D. (1996). The evidence for radiation hormesis. *21st Century Sci. Technol.*, **9**(3), 12–20.
Maisin, J. R., Wambersie, A., and Gerber, G. B. (1988). Life-shortening and disease incidence in C57B1 mice after single and fractionated g and high-energy neutron exposure. *Radiation Research*, **113**, 300–317.
Matanowski, G. M. (1991). Health effects of low-level radiation in shipyard workers final report, June, 1991. Report DOE DE–AC02–79 EV10095.
Mayneord, W. V. (1964). *Radiation and Health.* The Rock Carling Fellowship. The Nuffield Provincial Hospital Trust.
Mifune, M., Sobue, T., and Arimoto, H. (1992). Cancer mortality survey in a spa area (Misasa, Japan) with a high radon background. *Japanese J. Cancer Res.* **83**, 1–5.
Miller, A. B., Howe, G. R., Sherman, G. J et al. (1989). Mortality from breast cancer after irradiation during fluoroscopic examinations in patients being treated for tuberculosis. *New Engl. J. Med.* **321**, 1285–1289.
Modan, B. (1993). Low dose radiation carcinogenesis – issues and interpretation: The 1993 G. William Morgan Lecture. *Health Physics*, **6**, 475–480.
Mossman, K. L., Goldman, M.,. Masse, F. et al. (1996). Radiation Risk in Perspective, Health Physics Society Position Statement, 1996. Internet, March 12.
Neel, J. V., Satoh, C., Goriki, K. et al. (1988). Search for mutations altering protein change and/or function in children of atomic bomb survivors: final report. *Am. J. Hum. Genet.*, **42**, 663–676.
Planel, H., Soleilhavoup, J. P., Taxidor, R. et al. (1987). Influence on cell proliferation of background radiation or exposure to very low, chronic g radiation. *Health Physics*, **52**(5), 571–578.
Robinette, C. D., Jablon, S., and Preston, T. L. (1985). Mortality studies in nuclear weapon tests participants. In *Medical Follow-up Agency*, National Research Council. National Academy Press, Washington DC, pp. 1–47.
Rolevich, I. V., Feniks, I. A., Baboshov, E. M., and Lych, G. M. (1996). Social, Economic, institutional and political impact of the Chernobyl accident in Belarus. Statement of the Head of National Delegation of Belarus, 6 Session. International Conference One Decade After Chernobyl: Summing up the Consequences of the Accident. IAEA, Vienna (in Russian).
Schiager, K. J. (1992). Radon- risk and reason. In J. H. Lehr (ed.) *Rational Readings on Environmental Concerns*, Van Nostrand Reinhold, New York, pp. 619–626.
Sohrabi, M. (1990). Recent radiological studies of high level natural radiation areas of Ramsar. *Proc. International Conference on High Levels of Natural Radiation*. IAEA, Vienna, pp. 39–45.

Sunta, C. M. (1990). A review of the studies of high background areas of the S-W coast of India. *Proc. International Conference on High Levels of Natural Radiation.* IAEA, Vienna, pp. 71–86.
Taylor, L. S. (1980). Some non-scientific influences on radiation protection standards and practice. *Proc. 5th International Congress of the International Radiation Protection Association* – Vol. 1. The Israel Health Physics Society, Jerusalem, pp. 307–319.
Tirmarche, N., Rannou, A. Mollie A. and Sauve A. (1988). Epidemiological study of regional cancer mortality in France and natural radiation. *Radiation Protection Dosimetry*, **24**, 479–482.
Ullrich, R. L.,. Jernigan, M. C., and Cosgrove. G. E. (1976). The influence of dose and dose rate on the incidence of neoplastic disease in RFM mice after neutron irradiation. *Radiation Research*, **68**, 115–131.
UNSCEAR (1958). Report of the United Nations Scientific Committee on the Effects of Atomic Radiation. United Nations, General Assembly Official Records: Thirteenth Session, Supplement No. 17 (A/3838).
UNSCEAR (1977). Sources and Effects of Ionizing Radiation. United Nations Scientific Committee on the Effects of Atomic Radiation. New York.
UNSCEAR (1993). Sources and Effects of Ionizing Radiation. United Nations Scientific Committee on the Effects of Atomic Radiation. New York.
UNSCEAR (1994). Sources and Effects of Ionizing Radiation. United Nations Scientific Committee on the Effects of Atomic Radiation. New York.
Waligórski, M. (1997). On the present paradigm of radiation protection – A track structure perspective. Nukleonika, submitted.
Walinder, G. (1987). Epistemological problems in assessing cancer risks at low radiation doses. *Health Physics*, **52**(5), 675–678.
Walinder, G. (1995). *Has radiation protection become a health hazard?* Kärnkraftsäkerhet and Utbildning AB, Nyköping.
Wei, L., Zha, Y., Tao, Z. et al. (1990). Epidemiological investigation of radiological effects in high background radiation areas of Yangjiang, *China. J. Radiat. Res.*, **31**, 119–136.
Weinberg, A. M. (1972). Science and Trans–Science. *Minerva*, **10**, 209–222.
Yoshimoto, Y., Schull, W. J., Kato, H., Neel, J. V. (1991). Mortality among the offspring (F1) of atomic bomb survivors, 1946-85. Radiation Effects Research Foundation, Hiroshima. Report No. RERF TR 1–91, pp. 1–27.

7 Pollution, pesticides and cancer misconceptions

Bruce N. Ames and Lois S. Gold

Summary

The major causes of cancer are:

(a) Smoking, which is responsible for about a third of US cancer (and 90 per cent of lung cancer).
(b) Dietary imbalances, for example, lack of dietary fruits and vegetables: The quarter of the population eating the least fruits and vegetables has double the cancer rate for most types of cancer compared with the quarter eating the most.
(c) Chronic infections: mostly in developing countries.
(d) Hormonal factors influenced by lifestyle.

There is no epidemic of cancer, except for lung cancer from smoking. Cancer mortality rates have declined 15 per cent since 1950 (excluding lung cancer and adjusted for the increased lifespan of the population).

Regulatory policy focused on traces of synthetic chemicals is based on misconceptions about animal cancer tests. Recent research contradicts these ideas:

(a) Rodent carcinogens are not rare. Half of all chemicals tested in standard high dose animal cancer tests, whether occurring naturally or produced synthetically, are 'carcinogens'.
(b) There are high dose effects in these rodent cancer tests that are not relevant to low dose human exposures and which can explain the high proportion of carcinogens.
(c) Though 99.9 per cent of the chemicals humans ingest are natural, the focus of regulatory policy is on synthetic chemicals.

Over 1000 chemicals have been described in coffee: 27 have been tested and 19 are rodent carcinogens.

Plants we eat contain thousands of natural pesticides, which protect plants from insects and other predators: 64 have been tested and 35 are rodent carcinogens.

There is no convincing evidence that synthetic chemical pollutants are important for human cancer. Regulations that try to eliminate minuscule levels of synthetic chemicals are enormously expensive: EPA estimates its regulations cost $140 billion/year. The US spends 100 times more to prevent one hypothetical, highly uncertain, death from a synthetic chemical than it spends to save a life by medical intervention. Attempting to reduce tiny hypothetical risks also has costs; for example, if reducing synthetic pesticides makes fruits and vegetables more expensive, thereby decreasing consumption, then cancer will be increased, particularly for the poor.

Improved health comes from knowledge due to biomedical research, and from lifestyle changes by individuals. Little money is spent on biomedical research or on educating the public about lifestyle hazards, compared with the costs of regulations.

Myths and facts about synthetic chemicals and human cancer

Various misconceptions about the relationship between environmental pollution and human disease, particularly cancer, drive regulatory policy. Here, we highlight nine such misconceptions and briefly present the scientific evidence that undermines each.

Misconception No. 1: cancer rates are soaring

Cancer death rates overall in the US (after adjusting for age and excluding lung cancer due to smoking) have declined 15 per cent since 1950 (Kosary et al. 1995, Doll and Peto 1981). The particular types of cancer deaths that have decreased since 1950 are stomach, cervical, uterine, and rectal. The types that have increased are primarily lung cancer (90 per cent is due to smoking, as are 35 per cent of all cancer deaths in the US), melanoma (probably due to sunburns), and non-Hodgkin's lymphoma. (Cancer incidence rates are also of interest, but they should not be taken in isolation, because trends in the recorded incidence rates are biased by improvements in registration and diagnosis (Doll and Peto 1981; Devesa et al. 1995).)

Cancer is one of the degenerative diseases of old age, increasing exponentially with age in both rodents and humans. External factors, however, can markedly increase cancer rates (e.g. cigarette smoking in humans) or decrease them (e.g. caloric restriction in rodents). Life expectancy has continued to rise since 1950. The increases in cancer deaths are due to the delayed effect of increases in smoking and to increasing life expectancy (Doll and Peto 1981, Devesa et al. 1995).

Misconception No. 2: environmental synthetic chemicals are an important cause of human cancer

Neither epidemiology nor toxicology supports the idea that synthetic industrial chemicals are important for human cancer. Epidemiological studies have identified the factors that are likely to have a major effect on reducing rates of cancer: reduction of smoking, improving diet (e.g. increased consumption of fruits and vegetables), and control of infections (Ames et al. 1995). Although some epidemiologic studies find an association between cancer and low levels of industrial pollutants, the associations are usually weak, the results are usually conflicting, and the studies do not correct for potentially large confounding factors like diet. Moreover, the exposure to synthetic pollutants are tiny and rarely seem plausible as a causal factor when compared with the background of natural chemicals that are rodent carcinogens (Gold et al. 1992) . Even assuming that the EPA's worst-case risk estimates for synthetic pollutants are true risks, the proportion of cancer that EPA could prevent by regulation would be tiny (Gough 1990). Occupational exposure to carcinogens can cause cancer, though how much has been a controversial issue: a few percent seems a reasonable estimate (Ames et al. 1995). The main contributor has been asbestos in smokers. Exposures to substances in the workplace can be high in comparison with other chemical exposures in food, air, or water. Past occupational exposures have sometimes been high and therefore comparatively little quantitative extrapolation from high dose rodent tests to high dose occupational exposures may be required for risk assessment. Since occupational cancer is concentrated among small groups exposed at high levels, there is an opportunity to control or eliminate risks once they are identified. We estimate that diet accounts for about one-third of cancer risk (Ames et al. 1995) in agreement with the earlier estimate of Doll and Peto (1981). Other factors are lifestyle influencing hormones, avoidance of intense sun exposure, increased physical activity, and reduced consumption of alcohol.

Since cancer is due, in part, to normal ageing, the influence of the major external risk factors for cancer must be diminished (smoking, unbalanced diet, chronic infection and hormonal factors). Cancer will occur at a later age, and the proportion of cancer caused by normal metabolic processes will increase. Ageing and its degenerative diseases appear to be due in good part to the accumulation of oxidative damage to DNA and other macromolecules (Ames et al. 1993b). By-products of normal metabolism – superoxide, hydrogen peroxide, and hydroxyl radical – are the same oxidative mutagens produced by radiation. Oxidative lesions in DNA accumulate with age, so that by the time a rat is old it has about a million oxidative DNA lesions per (Ames et al. 1993b). Mutations also accumulate with age. DNA is oxidized in normal metabolism because antioxidant defences, though numerous, are not perfect. Antioxidant defences against oxidative damage include vitamins C and E and carotenoids, most of which come from dietary fruits and vegetables.

Smoking contributes to about 35 per cent of US cancer, about one-quarter of heart disease, and about 400 000 premature deaths per year in the United States (Peto et al. 1994). Tobacco is a known cause of cancer of the lung, bladder, mouth, pharynx, pancreas, stomach, larynx, oesophagus and possibly colon. Tobacco causes even more deaths by diseases other than cancer. Smoke contains a wide variety of mutagens and rodent carcinogens. Smoking is also a severe oxidative stress and causes inflammation in the lung. The oxidants in cigarette smoke – mainly nitrogen oxides – deplete the body's antioxidants.

Thus, smokers must ingest two to three times more Vitamin C than non-smokers to achieve the same level in blood, but they rarely do. Inadequate concentration of Vitamin C in plasma is more common among single males, the poor, and smokers (Ames et al. 1993b). Men with inadequate diets or who smoke may damage both their somatic DNA and the DNA of their sperm. When the level of dietary Vitamin C is insufficient to keep seminal fluid Vitamin C at an adequate level, oxidative lesions in sperm DNA are increased 250 per cent (Fraga et al. 1991; Ames et al. 1994; Fraga et al. 1996) . Paternal smoking, therefore, may plausibly increase the risk of birth defects and appears to increase childhood cancer in offspring (Fraga et al. 1991; Ames et al. 1994; Ji et al. 1997).

Chronic inflammation from chronic infection results in release of oxidative mutagens from phagocytic cells and is a major contributor to cancer (Ames et al. 1995; Christen et al. 1997) White cells and other phagocytic cells of the immune system combat bacteria, parasites, and virus-infected cells by destroying them with potent, mutagenic oxidizing agents. The oxidants protect humans from immediate death from infection, but they also cause oxidative damage to DNA, mutation, and chronic cell killing with compensatory cell division (Shacter et al. 1988,Yamashina et al. 1986) and thus contribute to the carcinogenic process. Antioxidants appear to inhibit some of the pathology of chronic inflammation. We estimate that chronic infections contribute to about one-third of the world's cancer, mostly in developing countries.

Endogenous reproductive hormones play a large role in cancer, including breast, prostate, ovary and endometrium (Henderson et al. 1991; Feigelson and Henderson 1996), contributing to as much as 20 per cent of all cancer. Many lifestyle factors such as lack of exercise, obesity and reproductive history influence hormone levels and therefore risk.

Genetic factors play a significant role in cancer and interact with lifestyle and other risk factors. Biomedical research is uncovering important genetic variation in humans.

Misconception No. 3: reducing pesticide residues is an effective way to prevent diet-related cancer

On the contrary, fruits and vegetables are of major importance for reducing cancer: if they become more expensive by reducing the use of synthetic pesticides, cancer is likely to increase. People with low incomes eat fewer fruits and vegetables and spend a higher percentage of their income on food.

Dietary fruits and vegetables and cancer prevention. Consumption of adequate fruits and vegetables is associated with a lowered risk of degenerative diseases including cancer, cardiovascular disease, cataracts, and brain dysfunction (Ames et al. 1993b). Over 200 studies in the epidemiological literature have been reviewed that show, with great consistency, an association between a lack of adequate consumption of fruits and vegetables and cancer incidence (Block et al. 1992; Steinmetz and Potter 1991; Hill et al. 1994) (Table 7.1). The quarter of the population with the lowest dietary intake of fruits and vegetables compared with the quarter with the highest intake has roughly twice the cancer rate for most types of cancer (lung, larynx, oral cavity, oesophagus, stomach, colon and rectum, bladder, pancreas, cervix, and ovary). Only 22 per cent of Americans meet the intake recommended by the National Cancer Institute (NCI) and the National Research Council (Hunter and Willett 1993; Block 1992; Patterson et al. 1990): 5 servings of fruits and vegetables per day. When the public is told about hundreds of minor hypothetical risks, they lose perspective on what is important: half the public does not know that fruits and vegetables protect against cancer (NCI 1996).

Micronutrients in fruits and vegetables are anti-carcinogens, and their antioxidants may account for some of their beneficial effect as discussed in Misconception No. 2. However, the effects of dietary antioxidants are difficult to disentangle by epidemiological studies from other important vitamins and ingredients in fruits and vegetables (Steinmetz and Potter 1991; Hill et al. 1994; Block 1992; Steinmetz and Potter 1996).

Folate deficiency, one of the most common vitamin deficiencies, causes extensive chromosome breaks in human genes (Blount et al. 1997) Approximately 10 per cent of the US population (Senti and Pilch 1985) is deficient at the level causing chromosome breaks. In two small studies of low income (mainly African-American) elderly people (Bailey et al. 1979) and adolescents (Bailey et al. 1982), nearly half were folate deficient to this level. The mechanism is deficient methylation of uracil to thymine, and subsequent incorporation of uracil into human DNA (4 million/cell) (Blount et al. 1997).

Table 7.1. Review of epidemiological studies on cancer showing protection by consumption of fruits and vegetables

Cancer site	*Fraction of studies showing significant cancer protection*	*Relative risk (median) low versus high quartile) of consumption*
Epithelial		
Lung	24/25	2.2
Oral	9/9	2.0
Larynx	4/4	2.3
Oesophagus	15/16	2.0
Stomach	17/19	2.5
Pancreas	9/11	2.8
Cervix	7/8	2.0
Bladder	3/5	2.1
Colorectal	20/35	1.9
Miscellaneous	6/8	—
Hormone-dependent		
Breast	8/14	1.3
ovary/endometrium	3/4	1.8
Prostate	4/14	1.3
Total	129/172	

Source: Block et al. (1992)

During repair of uracil in DNA, transient nicks are formed; two opposing nicks causes a chromosome break. Both high DNA uracil levels and chromosome breaks in humans are reversed by folate administration (Blount et al. 1997). Chromosome breaks could contribute to the increased risk of cancer and cognitive defects associated with folate deficiency in humans (Blount et al. 1997). Folate deficiency also damages human sperm (Wallock et al. 1997), causes neural tube defects in the foetus, and 10 per cent of US heart disease (Blount et al. 1997).

Other micronutrients are likely to play a significant role in the prevention and repair of DNA damage, and thus are important to the maintenance of long-term health. Deficiency of vitamin B12 causes a functional folate deficiency, accumulation of homocysteine (a risk factor for heart disease) (Wickramasinghe and Fida 1994), and misincorporation of uracil into DNA (Wickramasinghe and Fida 1994). Strict vegetarians are at increased risk of developing a Vitamin B12 deficiency (Herbert 1996). Niacin contributes to the repair of DNA strand breaks by maintaining nicotinamide adenine dinucleotide levels for the poly ADP-ribose protective response to DNA damage (Zhang et al. 1993). As a result, dietary insufficiencies of niacin (15 per cent of some populations are deficient (Jacobson 1993)), folate, and antioxidants may act synergistically to adversely affect DNA synthesis and repair. Diets deficient in fruits and vegetables are commonly low in folate, antioxidants, (e.g. Vitamin C) and many other micronutrients, and result in significant amounts of DNA damage and higher cancer rates (Ames et al. 1995; Block et al. 1992; Subar et al. 1989).

Optimizing micronutrient intake can have a major impact on health. Increasing research in this area and efforts to improve micronutrient intake and balanced diet should be a high priority for public policy.

Misconception No. 4: human exposures to carcinogens and other potential hazards are nearly all exposures to synthetic chemicals

On the contrary, 99.9 per cent of the chemicals humans ingest are natural. The amounts of synthetic pesticide residues in plant foods are insignificant compared with the amount of natural pesticides produced by plants themselves (Ames et al. 1990a; Ames et al. 1990b). Of all dietary pesticides that humans eat, 99.99 per cent are natural: they are chemicals produced by plants to defend themselves against fungi, insects, and other animal predators (Ames et al. 1990a; Ames et al. 1990b). Each plant produces a different array of such chemicals. On average, Americans ingest roughly 5 000 to 10 000 different natural pesticides and their breakdown products. Americans eat about 1 500 mg of natural pesticides per person per day, which is about 10 000 times more than they consume of synthetic pesticide residues.

Even though only a small proportion of natural pesticides has been tested for carcinogenicity, half of those tested (35/64) are rodent carcinogens, and naturally occurring pesticides that are rodent carcinogens are ubiquitous in fruits, vegetables, herbs, and spices (Gold et al. 1997b) (Table 7.2).

Cooking foods produces about 2 000 mg per person per day of burnt material that contains many rodent carcinogens and many mutagens. By contrast, the residues of 200 synthetic chemicals measured by FDA, including the synthetic pesticides thought to be of greatest importance, average only about 0.09 mg per person per day (Ames et al. 1990a; Gold et al. 1997). The known natural rodent carcinogens in a single cup of coffee are about equal in weight to an entire year's worth of carcinogenic synthetic pesticide residues, even though only 3 per cent of the natural chemicals in roasted coffee have been tested for carcinogenicity (Gold et al. 1992) (Table 7.3). This does

Table 7.2. Carcinogenicity of natural plant pesticides tested in rodents (fungal toxins are not included)

Carcinogens: N = 35	acetaldehyde methylformylhydrazone, allyl isothiocyanate, arecoline.HCI, benzaldehyde, benzyl acetate, caffeic acid, catechol, clivorine, coumarin, crotonaldehyde, cycasin and methylazoxymethanol acetate, 3,4-dihydrocoumarin, estragole, ethyl acrylate, N2-g-glutamyl-p-hydrazinobenzoic acid, hexanal methylformylhydrazine, p-hydrazinobenzoic acid.HCI, hydroquinone, 1-hydroxyanthraquinone, lasiocarpine, d-limonene, 8-methoxypsoralen, N-methyl-N-formylhydrazine, a-methylbenzyl alcohol, 3-methylbutanal methylformylhydrazone, methylhydrazine, monocrotaline, pentanal methyl-formylhydrazone, petasitenine, quercetin, reserpine, safrole, senkirkine, sesamol, symphytine
Noncarcinogens: N = 29	atropine, benzyl alcohol, biphenyl, d-carvone, deserpidine, disodium glycyrrhizinate, emetine.2HCI, ephedrine sulphate, eucalyptol, eugenol, gallic acid, geranyl acetate, b-N-[g-l(+)-glutamyl]-4-hydroxy-methylphenylhydrazine, glycyrrhetinic acid, glycyrrhizinate, disodium, p-hydrazinobenzoic acid, isosafrole, kaempferol, d-menthol, nicotine, norharman, pilocarpine, piperidine, protocatechuic acid, rotenone, rutin sulfate, sodium benzoate, turmeric oleoresin, vinblastine
These rodent carcinogens occur in:	absinthe, allspice, anise, apple, apricot, banana, basil, beet, broccoli, Brussels sprouts, cabbage, cantaloupe, caraway, cardamom, carrot, cauliflower, celery, cherries, chilli pepper, chocolate milk, cinnamon, cloves, cocoa, coffee, collard greens, comfrey herb tea, coriander, currants, dill, eggplant, endive, fennel, garlic, grapefruit, grapes, guava, honey, honeydew melon, horseradish, kale, lemon, lentils, lettuce, licorice, lime, mace, mango, marjoram, mushrooms, mustard, nutmeg, onion, orange, paprika, parsley, parsnip, peach, pear, peas, black pepper, pineapple, plum, potato, radish, raspberries, rhubarb, rosemary, rutabaga, sage, savory, sesame seeds, soybean, star anise, tarragon, tea, thyme, tomato, turmeric, and turnip.

Source: Gold et al

Table 7.3. Carcinogenicity in rodents of natural chemicals in roasted coffee

Positive. N=19	acetaldehyde, benzaldehyde, benzene, benzofuran, benzo(a)pyrene, caffeic acid, catechol, 1,2,S,6-dibenzanthracene, ethanol, ethylbenzene, formaldehyde, furan, furfural, hydrogen peroxide, hydroquinone, limonene, styrene, toluene, xylene
Not positive: N=8	acrolein, biphenyl, choline, eugenol, nicotinamide, nicotinic acid, phenol, piperidine
Uncertain:	caffeine
Yet to test:	~ 1000 chemicals

Source: Gold et al.

not mean that coffee is dangerous, but rather that assumptions about high dose animal cancer tests for assessing human risk at low doses need re-examination. No diet can be free of natural chemicals that are rodent carcinogens (Gold et al. 1997a).

Misconception No. 5: cancer risks to humans can be assessed by standard high dose animal cancer tests

Approximately half of all chemicals – whether natural or synthetic – that have been tested in standard animal cancer tests are rodent carcinogens (Ames et al. 1996; Gold et al. 1997b) (Table 7.4). What are the explanations for the high positivity rate? In standard cancer tests rodents are given chronic, near-toxic doses, the maximum tolerated dose (MTD). Evidence is accumulating that it may be cell division caused by the high dose itself, rather than the chemical *per se*, that is increasing the cancer rate. High doses can cause chronic wounding of tissues, cell death, and consequent chronic cell division of neighbouring cells, which is a risk factor for cancer (Ames et al. 1996). Each time a cell divides it increases the probability that a mutation will occur, thereby increasing the risk for cancer. At the low levels to which humans are usually exposed, such increased cell division does not occur. Therefore, the very low levels of chemicals to which humans are exposed through water pollution or synthetic pesticide residues are likely to pose no or minimal cancer risks.

It seems likely that a high proportion of all chemicals, whether synthetic or natural, might be 'carcinogens' if run through the standard rodent bioassay at the MTD, but this will be primarily due to the effects of high doses for the non-mutagens, and a synergistic effect of cell division at high doses with DNA damage for the mutagens (Butterworth et al. 1995; Ames et al. 1993a; Ames and Gold 1990). Without additional data on the mechanism of carcinogenesis for each chemical, the interpretation of a positive result in a rodent bioassay is highly uncertain. The carcinogenic effects may be limited

Table 7.4. Proportion of chemicals evaluated as carcinogenic

Chemicals tested in both rats and mice	330/559	(59%)
Naturally-occurring chemicals	73/127	(57%)
Synthetic chemicals	257/432	(59%)
Chemicals tested in rats and/or mice		
Natural pesticides	35/64	(55%)
Mold toxins	14/23	(61%)
Chemicals in roasted coffee	19/28	(68%)
Innes negative chemicals retested[a]	16/34	(47%)
Drugs in the Physician's Desk Reference	117/241	(49%)

a The 1969 study by Innes et al. is frequently cited as evidence that the proportion of carcinogens is low, as only 9% of 119 chemicals tested (primarily pesticides) were positive in cancer tests on mice. However, these tests lacked the power of modern tests. We have found 34 of the Innes negative chemicals that have been retested using modern protocols: 16 were positive again about half.

Source: Gold et al.

to the high dose tested. The recent report of the National Research Council, *Science and Judgement in Risk Assessment* (NCI 1994) supports these ideas. The EPA's draft document *Working Paper for Considering Draft Revisions to the US EPA Guidelines for Cancer Risk Assessment* (NCI 1994) is a step toward improvement in the use of animal cancer test results.

In regulatory policy, the 'virtually safe dose' (VSD), corresponding to a maximum, hypothetical cancer risk of one in a million, is estimated from bioassay results using a linear model. To the extent that carcinogenicity in rodent bioassays is due to the effects of high doses for the non-mutagens, and a synergistic effect of cell division at high doses with DNA damage for the mutagens, then this model is inappropriate. Moreover, as currently calculated, the VSD can be known without ever conducting a bioassay: for 96 per cent of the NCI/NTP rodent carcinogens, the VSD is within a factor of 10 of the ratio MTD/740,000 (Gaylor and Gold 1995). This is about as precise as the estimate obtained from conducting near-replicate cancer tests of the same chemical (Gaylor and Gold 1995).

Misconception No. 6: synthetic chemicals pose greater carcinogenic hazards than natural chemicals

Gaining a broad perspective about the vast number of chemicals to which humans are exposed can be helpful when setting research and regulatory priorities (Gold et al. 1992; Ames et al. 1990b; Gold et al. 1993; Ames et al. 1987). Rodent bioassays provide little information about mechanisms of carcinogenesis and low dose risk. The assumption that synthetic chemicals are particularly hazardous has led to a bias in testing, such that synthetic chemicals account for 77 per cent of the 559 chemicals tested chronically in both rats and mice (Table 7.4). The natural world of chemicals has never been tested systematically. One reasonable strategy is to use a rough index to compare and rank possible carcinogenic hazards from a wide variety of chemical exposures at levels that humans typically receive, and then to focus on those that rank highest (Gold et al. 1992; Ames et al. 1987; Gold et al. 1994b). Ranking is a critical first step that can help to set priorities for selecting chemicals for chronic bioassay or mechanistic studies, for epidemiological research, and for regulatory policy. Although one cannot say whether the ranked chemical exposures are likely to be of major or minor importance in human cancer, it is not prudent to focus attention on the possible hazards at the bottom of a ranking if, using the same methodology to identify hazard, there are numerous common human exposures with much greater possible hazards. Our analyses are based on the HERP index (Human Exposure/Rodent Potency), which indicates what percentage of the rodent carcinogenic potency (TD_{50} in mg/kg/day) a human receives from a given daily lifetime exposure (mg/kg/day). TD_{50} values in our Carcinogenic Potency Database span a 10-million-fold range across chemicals (Gold et al. 1997c) (Table 7.5).

Overall, our analyses have shown that HERP values for some historically high exposures in the workplace and some pharmaceuticals rank high, and that there is an enormous background of naturally occurring rodent carcinogens in typical portions of common foods that cast doubt on the relative importance of low dose exposures to residues of synthetic chemicals such as pesticides (Gold et al. 1992; Ames et al. 1987; Gold et al. 1994a). A committee of the National Research Council/National Academy of Sciences recently reached similar conclusions about natural as opposed to synthetic chemicals in the diet, and called for further research on natural chemicals (NRC 1996).

The possible carcinogenic hazards from synthetic pesticides (at average exposures) are minimal compared with the background of nature's pesticides, though neither may be a hazard at the low doses consumed (Table 7.5). Table 7.5 also indicates that many ordinary foods would not pass the regulatory criteria used for synthetic chemicals. For many natural chemicals the HERP values are in the top half of the table, even though natural chemicals are markedly underrepresented because so few have been tested in rodent bioassays. Caution is necessary in drawing conclusions from the occurrence in the diet of natural chemicals that are rodent carcinogens. It is not argued here that these dietary exposures are necessarily of much relevance to human cancer. Our results call for a re-evaluation of the utility of animal cancer tests in protecting the public against minor hypothetical risks.

Misconception No. 7: the toxicology of synthetic chemicals is different from that of natural chemicals

It is often assumed that because natural chemicals are part of human evolutionary history, whereas synthetic chemicals are recent, the mechanisms

Table 7.5. Ranking possible carcinogenic hazards from average US exposures

Possible hazard: HERP (%)	*Average daily US exposure*	*Human dose of rodent carcinogen*	*Potency TD_{50} (mg/kg/day)*[a]	
			Rats	*Mice*
140	EDB: workers (high exposure) (before 1977)	Ethylene dibromide, 150 mg	1.52	(7.45)
17	Clofibrate	Clofibrate, 2 g	169	
14	Phenobarbital, 1 sleeping pill	Phenobarbital, 60 mg	(+)	6.09
6.8	1,3-Butadiene: rubber workers (1978-86)	1,3-Butadiene, 66.0 mg	(261)	13.9
6.1	Tetrachloroethylene: dry cleaners with dry-to-dry units (1980-90)[b]	Tetrachloroethylene, 433 mg	101	(126)
4.0	Formaldehyde: workers	Formaldehyde, 6.1 mg	2.19	(43.9)
2.1	**Beer, 257 g**	**Ethyl alcohol, 13.1 ml**	9110	(—)
1.4	Mobile home air (14 hours/day)	Formaldehyde, 2.2 mg	2.19	(43.9)
0.9	Methylenechloride: workers (1940s-80s)	Methylenechloride, 471 mg	724	(918)
0.5	**Wine, 28.0 g**	**Ethyl alcohol, 3.36 ml**	91 I0	(—)
0.4	Conventional home air (14 hours/day)	Formaldehyde, 598 mg	2.19	(43.9)
0. 1	**Coffee, 13.3 g**	**Caffeic acid, 23.9 mg**	297	(4900)
0.04	**Lettuce, 14.9 g**	**Caffeic acid, 7.90 mg**	297	(4900)
0.03	**Safrole in spices**	**Safrole, 1.2 mg**	(441)	51.3
0.03	**Orange juice, 138 g**	**d-Limonene, 4.28 mg**	204	(—)
0.03	**Pepper, black, 446 mg**	**d-Limonene, 3.S7 mg**	204	(—)
0.02	**Mushroom (*Agaricus bisporus* 2.55 g)**	**Mixture of hydrazines, etc. (whole mushroom)**	—	20,300
0.02	**Apple, 32.0 g**	**Caffeic acid, 3.40 mg**	297	(4900)
0.02	**Coffee, 13.3 g**	**Catechol, 1.33 mg**	118	(244)
0.02	**Coffee, 13.3 g**	**Furfural, 2.09 mg**	(683)	197
0.009	BHA: daily US avg (1975)	BHA, 4.6 mg	745	(5530)
0.008	**Beer (before 1979), 257 g**	**Dimethylnitrosamine, 726 ng**	0.124	(0.189)
0.008	**Aflatoxin: daily US avg (1984-89**	**Aflatoxin, 18 ng**	0.0032	(+)
0.007	**Cinnamon, 21.9 mg**	**Coumarin, 65.0 mg**	13.9	(103)
0.006	**Coffee, 13.3 g**	**Hydroquinone, 333 mg**	82.8	(225)
0.005	Saccharin: daily US avg (1977)	Saccharin, 7 mg	2140	(—)
0.005	**Carrot, 12.1 g**	**Aniline, 624 mg**	194C	(—)
0.004	**Potato, 54.9 g**	**Caffeic acid, 867 mg**	297	(4900)
0.004	**Celery, 7.95 g**	**Caffeic acid, 858 mg**	297	(4900)
0.004	**White bread, 67.6 g**	**Furfural, 500 mg**	(683)	197
0.003	**Nutmeg, 27.4 mg**	**d-Limonene, 466 mg**	204	(—)
0.003	Conventional home air (14 hour/day)	Benzene, 155 mg	(169)	77.5
0.002	**Carrot, 12.1 g**	**Caffeic acid, 374 mg**	297	(4900)
0.002	Ethylene thiourea: daily US avg (1990)	Ethylene thiourea, 9.51 mg	7.9	(23.5)
0.002	[DDT: daily US avg (before 1972 ban)]	[DDT, 13.8 mg]	(84.7)	12.3
0.001	**Plum, 2.00 g**	**Caffeic acid, 276 mg**	297	(4900)
0.001	BHA: daily US avg (1987)	BHA, 700 mg	745	(5530)
0.001	**Pear, 3.29 g**	**Caffeic acid, 240 mg**	297	(4900)
0.00I	[UDMH: daily US avg (1988)]	[UDMH, 2.82 mg (from Alar)]	(—)	3.96
0.0009	**Brown mustard, 68.4 mg**	**Allyl isothiocyanate, 62.9 mg**	96	(—)
0.0008	[DDE: daily US avg (before 1972 ban)]	[DDE, 6.91 mg]	(—)	12.5

Table 7.5. Ranking possible carcinogenic hazards from average US exposures (cont'd)

0.0007	TCDD: daily US avg (1994)	TCDD, 12.0 pg	0.0000235	(0.000156)
0.0007	Bacon, 11.5 g	Diethylnitrosamine, 11.5 ng	0.0237	(+)
0.0006	**Mushroom (*Agaricus bisporus* 2.55 g)**	**Glutamyl-p-hydrazino-benzoate, 107 mg**	.	277
0.0005	**Jasmine tea, 2.19 g**	**Benzyl acetate, 504 mg**	(—)	1440
0.0004	**Bacon, 11.5 g**	**N-Nitrosopyrrolidine, 196 ng**	(0.799)	0.679
0.0004	**Bacon, 11.5 g**	**Dimethylnitrosamine, 34.5 ng**	0.124	(0.189)
0.0004	[EDB: Daily US avg (before 1984 ban)]	[EDB, 420 ng]	1.52	(7.45)
0.0004	Tap water, 1 litre (1987-92)	Bromodichloromethane, 13 mg	(72.5)	47.7
0.0003	**Mango, 1.22 g**	**d-Limonene, 48.8 mg**	204	(—)
0.0003	**Beer, 257 g**	**Furfural, 39.9 mg**	(683)	197
0.0003	Tap water, 1 litre (1987-92)	Chloroform, 17 mg	(262)	90.3
0.0003	Carbaryl: daily US avg (1990)	Carbaryl, 2.6 mg	14.1	(—)
0.0002	**Celery, 7.95 g**	**8-Methoxypsoralen, 4.86 mg**	32.4	(—)
0.0002	Toxaphene: daily US avg (1990)	Toxaphene, 595 ng	(—)	5.57
0.00009	**Mushroom (*Agaricus bisporus*, 2.55 g)**	**p-Hydrazinobenzoate, 28 mg**	.	45
0.00008	PCBs: daily US avg (1984-86)	PCBs, 98 ng	1.74	(9.58)
0.00008	DDE/DDT: daily US avg (1990)	DDE, 659 ng	(—)	12.5
0.00007	**Parsnip, 54.0 mg**	**8-Methoxypsoralen, 1.57 mg**	32.4	(—)
0.00007	**Toast, 67.6 g**	**Urethane, 811 ng**	(41.3)	16.9
0.00006	**Hamburger, pan fried, 85 g**	**PhIP, 176 ng**	4.29c	(28.6c)
0.00005	**Estragole in spices**	**Estragole, 1.99 mg**	.	51.8
0.00005	**Parsley, fresh, 324 mg**	**8-Methoxypsoralen, 1.17 mg**	32.4	(—)
0.00003	**Hamburger, pan fried, 85 g**	**MeIQx, 38.1 ng**	1.99	(24.3)
0.00002	Dicofol: daily US avg (1990)	Dicofol, 544 ng	(—)	32.9
0.00001	**Cocoa, 3.34 g**	**a-Methylbenzyl alcohol, 4.3 mg**	458	(—)
0.00001	**Beer, 257 g**	**Urethane, 115 ng**	(41.3),	16.9
0.000005	**Hamburger, pan fried, 85 g**	**IQ, 6.38 ng**	1.89c	(19.6)
0.000001	Lindane: daily US avg (1990)	Lindane, 32 ng	(—)	30.7
0.0000004	PCNB: daily US avg (1990)	PCNB (Quintozene), 19.2 ng	(—)	71.1
0.0000001	Chlorobenzilate: daily US avg (1989)	Chlorobenzilate, 6.4 ng	(—)	93.9
<0.00000001	Chlorothalonil: daily US avg (1990)	Chlorothalonil, <6.4 ng	828d	(—)
0.000000008	Folpet: daily US avg (1990)	Folpet, 12.8 ng		2280d
0.000000006	Captan: daily US avg (1990)	Captan, 11.5 ng	269d	(2730d)

a = no data in CPDB; (—) = negative in cancer test; (+) = positive cancer test(s) not suitable for calculating a TD_{50}.

b = This is not an average, but a reasonably large sample (1027 workers).

c TD_{50} harmonic mean was estimated for the base chemical from the hydrochloride salt.

d Additional data from EPA that is not in the CPDB were used to calculate these TD_{50} harmonic means.

[Chemicals that occur naturally in foods are in bold.] Daily human exposure: Reasonable daily intakes are used to facilitate comparisons. The calculations assume a daily dose for a lifetime. Possible hazard: The human dose of rodent carcinogen is divided by 70 kg to give a mg/kg/day of human exposure, and this dose is given as the percentage of the TD_{50} in the rodent (mg/kg/day) to calculate the Human Exposure/Rodent Potency index (HERP), i.e. 100% means that the human exposure in mg/kg/day is equal to the dose estimated to give 50% of the rodents tumours. TD_{50} values used in the HERP calculation are averages calculated by taking the harmonic mean of the TD_{50}s of the positive tests in that species from the Carcinogenic Potency Database. Average TD_{50} values have been calculated separately for rats and mice, and the more potent value is used for calculating possible hazard.

Source. Gold et al.

that have evolved in animals to cope with the toxicity of natural chemicals will fail to protect against synthetic chemicals. This assumption is flawed for several reasons (Ames et al. 1990b; Ames et al. 1996).

(a) Humans have many natural defences that make us well buffered against normal exposures to toxins (Ames, Profet, Gold 1990b), and these are usually general, rather than tailored for each specific chemical. Thus they work against both natural and synthetic chemicals. Examples of general defences include the continuous shedding of cells exposed to toxins – the surface layers of the mouth, oesophagus, stomach, intestine, colon, skin, and lungs are discarded every few days; DNA repair enzymes, which repair DNA damaged from many different sources; and detoxification enzymes of the liver and other organs which generally target classes of toxins rather than individual toxins. That defences are usually general, rather than specific for each chemical, makes good evolutionary sense. The reason that predators of plants evolved general defences is presumably to be prepared to counter a diverse and ever-changing array of plant toxins in an evolving world; if a herbivore had defences against only a set of specific toxins, it would be at a great disadvantage in obtaining new food when favoured foods became scarce or evolved new toxins.

(b) Various natural toxins, which have been present throughout vertebrate evolutionary history, nevertheless cause cancer in vertebrates (Ames et al. 1990b; Gold et al. 1997). Mould toxins, such as aflatoxin, have been shown to cause cancer in rodents and other species including humans (Table 7.4). Many of the common elements are carcinogenic to humans at high doses (e.g. salts of cadmium, beryllium, nickel, chromium, and arsenic) despite their presence throughout evolution. Furthermore, epidemiological studies from various parts of the world show that certain natural chemicals in food may be carcinogenic risks to humans; for example, the chewing of betel nuts with tobacco has been correlated with oral cancer world-wide.

(c) Humans have not had time to evolve a 'toxic harmony' with all of their dietary plants. The human diet has changed dramatically in the last few thousand years. Indeed, very few of the plants that humans eat today (e.g. coffee, cocoa, tea, potatoes, tomatoes, corn, avocados, mangoes, olives, and kiwi fruit), would have been present in a hunter-gatherer's diet. Natural selection works far too slowly for humans to have evolved specific resistance to the food toxins in these newly introduced plants.

(d) DDT is often viewed as the quintessentially dangerous synthetic pesticide because it concentrates in the tissues and persists for years, being slowly released into the bloodstream. DDT, the first synthetic pesticide, eradicated malaria from many parts of the world, including the US. It was effective against many vectors of disease such as mosquitoes, tsetse flies, lice, ticks, and fleas. DDT was also lethal to many crop pests, and significantly increased the supply and lowered the cost of food, making fresh nutritious foods more accessible to poor people. It was also remarkably non-toxic to humans. A 1970 National Academy of Sciences report concluded: 'In little more than two decades DDT has prevented 500 million deaths due to malaria, that would otherwise have been inevitable (NAS 1970).' There is no convincing epidemiological evidence, nor is there much toxicological plausibility, that the levels normally found in the environment are likely to be a significant contributor to cancer. DDT was unusual with respect to bioconcentration, and because of its chlorine substituents it takes longer to degrade in nature than most chemicals; however, these are properties of relatively few synthetic chemicals. In addition, many thousands of chlorinated chemicals are produced in nature and natural pesticides also can bioconcentrate if they are fat soluble. Potatoes, for example, naturally contain the fat soluble neurotoxins solanine and chaconine, which can be detected in the bloodstream of all potato eaters. High levels of these potato neurotoxins have been shown to cause birth defects in rodents (Ames et al. 1990b).

(e) Since no plot of land is immune to attack by insects, plants need chemical defences – either natural or synthetic – in order to survive pest attack. Thus, there is a trade-off between naturally occurring pesticides and synthetic pesticides. One consequence of disproportionate concern about synthetic pesticide residues is that some plant breeders develop plants to be more insect-resistant by making them higher in natural toxins. A recent case illustrates the potential hazards of this approach to pest control: when a major grower introduced a new variety of highly insect-resistant celery people who handled the celery developed rashes when they were subsequently exposed to sunlight. Some detective work found that the pest-resistant celery contained 6,200 parts per billion (ppb) of carcinogenic (and mutagenic) psoralens instead of the 800 ppb present in common celery (Ames et al. 1990b).

Misconception No. 8: pesticides and other synthetic chemicals are disrupting our hormones

Synthetic hormone mimics are likely to be the next big environmental issue, with accompanying large expenditures. Hormonal factors are important in cancer (Misconception No. 2). A recent book (Colburn et al. 1996), holds that traces of synthetic chemicals, such as pesticides with weak hormonal activity, may contribute to cancer and reduce sperm counts. The book ignores the fact that our normal diet contains natural chemicals that have oestrogenic activity millions of times higher than that due to the traces of synthetic oestrogenic chemicals (Safe 1994, 1995) and that lifestyle factors can markedly change the levels of endogenous hormones (Misconception No. 2). The low levels of exposure to residues of industrial chemicals in humans are toxicologically implausible as a significant cause of cancer or of reproductive abnormalities, especially when compared with the natural background (Safe 1994, 1995; Reinli and Block 1996). In addition, it has not been satisfactorily shown that sperm counts really are declining (Kolata 1996), and, even if they were, there are many more likely causes, such as smoking and diet (Misconception No. 2).

Misconception No. 9: regulation of low hypothetical risks advances public health

There is no risk-free world, and resources are limited; therefore, society must set priorities based on which risks are most important in order to save the most lives. The EPA reports that its regulations cost $140 billion per year. It has been argued that overall these regulations harm public health (Viscusi 1992), because 'wealthier is not only healthier but highly risk reducing'. One estimate indicates 'that for every 1 per cent increase in income, mortality is reduced by 0.05 per cent' (Shanahan and Thierer 1996). In addition, the median toxin control programme costs 58 times more per life-year saved than the median injury prevention programme and 146 times more than the median medical programme (Tengs et al. 1995). It has been estimated that the US could prevent 60 000 deaths a year by redirecting resources to more cost-effective programems (Tengs and Graham 1996). The discrepancy is likely to be greater because cancer risk estimates used for toxin control programmes are worst-case, hypothetical estimates, and the true risks at low dose are often likely to be zero (Gold et al. 1992; Gold et al. 1997; Graham and Wiener 1995) (Misconception No. 5).

Regulatory efforts to reduce low-level human exposures to synthetic chemicals are expensive because they aim to eliminate minuscule concentrations that now can be measured with improved techniques. These efforts are distractions from the major task of improving public health through increasing knowledge, public understanding of how lifestyle influences health, and effectiveness in incentives and spending to maximize health. Basic biomedical research is the basis for improved public health and longevity, yet its cost is less than 10 per cent the cost to society of EPA regulations.

Of course, rules on air and water pollution are necessary (e.g. it was a public health advance to phase lead out of gasoline) and, clearly, cancer prevention is not the only reason for regulations. But worst-case scenarios, with their associated large costs to the economy, are not in the interest of public health and can be counterproductive.

References

Ames, B. N., Magaw, R., and Gold, L. S. (1987). Ranking possible carcinogenic hazards. *Science* **236**, 271-280.
Ames, B. N., and Gold, L. S. (1990). Chemical carcinogenesis: Too many rodent carcinogens. *Proc. Natl. Acad. Sci. USA* **87**, 7772-7776.
Ames, B. N., Profet, M., and Gold, L. S. (1990a). Dietary pesticides (99. 99 per cent all natural). *Proc. Natl. Acad. Sci. USA* **87**, 7777-7781.
Ames, B. N., Profet, M., and Gold, L. S. (1990b). Nature's chemicals and synthetic chemicals: Comparative toxicology. *Proc. Natl. Acad. Sci. USA* **87**, 7782-7786.
Ames, B. N., Shigenaga, M. K., and Gold, L. S. (1993a). DNA lesions, inducible DNA repair, and cell division: Three key factors in mutagenesis and carcinogenesis. *Environ. Health Perspect.* **101**(Suppl 5),35-44.
Ames, B. N., Shigenaga, M. K., and Hagen, T. M. (1993b). Oxidants, antioxidants, and the degenerative diseases of ageing. *Proc. Natl. Acad. Sci. USA* **90**, 7915-7922.
Ames, B. N., Motchnik, P. A., Fraga, C. G. et al. (1994). Antioxidant prevention of birth defects and cancer. In *Male-Mediated Developmental Toxicity*. (D. R. Mattison, and A. Olshan, eds). New York: Plenum Publishing Corporation, pp. 243-259.
Ames, B. N., Gold, L. S., and Willett, W. C. (1995). The causes and prevention of cancer. *Proc. Natl. Acad. Sci. USA* **92**, 5258-5265.
Ames, B. N., Gold, L. S., and Shigenaga, M. K. (1996). Cancer prevention, rodent high dose cancer tests and risk assessment. *Risk Anal.* **16**, 613-617.
Bailey, L. B., Wagner, P. A., Christakis, G. J. et al. (1979). Folacin and iron status and haematological findings in predominately black elderly persons from urban low-income households. *Am. J. Clin. Nutr.* **32**, 2346-2353.
Bailey, L. B., Wagner, P. A., Christakis, G. J. et al. (1982). Folacin and iron status and haematological findings in black and Spanish-American adolescents from urban low-income households. *Am. J. Clin. Nutr.* **35**, 1023-1032.
Block, G. (1992). The data support a role for antioxidants in reducing cancer risk. *Nutr. Reviews* **50**, 207-213.
Block, G., Patterson, B., and Subar, A. (1992). Fruit, vegetables and cancer prevention: A review of the epidemiologic evidence. *Nutr. and Canc.* **18**, 1-29.
Blount, B. C., Mack, M. M., Wehr, C. et al. (1997). Folate deficiency causes uracil misincorporation into human DNA and chromosome breakage: Implications for cancer and neuronal damage. *Proc. Natl. Acad. Sci. USA* **94**, in press.
Butterworth, B., Conolly, R., and Morgan, K. (1995). A strategy for establishing mode of action of chemical carcinogens as a guide for approaches to risk assessment. *Canc. Lett.* **93**, 129-146.
Christen, S., Hagen, T. M., Shigenaga, M. K. et al. (1997). Chronic infection and inflammation lead to mutation and cancer. In *Microbes and Malignancy: Infection as a Cause of Cancer.* (J. Parsonnet, S. Horning, eds) Oxford: Oxford University Press, in press.
Colburn, T., Dumanoski, D., and Myers, J. P. (1996). *Our Stolen Future: Are we Threatening our Fertility, Intelligence, and Survival?: A Scientific Detective Story.* New York: Dutton.
Devesa, S. S., Blot, W. J., Stone, B. J. et al. (1995). Recent cancer trends in the United States. *J. Natl. Canc. Inst.* **87**, 175-182.
Doll, R., and Peto, R. (1981). The causes of cancer. Quantitative estimates of avoidable risks of cancer in the United States today. *J. Natl. Canc. Inst.* **66**, 1191-1308.
Feigelson, H. S., and Henderson, B. E. (1996). Oestrogens and breast cancer. *Carcinogenesis* **17**, 2279-2284.
Fraga, C. G., Motchnik, P. A., Shigenaga, M. K. et al. (1991). Ascorbic acid protects

against endogenous oxidative damage in human sperm. *Proc. Natl. Acad. Sci. USA* **88**, 11003-11006.

Fraga, C. G., Motchnik, P. A., Wyrobek, A. J. et al. (1996). Smoking and low antioxidant levels increase oxidative damage to sperm DNA. *Mutat. Res.* **351**, 199-203.

Gaylor, D. W., and Gold, L. S. (1995). Quick estimate of the regulatory virtually safe dose based on the maximum tolerated dose for rodent bioassays. *Regul. Toxicol. Pharmacol.* **22**, 57-63.

Gold, L. S., Slone, T. H., Stern, B. R. et al. (1992). Rodent carcinogens, Setting priorities. *Science* **258**, 261-265.

Gold, L. S., Slone, T. H., Stern, B. R. et al. (1993). Possible carcinogenic hazards from natural and synthetic chemicals: Setting priorities. In Cothern, C. R. (ed.), *Comparative Environmental Risk Assessment.* Boca Raton, FL: Lewis Publishers, pp. 209-235.

Gold, L. S., Garfinkel, G. B., and Slone, T. H. (1994a). Setting priorities among possible carcinogenic hazards in the workplace. In *Chemical Risk Assessment and Occupational Health, Current Applications, Limitations, and Future Prospects.* (C. Smith, D. C. Christiani, K. T. Kelsey, eds). Westport, CT: Greenwood Publishing Group, pp. 91-103.

Gold, L. S., Slone, T. H., Manley, N. B., and Ames, B. N. (1994b). Heterocyclic amines formed by cooking food: Comparison of bioassay results with other chemicals in the Carcinogenic Potency Database. *Canc. Lett.* **83**, 21-29.

Gold, L. S., Slone, T. H., and Ames, B. N. (1997a). Prioritization of possible carcinogenic hazards in food. In Tennant, D. (ed.), *Food Chemical Risk Analysis.* London: Chapman and Hall, in press.

Gold, L. S., Slone, T. H., and Ames, B. N. (1997b). Overview and update analyses of the carcinogenic potency database. In *Handbook of Carcinogenic Potency and Genotoxicity Databases.* (L. S. Gold, and E. Zeiger, eds) Boca Raton, FL: CRC Press, 661-685.

Gold, L. S., Slone, T. H., Manley, N. B. et al. (1997c). Carcinogenic potency database. In *Handbook of Carcinogenic Potency and Genotoxicity Databases* (L. S. Gold, and E. Zeiger, eds) Boca Raton, FL: CRC Press, pp. 1-605.

Gough, M. (1990). How much cancer can EPA regulate anyway? *Risk Anal.* **10**, 1-6.

Graham, J., and Wiener, J., eds (1995). *Risk versus Risk: Trade-offs in Protecting Health and the Environment.* Cambridge, Massachusetts: Harvard University Press.

Hahn, R. W., ed. (1996). *Risks, Costs, and Lives Saved: Getting Better Results from Regulation.* New York: Oxford University Press.

Henderson, B. E., Ross, R. K., and Pike, M. C. (1991). Toward the primary prevention of cancer. *Science* **254**, 1131-1138.

Herbert, V. (1996). Vitamin B-12. In *Present Knowledge in Nutrition* (Ziegler, E. E., and Filer, L. J., eds). Washington, DC: ILSI Press, pp. 191-205.

Hill, M. J., Giacosa, A., and Caygill, C. P. J. (1994). *Epidemiology of Diet and Cancer.* Chichester: Ellis Horwood.

Hunter, D. J., and Willett, W. C. (1993). Diet, body size, and breast cancer. *Epidemiol. Rev.* **15**, 110-132.

Innes, J. R. M., Ulland, B. M., Valerio, M. G. et al. (1969). Bioassay of pesticides and industrial chemicals for tumourigenicity in mice: A preliminary note. *J. Natl. Canc. Inst.* **42**, 1101-1114.

Jacobson, E. L. (1993). Niacin deficiency and cancer in women. *J. Am. Coll. Nutr.* **12**, 412-6.

Ji, B. -T., Shu, X. -O., Linet, M. S. et al. (1997). Paternal cigarette smoking and the risk of childhood cancer among offspring of non-smoking mothers. *J. of Natl. Canc. Inst.* **89**, 238-244.

Kolata, G. (1996). Measuring men up, sperm by sperm. *New York Times*, May 4 E4(N), E4(L), (col. 1).

Kosary, C. L., Ries, L. A. G., Miller, B. A. et al. (eds) (1995). *SEER Cancer Statistics Review, 1973-1992.* National Cancer Institute, Bethesda, MD.
National Academy of Sciences (US). Committee on Research in the Life Sciences. (1970). *The Life Sciences: Recent Progress and Application to Human Affairs, the World of Biological Research, Requirement for the Future.* National Academy of Sciences, Washington.
National Research Council (1994). *Science and Judgement in Risk Assessment.* Committee on Risk Assessment of Hazardous Air Pollutants, Washington, DC.
National Research Council (1996). *Carcinogens and Anti-carcinogens in the Human Diet: A Comparison of Naturally Occurring and Synthetic Substances.* Washington, DC: National Academy Press.
National Cancer Institute Graphic (A). (1996). Why eat five? *J. Natl. Canc. Inst.* **88**, 1314.
Patterson, B. H., Block, G., Rosenberger, W. F. et al. (1990). Fruit and vegetables in the American diet: Data from the NHANES II survey. *Am. J. Public Health* **80**, 1443-1449.
Peto, R., Lopez, A. D., Boreham, J. et al. (1994). *Mortality from Smoking in Developed Countries 1950-2000.* Oxford: Oxford University Press.
Reinli, K., and Block, G. (1996). Phyto-oestrogen content of foods – A compendium of literature values. *Nutr. Canc.* **26**, 1996.
Safe, S. H. (1994). Dietary and environmental oestrogens and anti-oestrogens and their possible role in human disease. *Environ. Sci. Pollution Res.* **1**, 29-33.
Safe, S. H. (1995). Environmental and dietary oestrogens and human health: Is there a problem? *Env. Health Persp.* **103**, 346-351.
Senti, F. R., and Pilch, S. M. (1985). Analysis of folate data from the second National Health and Nutrition Examination Survey (NHANES II). *J. Nutr.* **115**, 1398-402.
Shacter, E., Beecham, E. J., Covey, J. M. et al. (1988). Activated neutrophils induce prolonged DNA damage in neighbouring cells [published erratum appears in Carcinogenesis 1989 Mar 10(3),628]. *Carcinogenesis* **9**, 2297-2304.
Shanahan, J. D., and Thierer, A. D. (1996). How to talk about risk: How well-intentioned regulations can kill: TP13, Washington, DC, Heritage Foundation.
Steinmetz, K. A., and Potter, J. D. (1991). Vegetables, fruit, and cancer. I. Epidemiology. *Cancer Causes Control* **2**, 325-357.
Steinmetz, K. A., and Potter, J. D. (1996). Vegetables, fruit, and cancer prevention: A review. *J. Am. Diet Assoc.* **96**, 1027-1039.
Subar, A. F., Block, G., and James, L. D. (1989). Folate intake and food sources in the US population. *Am. J. Clin. Nutr.* **50**, 508-16.
Tengs, T. O., Adams, M. E., Pliskin, J. S. et al. (1995). Five-hundred life-saving interventions and their cost-effectiveness. *Risk Anal.* **15**, 369-389.
Tengs, T. O., and Graham, J. D. (1996). The opportunity costs of haphazard social investments in life-saving. In: *Risks, Costs, and Lives Saved: Getting Better Results from Regulation* (R. Hahn, ed.). New York: Oxford University Press, pp. 165-173.
Viscusi, W. K. (1992). *Fatal Trade-offs.* Oxford: Oxford University Press.
Wallock, L., Woodall, A., Jacob, R., and Ames, B. (1997). Nutritional status and positive relation of plasma folate to fertility indices in non-smoking men. *FASEB J. (Abstract)* in press.
Wickramasinghe, S. N., and Fida, S. (1994). Bone marrow cells from vitamin B12- and folate-deficient patients misincorporate uracil into DNA. *Blood* **83**, 1656-61.
Yamashina, K., Miller, B. E., and Heppner, G. H. (1986). Macrophage-mediated induction of drug-resistant variants in a mouse mammary tumour cell line. *Cancer Res.* **46**, 2396-2401.
Zhang, J. Z., Henning, S. M., and Swendseid, M. E. (1993). Poly(ADP-ribose) polymerase activity and DNA strand breaks are affected in tissues of niacin-deficient rats. *J. Nutr.* **123**, 1349-55.

8 Interpretation of epidemiological studies with modestly elevated relative risks

Göran Pershagen

Summary

Epidemiological studies showing strong effects are rarely controversial and biases are unlikely to alter the fundamental conclusions of such studies. However, for studies that show a relatively weak association – a relative risk of 2 or lower – as is typical for exposures to environmental hazards, problems caused by biases must be assessed carefully. Indeed, such biases may explain all or part of the sometimes observed increase in relative risk of lung cancer in cases exposed to environmental tobacco smoke (ETS). This paper reviews the literature on problems caused by biases in such studies.

Epidemiological bias is subdivided into three categories: selection, misclassification and confounding.

Selection bias occurs because participants in a study are not drawn at random from the population. The problem tends to be more severe in case–control studies because the type of selection bias is likely to vary between cases and controls.

Misclassification of exposure and health effects is very common, occurring both randomly and systematically, and introducing bias into most epidemiological studies.

Confounding occurs when there is a correlation between the health effect of concern and an unaccounted tendency in cases which is also correlated to the suspected causal agent.

New developments in molecular biology may make it possible to enhance the resolution power of epidemiological studies by increasing the precision of exposure information, enhancing the specificity of outcome measures and identifying susceptible groups.

In order for a researcher to draw a negative conclusion (rather than an inconclusive one) from a study, the study must be large and the exposure data precise. Unsurprisingly, negative results are rarely reported.

Introduction

Epidemiological evidence is increasingly used in risk assessment. When risks are substantial, such as lung cancer from tobacco smoking and from some occupational exposures, there is generally little controversy in the interpretation of epidemiological findings. However, when associations are weak, i.e. relative risks of the order of two or lower, the consequences of bias have to be assessed carefully. Most studies of risk factors in the general environment show only modestly elevated risks, but these effects may still be important from a public health point of view if large numbers of people are exposed.

The understanding of bias and its consequences is central to epidemiology. This is particularly important in the evaluation of epidemiological evidence showing weak associations. The aim of this review is to discuss qualitative and quantitative implications of bias for the interpretation of epidemiological studies with modestly elevated relative risks. Both risk identification and risk estimation will be addressed. Examples are primarily taken from studies on environmental causes of lung cancer, such as environmental tobacco smoke (ETS) and residential radon exposure. However, it is not the intention to evaluate comprehensively the scientific evidence on certain environmental exposures but to illustrate methodological problems in the interpretation of data.

Role of bias

Several definitions of bias in epidemiological studies have been proposed (Miettinen 1985; Rothman 1986; Norell 1992; Steineck and Ahlbom 1992). The definition used here is based on the concept of the study base, i.e. the person-time experience in which an epidemiological study is conducted (Miettinen 1985). Consequently, bias is subdivided into selection, misclassification and confounding pertaining to the conditions in the study base (Norell 1992). Since a longitudinal design generally provides the most conclusive epidemiological evidence, the discussion is focused on cohort and case–control studies.

Selection

Bias may result from the selection of study subjects for an epidemiological study. In most epidemiological investigations all individuals cannot be followed up and classified in respect of disease. This can result in an over- or under-estimation of the relative risk. For example, in a hospital-based case–control study on residential radon exposure and lung cancer it was indicated that only about half of the cases in the population were included (Svensson 1988). Furthermore, referral to the hospitals was associated with urbanization, which showed some relation to a residential radon level. A potential bias was avoided by the use of hospital controls experiencing a similar type of selection.

Particular problems arise in the selection of controls for case–control studies. Ideally, controls should reflect the exposure situation in the study base (Miettinen 1985). They may be sampled from the population or selected from among hospital patients. It is generally easier to achieve representativeness for population controls, but the non-response is often higher than for hospital controls. Both hospitals and population controls are sometimes used in the same study (Pershagen et al. 1992), and findings are strengthened if results using either type of control group show consistency.

In retrospective studies of lung cancer, and other diseases with a high lethality, it is often impossible to obtain exposure information directly from the study subjects. To enhance comparability of the exposure information, deceased controls are sometimes used (Pershagen et al. 1987; Pershagen et al. 1994). If the exposures of interest affect mortality, a misrepresentation will result, leading to biased estimates of relative risks (McLoughlin et al. 1985). For example, smoking habits are over-represented among dead controls, which gives rise to an under-estimation of risks associated with smoking (Järup and Pershagen 1991). If the only purpose is to control confounding from smoking, unbiased estimates of relative risks associated with other exposures may be achieved if these do not influence mortality (Howe 1991). Over-representation of smoking among deceased controls is avoided if those with smoking-related causes of death are excluded (Pershagen et al. 1994).

Misclassification

Most epidemiological investigations suffer from misclassification of exposure and health effects. The measurements of exposure and health effects can be afflicted with random and systematic errors. Both types of error result in biased estimates of the relative risk. It is essential to assess how misclassification can affect the direction and magnitude of this bias.

Imprecise exposure data will generally lead to dilution effects, i.e. the association between exposure and disease is weakened (Armstrong 1990). For example, in the studies on ETS and lung cancer, the exposure data were obtained from interviews or questionnaires, and, with few exceptions, no validation of the information was attempted (EPA 1992). The most common measure of exposure was spouses who smoked, which may be a more accurate predictor of ETS exposure for women than for men, where other sources predominate (Cummings et al. 1989; Sandler et al. 1989). It is clear, however, that there are many sources of ETS exposure which should be considered if the total exposure is to be assessed and that this depends also on cultural factors (Coughlin et al. 1989; Cummings et al. 1989; Riboli et al. 1990). On the other hand, the failure in some studies to obtain a clearer association between ETS exposure and lung cancer risk using more complex exposure models than smoking by spouses, may partly be explained by less reliable reporting of such exposures (Pron et al. 1988).

Attempts have been made to assess quantitatively the influence of non-differential misclassification of exposure on the risk estimates for ETS and lung cancer (Wald et al. 1986). Using data on urinary cotinine excretion, it is suggested that the relative risk comparing exposed with truly unexposed to ETS should be more in the order of 1.5 if the observed relative risk comparing non-smokers living with smokers and non-smokers is 1.3. Only recent exposure was included in this calculation and the imprecision will be even greater if lifetime exposure is of importance.

Misclassification of exposure may also be differential between cases and controls. A difference in the quality of exposure data provided by cases and controls is often referred to as recall bias. Some of the studies on ETS and lung cancer, which included deceased subjects, did not match on vital status in the selection of controls (EPA 1992). As a rule, this led to a higher percentage of surrogate respondents for the cases because of the high lethality of lung cancer. Qualitative differences in the information on ETS exposure between index and surrogate respondents could result in biased estimates for the relative risks, however, one study suggests that this would be unimportant (Cummings et al. 1989).

Non-differential misclassification of the health effects will also result in a dilution of the association. For example, secondary lung tumours and carcinomas with unknown primary site appeared in about one-sixth of reported cases of lung cancer on death certificates in the USA (Garfinkel 1981) and in central health registers in Sweden (Pershagen et al. 1987) among female non-smokers. When rare diseases are studied, the specificity of the measurement of disease is of greater importance for the attenuation effect than the sensitivity (Norell 1992).

Confounding

Confounding is unlikely to explain relative risks in the order of two or higher (Axelson 1978). However, an adequate control of confounding is crucial in epidemiological studies with modestly elevated relative risks. In the studies on ETS and lung cancer, probably the most important source of bias is confounding by unreported active smoking. There is a tendency for smokers to marry smokers, and in conjunction with some misclassification of smokers as non-smokers, this would tend to produce a spurious association between spousal smoking and lung cancer.

One crucial factor for estimation of the bias is the proportion of those classified as non-smokers who actually have been smokers. Using biological markers of nicotine exposure, it has been indicated that up to 3 per cent of subjects classifying themselves as non-smokers have levels consistent with regular smoking (Wald et al. 1986; Lee 1988; Riboli et al. 1990; Thompson et al. 1990). Studies involving multiple reports of smoking over several years as well as studies using surrogates suggest that in the order of about 5 per cent of ever-smokers may deny smoking (Lee 1988). Several investigations indicate surrogates can provide valid information to separate smokers from non-smokers (Pershagen 1984; McLoughlin et al. 1985; Metzner et al. 1989; US Surgeon General 1990; Nelson et al. 1994), which makes this source of information useful for validations. Very limited and inconclusive data are available for smoking misclassification at interview with newly diagnosed lung cancer cases, which was the source of information in most epidemiological studies on ETS exposure and lung cancer.

A few attempts have been made to quantify the influence of confounding by smoking on the risk estimates for lung cancer in non-smokers living with smokers. Using different assumptions, quite divergent conclusions were reached. Wald et al. (1986) claimed that this source of bias would explain only about 15 per cent of the observed risk, while Lee (1988) argued that most of the association was due to this factor. It is likely that the extent of smoking confounding bias differs between studies because of differences in smoking habits, and risks associated with smoking and misclassification rates.

Negative confounding is probably as common as positive confounding, and also deserves attention in the interpretations of epidemiological evidence. For example, negative confounding by smoking has been observed in studies on residential radon exposure and lung cancer (Létourneau et al. 1994; Pershagen et al. 1994). This may lead to under-estimation of the true effect or that no association is observed. Imprecision in the information on confounders, such as smoking habits, may give rise to residual confounding, even if these factors are controlled in the analysis.

Risk identification

In general, it is not possible to draw conclusions on causality from a single epidemiological study. This is particularly relevant if only weak associations were observed. Instead, the interpretation must rely on evidence from different sources, including animal experiments and other epidemiological studies. The evaluation is facilitated if the findings are supported by knowledge regarding etiologic mechanisms. For example, the credibility of a causal relation between exposure to electromagnetic fields and cancer based on epidemiological observations has been challenged in light of the lack of evidence of plausible mechanisms. On the other hand, cancer induction by low-level exposure to tobacco smoke and radon progeny may be less unlikely.

In an effort to enhance their statistical power, data from different epidemiological studies are sometimes combined in meta- or pooled analyses. Meta-analyses use summary measures from different studies, such as point estimates of relative risk and variance, while individual data for each study subject are used in the pooled analysis (Friedenreich 1993). Pooled analyses are generally more informative that meta-analyses, particularly in relation to the possibility for studies of subgroups.

It is questionable to what extent meta- or pooled analyses of epidemiological studies are useful for assessment of causality. For example, meta-analysis of the epidemiological studies on ETS and lung cancer in non-smokers clearly shows that there is a statistically significant increase in the relative risk of about 2–30 per cent (EPA 1992; Lee 1992). As indicated above, relative risks in this range may be explained by bias. An evaluation of qualitative and quantitative aspects of bias is this central when causality is addressed in the interpretation. Unfortunately, important information on data quality and potential confounding factors is often lacking. Evidence of consistency between and within studies, such as exposure-response relationships, facilitates the evaluation.

A meta-analysis may sometimes complicate the assessment of causality. It is difficult to consider adequately quality aspects of different studies in the combined analysis. Furthermore, there may be real differences in the exposure-response relation between studies because of interactions. For example, a recent pooled analysis of three studies on residential radon exposure and lung cancer showed no association (Lubin et al. 1994). However, two of the studies (from New Jersey and Stockholm) indicated positive relation, but not the third study. This study was performed in Shenyang, China, where air pollution levels from both indoor and outdoor sources were extremely high, and this may have affected the radiation doses to the bronchial epithelium (James 1988). There was no clear heterogeneity between the studies upon statistical testing, but these tests generally have a low power (Friedrenreich 1993).

In the interpretation of studies with modestly elevated relative risks, it is essential to differentiate between inconclusive and negative evidence (Ahlbom et al. 1990). Several criteria have to be met if an investigation is to be interpreted as negative. The observed relative risk should lie close to unity, with a narrow confidence interval. This necessitates a large study base if exposure and/or disease occurrence is rare. Furthermore, the exposure information must have a high precision, and negative confounding effects should be adequately controlled. It is difficult to fulfil these requirements at the same time, which means that conclusively negative epidemiological studies are rare.

Using new developments in molecular biology, it may be possible to enhance the resolutions power of epidemiological studies. This can involve greater precision of the exposure information, enhanced specificity of outcome measures and identification of susceptible groups. For example, exposure to genotoxic agents can be monitored by measurements of DNA-adducts and mutations of marker genes, such as the hypoxantine–guanin–phosphoribosyltransferase (hprt) gene (Hemminki et al. 1990, Tates et al. 1991). Furthermore, specific mutational patterns in genes involved in cancer induction, such as the p-53 tumour-suppresser gene, may be more closely linked to certain exposures (Vähäkangas et al. 1992). Genetically determined variations in the ability to metabolize xenobiotics could make it possible to focus on highly susceptible subgroups of the population in epidemiological studies (Idle 1991).

Risk estimation

When a causal association in considered credible, epidemiological investigations are particularly useful for quantitative risk assessment. If studies are population-based, the proportion of disease attributed to the exposure (aetiologic fraction) may also be computed. This is essential for setting priorities in prevention, but should be used cautiously, with due attention to the problems of quality that affect epidemiological evidence. However, the uncertainties are usually even greater when risk estimates are based on data in experimental animals.

In view of the difficulties in determining risks with high accuracy at low exposures in epidemiological studies, risk estimation is sometimes based on extrapolation from high doses. For example, studies in miners have constituted the basis for quantitative risk assessment on residential radon exposure and lung cancer (NAS 1988; ICRP 1993). The linear relative risk model is used widely for risk estimation regarding carcinogens (EPA 1983; WHO 1897). It is supported to some extent by empirical data down to low doses for agents causing mutations (NAS 1988). On the other hand, it may be more relevant to assume threshold effects for agents operating at other stages in the cancer induction process (Paustenbach 1989). Risk estimation based on downward extrapolation is most uncertain at the lowest doses. The exposure to environmental agents, such as residential radon, is often skewed which implies that the population attributable proportion is highly dependent on the estimates in the low exposure range.

Imprecision in exposure estimation is likely to cause an attenuation of the exposure-response relation (Armstrong 1990). If the precision of the method for exposure measurement is known, it is possible to adjust the risk estimates for this bias. For example, there have been great efforts to assess the quality of different methods of dietary assessment in epidemiological studies, and corrections have sometimes been performed of the risk estimates (Rosner et al. 1989). Unfortunately, the exact quality of most methods for exposure estimation is not known, which makes it difficult to evaluate adequately the degree of attenuation of the exposure-response relations.

A recent epidemiological study from Sweden showed an increased lung cancer risk related to estimated time weighted residential radon exposure (Pershagen et al. 1994). The increase in risk per unit exposure appeared linear and multiplied the risk associated with smoking. Although great efforts were made to obtain detailed information on radon exposure, including measurement in close to 9,000 homes of the study subjects, it is clear that a substantial uncertainty remained in the exposure estimation. Several methods used to correct for imprecision in the exposure information indicated that the observed risks represented an underestimation of the true risk by a factor of about two. Obviously, this must have profound effects on the risk assessment.

Conclusions

There is an inherent conflict between the demands from regulatory agencies as well as the general public for data enabling precise risk estimation and the information science can deliver. Qualitative and quantitative risk assessment is often based on epidemiological evidence showing modestly elevated relative risks, i.e. in the order of two or lower. The consequences of bias may be particularly deleterious for the interpretation of relative risks in this range. Important potential sources of bias include selection effects, misclassification and confounding.

Risk identification based on epidemiological data with modestly elevated relative risks should focus more on the quality of the studies and less on the statistical properties of the associations. Consequently, meta- or pooled analyses of different studies are of limited value in this respect. Evidence of internal consistency, such as exposure-response relationships and coherence between studies in more important for the interpretation.

In evaluating epidemiological investigations showing modestly elevated relative risks, it is essential to differentiate between inconclusive and negative studies, i.e. studies indicating no association between exposure and disease. Several criteria have to be met if a study is to be regarded as negative, including a large size and precise exposure data. Negative epidemiological studies are undoubtedly rare.

When a causal association is considered credible, epidemiological studies may be useful for quantitative risk assessment. Imprecision in the exposure estimation is likely to cause attenuation of the exposure–response relations, and may lead to substantial under-estimation of attributable risks. Risk estimates can be adjusted for this bias if the precision of the exposure measures is known.

In view of the difficulties in determining risks with high accuracy at low exposures in epidemiological studies, risk estimation is often based on extrapolation from high doses. Obviously, such risk estimates are most uncertain at the lowest doses. The uncertainty in the estimation of population attributable risks is aggravated because exposure to environmental agents are often skewed, with a large fraction of the population in the low dose range.

Acknowledgement

An earlier version of this chapter originally appeared in *The Role of Epidemiology in Regulatory Risk Assessment* (John Graham ed. 1995). Elsevier, pp. 53–61.

References

Ahlbom A., Axelson O., Stöttrup Hansen E. et al. (1990). Interpretation of negative studies in occupational epidemiology. *Scand. J. Work Environ. Health.* **16**, 153–157.

Armstrong B. (1990). The effects of measurement errors on relative risk regressions. *Am. J. Epidemiol.* **132**, 1176–1184.

Axelson O. (1978). Aspects on confounding in occupational health epidemiology. *Scand. J. Work Environ. Health.* **4**, 98–102.

Coughlin J., Hammond S. K. and Gann P. H. (1989). Development of epidemiological tools for measuring environmental tobacco smoke exposure. *Am. J. Epidemiol.* **130**, 696–704.

Cummings K. M., Markello S. J., Mahoney M. C. and Marshall J. R. (1989). Measurement of lifetime exposure to passive smoke. *Am. J. Epidemiol.* **130**, 122–132.
EPA (1983). Health assessment document for acrylonitrile. Washington DC: US Environmental Protection Agency.
EPA (1992). Respiratory health effects of passive smoking: lung cancer and other disorders. Washington DC: US Environmental Protection Agency.
Friedrenreich CM. (1993). Methods for pooled analyses of epidemiological studies. *Epidemiology* **4**, 295–302.
Garfinkel L. (1981Time trends in lung cancer mortality among non-smokers and a note on passive smoking. *J. Natl. Cancer. Inst.* **66**, 1061–1066.
Hemminki K., Grzybowska E., Chorazy M. et al. (1990). DNA adducts in humans environmentally exposed to aromatic compounds in an industrial area of Poland. *Carcinogenesis* **11** ,1229–1231.
Howe G. R. (1991). Using dead controls to adjust for confounders in case–control studies. *Am. J. Epidemiol.* **134**, 689–690.
Idle J. R. (1991). Is environmental carcinogenesis modulated by host polymorphism? *Mutation Res.* **247**, 259.
International Commission on Radiological Protection. (1993). Protection against Radon–222 at home and at work. ICRP Publication 65. *Ann. ICRP* **23**, 1–45.
James AC. (1988). Radon and its decay products in indoor air. In Nazaroff W. W. and Nero A. V. (eds) *Lung Dosimetry.* New York: John Wiley and Sons.
Järup L. and Pershagen G. (1991). Arsenic exposure, smoking and lung cancer in smelter workers – a case–control study. *Am. J. Epidemiol.* **134**, 545–551.
Lee P. N. (1992).*Environmental Tobacco Smoke and Mortality.* Basel: Karger.
Lee P. N. (1988). *Misclassification of Smoking Habits and Passive Smoking.* Berlin: Springer.
Létourneau E. G., Krewski D., Choi N. W. et al. (1994). A case–control study of residential radon and lung cancer in Winnipeg, Manitoba. *Am. J. Epidemiol.* **140**, 310–322.
Lubin J. H., Liang Z., Hrubec Z. et al. (1994). Radon exposure in residences and lung cancer among women: Combined analysis of three studies. *Cancer Causes Control* **5**, 114–128.
McLoughlin J. K., Dietz M. S., Mehl E. S. and Blot W. J. (1985). Reliability of surrogate information on cigarettes smoking by type of informant. *Am. J. Epidemiol.* **126**, 144–146.
Metzner H. L., Lamphiear D. E., Thompson F. E. et al. (1989). Comparison of surrogate and subject reports of dietary practices, smoking habits and weight among married couples in the Tecumseh diet methodology study. *J .Clin. Epidemiol.* **42**, 367–375.
Miettinen O. S. (1985). *Principles of Occurrence Research in Medicine.* New York: John Wiley and Sons.
National Academy of Sciences. (1988). Radon and other internally deposited alpha-emitters. BEIR IV. Washington DC: National Academy Press.
Nelson L. M., Longstreth W. T., Koepsell T. D. et al. (1994). Completeness and accuracy on interview data from proxy respondents: demographic, medical and lifestyle factors. *Epidemiology* **45**, 204–217.
Norell S. (1992). *A Short Course in Epidemiology.* New York: Raven Press.
Paustenbach D. J. (1989). Important recent advances in the practice of health risk assessment: implications for the 1990s. *Regul Toxicol Pharmacol.* **10**, 204–243.
Pershagen G. (1984). Validity of questionnaire data on smoking and other exposures, with special reference to environmental tobacco smoke. *Eur J Resp Dis.* **65**, 76–80.
Pershagen G., Hrubec Z. and Svensson C. (1987). Passive smoking and lung cancer in

Swedish women. *Am. J. Epidemiol.* **125**, 17–24.
Pershagen G., Liang Z. H., Hrubec Z., Svensson, C. and Boice J. D. Jr. (1992). Residential radon exposure and lung cancer in women. *Health Phys.* **63**, 179–186.
Pershagen G., Åkerblom G., Axelson O. et al. (1994). Residential radon exposure and lung cancer in Sweden. *N. Engl. J. Med.* **330**, 159–164.
Pron G. E., Burch D. J., Howe G. R. and Miller A. B. (1988).The reliability of passive smoking histories reported in a case–control study of lung cancer. *Am. J. Epidemiol.* **127**, 267–284.
Riboli E., Preston-Martin S., Saracci R. et al. (1990). Exposure of non-smoking women to environmental tobacco smoke: a ten-country collaborative study. *Cancer Causes Control* **1**, 243–252.
Rosner E, Willett W. C. and Spiegelman D. (1989). Correction of logistic regression relative risk estimates and confidence intervals for systematic within-person measurement error. *Stat. Med.* **8**, 1051–1069.
Rothman K. J. (1986). *Modern Epidemiology*. Boston: Little, Brown.
Sandler D. P., Helsing K. J., Comstock G. W. and Shore D. L. (1989). Factors associated with past household exposure to tobacco smoke. *Am. J. Epidemiol.* **129**, 380–387.
Steineck G and Ahlbom A. (1992). A definition of bias founded on the concept of the study base. *Epidemiology* **3**, 477–482.
Svensson C. Lung cancer etiology in women. (1988). Stockholm: Karolinska Institutet PhD thesis.
Tates A. D., Grummt T., Törnqvist M. et al. (1991). Biological and chemical monitoring of occupational exposure to ethylene oxide. *Mutation Res.* **250**, 483–497.
Thompson S. G., Stone R., Nanchahal K. and Wald N. J. (1990). Relation of urinary cotinine concentrations to cigarette smoking and to exposure to other people's smoke. *Thorax* **45**, 356–361.
US Surgeon General. (1990). The health benefits of smoking cessation. Washington DC: United States Department of Health and Human Services.
Vähäkangas K. H., Samet J. M., Metcalf R. A. et al. (1992). Mutations of p53 and ras genes in radon-associated lung cancer from uranium miners. *Lancet* **339**, 576–579.
Wald N. J., Nanchahal K., Thompson D. G. and Cuckle H. S. (1986). Does breathing other people's tobacco smoke cause lung cancer? *Br. Med. J.* **293**, 1217–1222.
World Health Organisation. (1987). Air quality guidelines for Europe. European series No 23. Copenhagen: WHO Regional Office for Europe.

9 The risks of dioxin to human health

Hans E. Müller

Summary

Dioxins are toxic by-products which are generated in small amounts by natural burning processes and technical synthesis of some chlorinated organic compounds. The accident in Seveso, Italy, in 1976 was viewed by the public as a tragedy of apocalyptic proportions. Many birds, chickens, and rabbits died, trees lost their leaves, and several thousand people were subsequently evacuated. Since then, dioxin has been seen as one of the most toxic chemicals known to man. Notwithstanding the fact that only a few people have ever died of acute dioxin poisoning, fear of dioxin has, for the past 20 years, been responsible for generating an enormous number of regulations.

The primary symptom of acute dioxin intoxication is chloracne, although hepatopathy, nervous disorders, gastrointestinal effects and skin lesions also appear occasionally. Chronic effects include mutagenic, carcinogenic and teratogenic effects, as well as adverse effects on immunity and reproduction. But epidemiological studies revealed that these consequent effects occur only for dioxin concentrations that are higher by a factor of 50 to 100 than that likely to occur in the general population following ingestion of dioxin from food (the primary source of exposure for humans). Since regulations limiting dioxin production were introduced, its environmental burden and concentration in food has been decreasing, so fears of imminent dioxin poisoning are now totally unjustified.

Introduction

So-called dioxins are often referred to as the most toxic man-made chemicals known. To be more exact, there are two groups of chemical compounds showing a similar pattern of toxicity: 75 structure isomer or congener chlorodibenzo-p-dioxins and its best known member, the 2,3,7,8-tetrachlorodibenzo-p-dioxin (TCDD), and 135 further congener chlorobenzofurans, from which the analogous substance 2,3,7,8-tetrachlorodibenzofuran (TCDF) is the most toxic. Regarding the polychlorinated dioxins (PCDDs) and polychlorinated dibenzofurans (PCDFs) from a chemical point of view, the molecular stability and symmetry corresponds to with the toxicity. In Table 9.1 is shown the rank of relative acute toxicity of some PCDDs and PCDFs (Müller 1993). The so-called toxic equivalents (TEQ) are values reduced to the toxicity of 2,3,7,8-TCDD expressed in terms of LD_{50} for the guinea pig.

Table 9.1. The different toxicity (LD_{50}) of PCDDs and PCDFs in relation to that of 2,3,7,8-TCDD and the appropriate toxic equivalents (TEQ)

Compound	*LD_{50} (mg/kg)*	*TEQ*
2,3,7,8-tetrachloro-DD[1]	1	1
1,2,3,7,8-pentachloro-DD	2	0.5
1,2,3,7,8,9-hexachloro-DD	10	0.1
1,2,3,4,6,7,8-heptachloro-DD	100	0.01
1,2,4,7,8-pentachloro-DD	1,125	0.00089
2,3,7-trichloro-DD	30,000	0.00003
2,3,4,7,8-pentachloro-DF[2]	2	0.5
2,3,7,8-tetrachloro-DF	10	0.1
1,2,3,7,8,9-hexachloro-DF	10	0.1
1,2,3,7,8-pentachloro-DF	20	0.05
1,2,3,4,7,8,9-heptachloro-DF	100	0.01

[1]DD = dibenzo-p-dioxin [2] DF = dibenzofuran

The mounting fears of dioxin and the accident at Seveso

Dioxins were synthesized for the first time in the last century; they have been produced as by-products in all chlorination processes of aromatic compounds for some decades. But the first observations of dioxin toxicity did not appear until the late 1940s, following accidents at Monsanto, BASF, and other herbicide manufactures (see Table 9.2).

Table 9.2 Chemical accidents releasing dioxins and numbers of people involved

Plant, situated	*Country*	*Year*	*Exposed*	*Dead by dioxin*
Monsanto, Nitro, West Virginia	USA	1949	884	0
Boehringer, Ingelheim	Germany	1952/3	60	0
BASF, Ludwigshafen	Germany	1953	247	1
Boehringer, Hamburg	Germany	1954	31	0
Diamond Alcali Corp. N.J.	USA	1956	73	0
Hooker	USA	1956	.	0
Rhone-Poulenc, Grenoble	France	1956	17	0
Philips-Duphar, Amsterdam (in question)	Netherlands	1963	141	4
Dow Chemicals, Midland	USA	1964	2192	0
Coalite & Chem.Prod., Bolsover	UK	1968	79	0
Stickstoffwerke Linz	Austria	1973	<100	0
Bayer Uerdingen	Germany	1974	<100	0
ICMESA, Seveso	Italy	1976	735 (zone A)	0
			4699 (zone B)	0

Although many people were exposed to relatively high concentrations of TCDD, chloracne was the only symptom presented in most cases (Kimmig and Schulz 1957). Notwithstanding the fact that during the next three decades only one death due to dioxin contamination was confirmed (four other deaths were possibly associated), dioxins were, by the late 1970s, perceived (especially by environmentalists) to be extremely toxic. From that time, the denouncement of dioxin began with an unscientific and moralizing attitude. For example, it was described as an 'insidious' substance (Bosshardt 1973).

The 1976 Seveso plant accident was a landmark case for dioxins for several reasons. In particular, the people directly affected were scared and that fear was transmitted by the media to the rest of the world, as the dramatic event unfolded. At noon on 10 July 1976, an explosion occurred during the production of 2,4,5-trichlorophenol in ICMESA's factory near Seveso, about 25 km north of Milan. A cloud of toxic material, estimated to contain between 159 g and 2 kg of TCDD, escaped into the environment and debris fell on an area of about 2.8 km^2. People soon detected skin lesions and, seeing many animals (including birds, chickens, and rabbits) dying and trees losing their leaves, they were frightened. The authorities divided the contaminated area into three zones (A, B, and R) and, within 20 days, evacuated 735 people from zone A. There were 4 699 people living in zone B, and 31 800 in zone R. All the residents underwent extensive medical examinations from 1976 to 1985, but chloracne of a small segment of the population was the only abnormal finding detected in the first instance (Anonymous 1988).

The accident at Seveso was more than just an unfortunate explosion in a chemical plant, it was confirmation of the fears that had been peddled by environmentalists about organochlorine compounds. Seveso became a menetekel – an icon for the pagan environmentalists – and a reason for banning the production of chlorine and all organic compounds containing it.

If Seveso represents the birth of the dioxin scare, Times Beach, a little town near St. Louis, represents the climax. Early in 1983, dioxin was detected in oil sprayed on local roads in traces of 50–100 ng/kg; in response all 2 240 residents were evacuated, despite the fact that only a few casualties had been caused by this allegedly potent toxin.

Subsequently, the hazard posed by dioxins was re-evaluated and it is this revised analysis that most people in the research community consider to be state-of-the-art science. But even today many environmentalists still see this revisionism as a sham (Wartenberg and Chess 1992). The accident at Seveso and the public attention it attracted motivated some scientists to examine the acute toxicity of TCDD and other PCDDs and PCDFs. But in order to attract research funding, the impacts of dioxins were over-stated and TCDD became known as 'the most toxic man-made chemical known'. It is true that a few dioxins, i.e. TCDD and TCDF, are very toxic, but only to the most susceptible animals. For example, Table 9.3 lists some toxins and demonstrates the eminent position of TCDD according to the minimum dose necessary to induce death (Forth et al. 1987).

Table 9.3. Rank of acute toxicity as minimum fatal dose of some substances

Substance	*Toxicity (mg/kg)*	*Mol wt.*
Botulism toxin A	0.00003	900 000
Tetanus toxin	0.0001	150 000
Ricin	0.02	63 000
Diphtheria toxin	0.3	72 000
Crototoxin	0.2	21 000
TCDD	1	320
Tetrodotoxin	10	319
Aflatoxin B1	10	312
Curarine I	500	597
Strychnine N-oxide	500	350
Nicotine	1000	162
Isoflurophate (DFP)	3000	184
Sodium cyanide	10 000	49
Sodium phenobarbital	100 000	232

But dioxin's rank in Table 9.3 should be compared with that in Table 9.4, which shows the LD_{50} of TCDD for different species (Müller 1991). Perhaps unsurprisingly, the toxicity of TCDD shows a difference of three orders of magnitude between the more susceptible guinea pigs and the more resistant hamsters or humans. TCDD is a substance comparable in acute toxicity to nicotine, for example, but probably no more toxic than that and therefore far less of an apocalyptic super-toxin than green pressure groups would have us believe. These dramatic exaggerations, without differentiation and reservation, have been well received by the public and so the hazard posed by dioxin is generally perceived to be greater than the evidence would justify.

Table 9.4. Acute toxicity in terms of LD_{50} of TCDD in different species

Species	*LD_{50} (mg/kg)*
Guinea pig	0.6–2.5
Mink	4
Rat	22–320
Chicken	50
Rhesus monkey	50–70
Dog	100–<3,000
Mouse	114–280
Rabbit	115–275
Hamster	1150–5,000
Humans	>1000–>5000

An impartial description of dioxin risks requires information on its generation and occurrence, on its path and mechanisms in the organism, and, finally, on epidemiological data.

Generation and occurrence

Dioxins originate in all combustion processes at temperatures of less than 1000°C that involve combinations of organic matter and chlorine-containing organic or inorganic substances, especially when catalysed by copper. For this reason, they are not always man-made – they are also generated by naturally burning wood (for example, in forest fires). Indeed, it has been estimated that on average as much as 60 kilograms of PCDDs are annually produced during forest fires in Canada alone (Abelson 1994). This is about 10 times more than the amount formed in the exceptional Seveso plant accident. Naturally generated TCDD/TCDFs are responsible for the fact that there were measurable background concentrations before humans began to produce chlorine and chlorinated organic compounds (Czuczwa et al. 1985, Hashimoto et al. 1990).

Man-made dioxins are often produced inadvertently during manufacturing processes and become components of waste emissions from industrial plants, especially those manufacturing organochlorines. Important sources are, or were, production of the antiseptic hexachlorophene, the preservative pentachlorophenol, the fungicide hexachlorobenzene, and the herbicide 2,4,5-trichlorophenoxyacetic acid (2,4,5-T), in which TCDD is produced as a contaminant (Kimmig and Schulz 1957; May 1973). Furthermore, dioxins are created by waste incinerators, power plants, and recycling processes for heavy metal, in particular by cable-burning factories, because of the combustion of PVC insulation together with copper.

One special case was the copper-smelting factory in Marsberg, Southern Westphalia, Germany, where, from 1938 to 1945, the poor copper ore was smelted together with coal and salt. The red clinker 'Kieselrot' was a by-product, and about 800 000 tons of it were utilized – until as recently as 1978 – as a surfacing material in thousands of leisure centres, playgrounds and sports fields in Germany, without perception of any sequelae by users over the decades. Its high PCDD/xPCDF-contamination up to 200 μg/kg was not detected until 1991, but after this public fears of dioxin led to the closure and decontamination of many such places at immense cost (Heudorf 1993).

Apart from these point sources of dioxin, there are many diffuse emissions, especially from cars using leaded fuel and by domestic heating. Dioxin emissions in the atmosphere have a half-life of about 3–5 days, but those deposited in earth degrade (slowly) and their half-life is about 10–20 years. PCDD/PCDFs are extremely insoluble in water (the solubility of TCDD is 2×10^{-7} g/l) and they continue in the surface without being dispersed by rainfall. For the same reason, they are also not re-absorbed by roots. Vegetables contain undetectable traces (if any), and dioxins do not play any significant role in most foods of plant origin. Two exceptions are kale and zucchini, which

Table 9.5. Contamination of the environment by dioxins, given in toxic equivalents (TEQ)

Material	*TEQ (ng/kg)*
Soil of farmland	2–8
Soil farmland after sludge deposition	7–14
Soil of woodland	20–30
Soil of motorway verge	< 2800
Soil in Seveso Italy	
Zone R (mean; maximum)	7; <79
Zone B (mean; maximum)	25; <360
Zone A (mean; maximum)	1900; <45 000
Soil in Times Beach, Mo, USA (mean; maximum)	50-100; <404 000
Red clinker, Marsberg, Germany ('Kieselrot')	<200 000
Soil in Rheinfelden, Germany (Dynamit Nobel)	<3 800 000
Human milk from The Netherlands and Germany	5–100
'Normal', not contaminated cow milk	<3
Meat (beef, mutton, pork)	<1
Sea fish	about 20–40
Fresh water fish	<5 000

accumulate them to a certain degree (Hülster 1994). Generally, the contamination of vegetables stems from dioxin emissions, and foliaceous vegetables are more contaminated, therefore, by dust than root vegetables. Table 9.5 shows the burden of some materials.

Intake of dioxins into the organism and elimination

In humans, respiratory and transcutaneous modes of intake of PCDDs and PCDFs represent only about two per cent of total intake. Two-thirds of dioxin contamination stems from foods of animal origin, the rest from dust on foods of plant origin. An average daily intake of dioxins via food consumption is given in Table 9.6.

The intake of TCDD/PCDFs leads to deposition in the adipose tissue of man. About 10 per cent of that concentration is found in the liver and one per cent in blood serum. There is a strong inverse correlation between acute toxicity of TCDD and relative proportion of fatty tissue of the whole body mass. The fat has, so to speak, a detoxification power – by depositing dioxins and keeping them away from susceptible tissues, especially from the liver. However, dioxins can be mobilized from the fatty tissue by starvation or

Table 9.6. Average daily intake of dioxins (pg per person) via food consumption in the early nineties in the Federal Republic of Germany (Beck 1994)

Food	*Consumption (g fat)*	*Intake TCDD (pg)*	*Total TEQ (pg)*	*Share*
Milk, milk products	27.6	4.6	41.7	32.0%
Meat, eggs + products	37.1	7.0	39.0	29.9%
Fish, fish products	1.0	5.1	33.9	26.0%
Vegetable	28.0	2.2	6.3	4.8%
Miscellaneous	6.9	1.1	9.4	7.3%
Total	100.6	20.0	130.3	100%

during lactation. Therefore, mother's milk contains relatively high concentrations of dioxins. Generally, they are slowly eliminated from the body (with a mean biological half-life of 7.1 years, 6–10 years, depending in the particular tissue in which they are stored). In the past, the rate of intake of dioxins was higher than the rate of elimination. The concentration of dioxins thus increased in proportion to age from <10 ng/kg fat at birth to about 70 ng/kg at the age of seventy (Beck 1994).

But as a result of decreasing environmental contamination, the accumulation rate is decreasing. For instance, in Germany TEQ concentrations of dioxin in human milk decreased from mean values of between 30 and 35 ng/kg fat in the mid-1980s to about 20 ng/kg fat in the early 1990s (Beck 1994).

Mode of action

There is a consensus that the first and essential step, before TCDD can cause any of its various toxic effects, be they cancer, birth defect or whatever, seems to be binding to and activation of the aromatic hydrocarbon receptor (AHR). It is activated by ligand binding and by dimerization with the AHR nuclear translocator (arnt). This complex interacts in the cell nucleus with the DNA and sets off a cascade of events by induction of genes encoding enzymes that catalyse the metabolism of foreign compounds (Birnbaum 1994; Hoffmann et al. 1991; Johnson 1991). In this way, they regulate the expression of enzymes, such as specific forms of cytochrome P450 that metabolizes lipid-soluble, foreign compounds and the enzymes provide a means for the rapid elimination of such xenobiotics.

The AHR is a member of the basic helix-loop-helix superfamily of DNA binding proteins and an essential protein – as an AH-deficient mouse line has shown (Fernandez-Salguero et al. 1995). Almost half of these mice died shortly after birth, although the survivors reached maturity and were fertile. However, they showed decreased accumulation of lymphocytes in the spleen and lymph nodes, although not in the thymus; their livers were reduced in size by 50 per cent and exhibited bile duct fibrosis. These effects point out that the AHR plays an important physiological role and its loss deprives the organism of its normal defence against some, as yet unidentified, endogenous toxic substances.

Therefore, the AHR is a general defence system. It is binding not only to more or less toxic aromatic endogenous compounds but also to exogenous substances. Apart from dioxin, there is especially in foods of plant origin also a broad spectrum of naturally occurring aromatic compounds, even anti-carcinogenic flavonoids, polyphenols, and indoles (Prochaska et al. 1992; Zhang et al. 1992 1994) which bind to and activate the AHR. This fact underlines the suggestion that many if not all of the toxic effects of dioxin mediated by the AHR may also be created by other substances in the same way and, for that reason, dioxin does not possess toxic effects in all cases. If this is so, low concentrations of dioxin well below the dissociation constant of dioxin from the AHR are unlikely to lead to toxicity. Current knowledge suggests that there is every reason to believe that the biological response to dioxin begins slowly at a concentration of about 1–10 ng/kg. But it shoots up after passing a critical concentration of about 100 ng/kg, reaching its maximum at about 10 mg/kg (Roberts 1991). In this case, selective alterations of P450 expression may be due to an imbalance between activation of toxic and/or carcinogenic compounds and its detoxification in tissue.

However, it is still unclear whether there are also other dioxin receptors beside AHR. There are some examples for such inconsistent effects caused by TCDD. Some hormone-like effects support this view.

Clinical manifestations of PCDD/PCDFs

Chloracne is the main symptom of intoxication by dioxins. However, it is an unspecific toxic effect, caused not only by dioxins but also by all chlorinated aromatic compounds (including DDT, PCB and PCP). Indeed, chloracne has been known since 1899 as Perna disease because workers were often contaminated with *per*chloro*nap*hthaline. Furthermore, chloracne is correlated with a deficiency of vitamin A, which may be depleted by dioxin. But it is unclear whether chloracne is actually due to that vitamin A deficiency. Finally, there are no clinical or histological differences between chloracne and acne vulgaris.

In Seveso, 447 persons fell ill with chloracne, but most of them recovered from it after a few weeks. Eleven years later, only three persons continued to suffer from strong chloracne type 4. The serum TCDD levels of five persons from Seveso zone A suffering from chloracne ranged from 826 to 27 821 ng/kg but most of them had serum levels higher than 15 000 ng/kg. However, others without chloracne showed TCDD levels of between 1 772 and 10 439 ng/kg, whereas the levels in the general uncontaminated population were less than 20 ng/kg (Anonymous 1988). The complex causes of chloracne suggest that it is not possible to determine an exact threshold dose of TCDD.

About one tenth as many people as suffered from chloracne were inflicted with various other skin problems following contamination by TCDD, including: itching, skin fragility, oiliness or thickening of skin, lack of perspiration, porphyria cutanea tarda, body hirsuteness, thinning of scalp hair, and growth of longer darker hairs in the eyebrows.

Furthermore, a wide variety of other conditions has been associated with exposure to dioxin: hepatic effects were described with hepatomegaly, elevated levels of liver enzymes and lipid metabolism disorder, manifesting in increased serum triglycerides, and hypo- as well as hypercholesterinemia. Even a high exposure to TCDD leads only to transient injuries (Mocarelli et al. 1986). But these are boosted by alcohol consumption. However, in no case was the wasting syndrome, showing hepatic oedema, starvation and cachexia, which affects rats intoxicated with TCDD, observed in humans.

Gastrointestinal effects after dioxin contamination primarily involve nausea and a decreased anal sphincter tone. Other gastrointestinal diseases claimed by TCDD contact, for example gastritis or gastrointestinal ulcer disease, showed no statistically significant association with any measure of TCDD exposure (Calvert et al. 1992).

Central and peripheral nervous disorders were reported, such as an elevated suicide rate of men suffering from severe chloracne, loss of hearing, sense of smell or taste, neurasthenia characterized by tendency to anger or irritability, sleep disorders, and emotional instability (Barbieri et al. 1988; Oliver 1975).

Chronic toxicity

Chronic toxicity of TCDD/TCDFs was purported to be due to mutagenic, carcinogenic and teratogenic effects as well as adverse effects on immunity and reproduction. This suspicion came from observations in animals that were fed with relatively high doses of dioxin. The question of whether humans exposed to lower concentrations are at risk can be answered only by exhaustive biostatistical studies on the most highly exposed populations.

The toxic effects of high doses of dioxin on reproduction in animals include an elevated rate of abortion, miscarriage, stillbirth, deformity, and malformation. But fertility disorders were not observed when rhesus monkeys were fed with doses of up to 25 ng TCDD/kg for two years. The observations from the Seveso collective are statistically consistent with the animal experiments (Bertazzi et al. 1993). Only the frequency of haemangiomas was somewhat higher than expected, but their spontaneous frequency is high in any case (Neubert 1985).

The significance of another animal experiment, in which rhesus monkeys developed endometriosis, is as yet unclear. In a group of monkeys fed daily doses of 25 ng dioxin/kg for 4 years, 71 per cent had moderate to severe endometriosis after 15 years, compared with 42 per cent of monkeys fed a lower dose of 5 ng/kg, and a control group not fed dioxin and without endometriosis (Rier et al. 1993). Overall, embryotoxic effects of dioxin do not seem to occur in humans below a (high) daily dose of between 1 and 10 ng TEQ/kg.

Preliminary observations show an alteration of the human sex ratio in the offspring of people exposed to high concentration of TCDD in zone A in Seveso. No males (only females) were born during the first biological half-life of TCDD, from 1976 to 1984, in families in which both father and mother had serum TCDD level greater than 100 ng/kg when measured in 1976. Altogether, there was an excess of 48 females versus 26 males. But this ratio declined to 64 females versus 60 males between 1985 and 1994 (Mocarelli et al. 1996).

However, the undoubted effects of dioxin on the organism seem to be balanced on a long-term basis by regulatory systems, as another example shows. The effect of dioxin on the function of thyroid gland was known from animal studies, so the thyroid hormone concentrations of new-born babies were investigated. After delivery, thyroxin (T4) and the T4/thyroid binding globulin (TBG) ratio were statistically significantly elevated in the high-exposure group of infants fed with mother's milk containing 38 (29–63) ng TEQ/kg compared with those getting milk containing only 19 (9–28) ng TEQ/kg. But eleven weeks after birth compensation had occurred in the T4 and T4/TBG ratio (Pluim et al. 1992).

Finally, there may exist acute toxic effects of TCDD/TCDFs on the cellular immune system (Vos et al. 1973) as confirmed in AHR-deficient mice by Fernandez-Salguero (1995). Long-term effects are claimed, for example increased inflammation and proneness of skin to infection, but they have not been confirmed. Of course, similar processes are seen in acne vulgaris and the consequence of such a discovery would be the necessary acceptance of an immunodeficiency in all humans suffering from any form of acne. In any case, single observations of depressed reactions of T lymphocytes from humans contaminated by TCDD could not be evaluated with statistical significance (Hoffman et al. 1986).

The conclusion from this is that there certainly is a possibility that impairment of immune and reproductive systems and the liver result from contamination with very high concentrations of dioxin, but that such concentrations are extremely seldom reached, even in exposed and highly contaminated humans.

Carcinogenicity

Regarding the carcinogenicity of TCDD, Kociba et al. (1978, 1979) performed an important animal experiment which showed that TCDD is potentially carcinogenic. This result induced terrible fears, but these were largely due to ignorance and possibly even conscious scaremongering. However, there are scientific as well as popular misunderstandings.

The scientific problem results from difficulties associated with transposing the results of animal experiments to humans (Rall 1988). In the experiments which found TCDD to be carcinogenic in animals, the doses of TCDD used were extraordinarily high. What was not taken into account was that at such large doses many naturally occurring chemicals induce increased mitogenesis. Because of the correlation between mitogenesis and carcinogenesis, each mitogenic substance inevitably possesses a certain carcinogenic potency (Ames and Gold 1990; Cohen and Ellwein 1990).

The popular misunderstandings concern the belief that cancer is a rare and irreversible one-step event caused by synthetic chemical carcinogens and that there is no safe dose or practical threshold below which no adverse effects occur. None of these perceptions are correct, although they seem to be responsible for many regulations in many countries.

In reality, cancer is always a complex multi-step process triggered off by mutagens, prevented or stopped by DNA repair enzymes as well as anticancer substances, and promoted by others – including misled hormonal and immunological control mechanisms of the organism. Therefore, cancer induction at the cell level happens frequently, particularly since about half of all known natural, as well as synthetic, chemicals are probably 'carcinogenic' (Huff et al. 1988). However, in contrast to the hundreds of chemicals that have been observed to possess carcinogenic activity in laboratory animals, less than three dozen are known to induce cancer in man. The reason that all the other chemicals are under suspicion is that the traditional toxicological risk assessment assumes that there is a linear relationship between dose and effect. However, the existence of a threshold for carcinogenesis seems rather a good explanation for why macroscopic cancers fortunately develop rather rarely. But if a threshold does exist, this means that many fears of carcinogens are unfounded.

Unlike most of the chemical carcinogens that have been identified in laboratory animals, TCDD does not appear to be directly genotoxic or to be converted to a genotoxic metabolite. In the sequence of events leading to cancer, TCDD is, rather, a promoter – as are many other compounds – inducing different enzymes and factors of cell growth. But it is by no means an initiating substance or a complete carcinogen showing an initiating plus promoting function. Indeed, the hormone-like properties of TCDD have quite the opposite effect. On the one hand, TCDD enhances the growth of some forms of cancer but, on the other hand, it also inhibits the growth of hormone-dependent tumours (Bertazzi et al. 1993). A further hormone effect is the dependence upon sex of tumour growth. Male animals generally are more resistant to cancer induction by TCDD than female. Altogether, TCDD is neither a typical anticancer substance nor a typical carcinogen but a co-carcinogen acting through epigenetic mechanisms.

As mentioned, dioxin has frequently been the subject of scaremongering by environmentalists and others, who claim that it induces cancer; equally, the producers of dioxin have sought to counter such fears. As a result, numerous disputes concerning the 'true' impact of dioxin have erupted. In response, epidemiological studies were undertaken in order to answer the question about the carcinogenicity of TCDD. However, perhaps inevitably, these often came under fire from one side or the other. They were charged with alleged falsification of data – as with, for example, the early Monsanto studies (Zack and Gaffey 1983; Roberts 1991), or with subjectively tendentious and unproven conclusions – this is the case with, for example, the paper on the Boehringer Hamburg accident (Manz et al. 1991; Carlo and Sund 1991; Triebig 1991).

Notwithstanding all this, the epidemiological studies by Bertazzi et al. (1993), Collins et al. (1993), Fingerhut et al. (1991), and Zober et al. (1990) show reliable results. They found that the mortality from several cancers previously associated with TCDD, i.e. cancers of stomach and liver, nasal cancers and Hodgkin's disease as well as non-Hodgkin lymphoma, were not significantly elevated in their cohorts in comparison with the general population. The mean serum TCDD levels, as adjusted for lipids, varied in the Fingerhut study in different sub-cohorts between 233 and 418 ng/kg, although all the workers had received their last occupational exposures 15 to 37 years earlier. Of course, the induced dioxin dose was higher by a factor of 5 to 35. A mean level of 7 ng/kg was found in the control group of unexposed people. Zober et al. (1990) described median values in blood of 8.4 ng/kg for the group without chloracne and 24.5 ng/kg for the group with chloracne. The highest value observed by Zober et al. (1990) was 553 ng/kg, which implies, extrapolating to the time of exposure 35 years before, a blood level of TCDD of more than 20,000 ng/kg, which is extremely high. Similar values were determined in Seveso, as mentioned above (Anonymous 1988). Table 9.7 shows the results of the four studies.

One further result of Fingerhut et al. (1991) was a statistically significant increased mortality for soft-tissue sarcoma (SMR 922; CI 190–2,695) as well as a weak increased mortality from cancers of the trachea, bronchus, and lung (SMR 139, CI 99–189). Collins et al. (1993) confirmed these findings but they showed that all soft-tissue sarcomas occurred among workers with 4-aminobiphenyl exposure, whereas no soft-tissue sarcomas were found among workers with TCDD exposure alone.

Finally, the excess mortality due to all malignancies, given by SMR values (baseline 100) of 120 115, and 117 in Collins et al. (1993), Fingerhut et al. (1991), and Zober et al (1990), respectively, may result, in part at least, from overestimation of the SMR. Two points are worth considering. First, the causes of death are well diagnosed in the verum groups, whereas the death certificates of the general population give less reliable diagnoses and are not so exact. Summary causes of death are overvalued in death certificates and, conversely, rare and exact diagnoses are undervalued (Höpker and Burkhardt 1984). For that reason, the figures of well diagnosed deaths, such as those from soft-tissue sarcoma, are underestimated in the general population and overestimated in the group of exposed workers. Second, the possible contribution of lifestyle factors as well as occupational exposure to other chemicals cannot be excluded. The excess mortality from soft-tissue sarcomas in the Fingerhut study was such a case, as Collins et al. (1993) showed.

In the end, the TCDD load resulting from occupational exposure (two

Table 9.7. Mortality follow-up studies of persons exposed to TCDD by several studies.

Specification	*Bertazzi et al.*		*Collins et al.*	*Fingerhut et al.*	*Zober et al.*
Exposed persons	724[1]	4 824[2]	754	5 172	247
Person-years of observation	2 313[1,3] 2 255[1,4]	15 503[2,3] 15 781[2,4]	23 198	116 748	8 398
Years of observation	9	9	38	15–37	34
All deaths	–	–	363	1 052	78
Deaths by all cancers	7[1,3] 7[1,4]	36[2,3] 76[2,4]	102 (= 28%)	265 (= 25%)	23 (= 29%)
Standardized mortality ratios (SMR)	100[1,3] [50–210][5] 70[1,4] [30–150]	80[2,3] [60–110] 110[2,4] [90–140]	120 [90–140]	115 [102–130]	117 [80–66]
Deaths by respiratory cancers	0[1,3][expected 0.2] 0[1,4][expected 1.8]	2[2,3] 24[2,4]	38	89[6] 40[7]	4
SMR	– –	80[2,3] [20–340] 110[2,4] [80–170]	110 [80–150]	111[6] [89–137] 139[7] [99–189]	201 [69–60]

[1] Zone A; [2] Zone B; [3] Female; [4] Male; [5] CI = Confidence Interval; [6] All Exposed; [7] Exposed >1 Year.

orders of magnitude higher than ambient concentrations) may result in a small increased cancer risk – of about the same order of magnitude as some lifestyle factors, such as alcohol consumption and smoking. However, the studies are not conclusive. On the one hand, it is obvious that high dioxin levels are linked to cancer in animals. On the other hand, it is unclear whether carcinogenicity of dioxin varies according to concentration in a similarly extreme manner as it does in acute toxicity. But the consensus is that there is no evidence that near-background exposure levels are likely to cause cancer.

Health risk assessments

The search for adverse health effects from dioxin has found no unequivocal epidemiological evidence sufficient to link dioxin to human cancers, suppression of immune function, or reproductivity, even amongst workers exposed to somewhat higher concentrations of dioxin. Chloracne is the only generally recognized result of dioxin exposure in humans. Therefore any purported risk to the general population of near-background exposure levels should be refuted. This risk assessment is supported by the known mode of action of dioxin and the fact that dioxin below its dissociation constant from the AHR is unlikely to lead to toxicity.

Nevertheless, some national agencies, for instance the US Environmental Protection Agency (EPA) and the German Federal Institute for Health Protection of Consumers and Veterinary Medicine (BgVV) perpetuate a false risk assessment by blurring the boundary between science and policy (Bradfield et al. 1994; Stone 1995). This obvious disparity between the concerns of scientists and that of the public and politicians is due to their different points of view. Most people find involuntary hazards, such as the health risk from dioxin exposure, far more abhorrent than voluntary ones, such as the cancer risk from smoking cigarettes. Moreover, people tend to perceive processes controlled by others – such as factory pollution – as more dangerous than actions under their own control, such as driving a car. As a result of this, relatively low threshold values were set. But these thresholds and the regulations they justify cannot remove the uncertainties if the existing risks are not identified clearly.

References

Abelson P. H. (1994). Chlorine and organochlorine compounds. *Science* **265**, 1155.

Ames B. N., and Gold L. S. (1990). Too many rodent carcinogens. Mitogenesis increases mutagenesis. *Science* **249**, 970–971.

Anonymous (1988). Preliminary report: 2,3,7,8-tetrachlorodibenzo-p-dioxin exposure to human – Seveso, Italy. *MMWR* **37**, 733–736.

Barbieri S., Pirovano C., Scarlato G., and Tarchini P. (1988). Long-term effects of 2,3,7,8-tetrachloro-dibenz-p-dioxin on the peripheral nervous system. *Neuroepidemiology* **7**, 29–37.

Beck H. (1995). Occurrence in food, human tissues and human milk, pp. 216–247. *Chlorinated Organic Chemicals – Their Effect on Human Health and the Environment.* The Toxicology Forum. Berlin Germany 19–21 September 1994.
Bertazzi P. A., Pesatori A. C., Consonni D. et al. (1993). Cancer incidence in a population accidentally exposed to 2,3,7,8-tetrachlorodibenzo-paradioxin. *Epidemiology* **4**, 398–406.
Birnbaum L. S. (1994). The mechanism of dioxin toxicity: relationship to risk assessment. *Environ. Health Perspect.* **102**, S157–S167.
Bosshardt H. P. (1973). *Dioxine – Heimtückische Gifte in unserer Umwelt. Entstehung, Eigenschaften und Verhalten von polychlorierten Dibenzo-p-Dioxinen.* Forschung und Technik, 70th edn.
Bradfield C. A., Gallo M. A., Gasiewicz T. A. et al. (1994). EPA dioxin reassessment. *Science* **266**, 1628–1629.
Calvert, G. M., Hornung R. W., Sweeney M. H. et al. (1992). Hepatic and gastrointestinal effects in an occupational cohort exposed to 2,3,7,8-tetrachlorodixbenzo-para-dioxin. *JAMA* **267**, 2209–2214.
Carlo G. L., and Sund K. G. (1991). Carcinogenicity of dioxin. *Lancet* **338**, 1393.
Cohen S. M., and Ellwein L. B. (1990). Cell proliferation in carcinogenesis. *Science* **249**, 1007–1011.
Collins J. J., Strauss M. E., Levinkas G. J., and Conner P. R. (1993). The mortality experience of workers exposed to 2,3,7,8-tetrachlorodibenzo-p-dioxin in a trichlorophenol process accident. *Epidemiology* **4**, 7–13.
Czuczwa J. M., Niessen F., and Hites R. A. (1985). Historical record of polychlorinated dibenzo-p-dioxins and dibenzofurans in Swiss lake sediments. *Chemosphere* **14**, 1175–1179.
Fernandez-Salguero P., Pineau T., Hilbert D. M., McPhail T., Lee S. S. T., Kimura S., Nebert D. W., Rudikoff S., Ward J. M., and Gonzalez F. J. (1995). Immune system impairment and hepatic fibrosis in mice lacking the dioxin-binding Ah receptor. *Science* **268**, 722–726.
Fingerhut M. A., Halperin W. E., Marlow D. A. et al. (1991). Cancer mortality in workers exposed to 2,3,7,8-tetrachlorodibenzo-p-dioxin. *New Engl. J. Med.* **324**, 212–218.
Forth W., Henschler D., and Rummel W. (1987). *Allgemeine und spezielle Pharmakologie und Toxikologie.* 5th edn. Bibliograph. Institut, Mannheim-Wien-Zürich.
Hashimoto S., Wakimoto T., and Tatsukawa R. (1990). PCDDs in the sediments accumulated about 8120 years ago from Japanese coastal areas. *Chemosphere* **21**, 825–835.
Heudorf U. (1993). Umgang mit Kieselrot auf Sport-, Spiel- und Freizeitflächen unter Berücksichtigung neuer toxikologischer Untersuchungsergebnisse. *Gesundh.-Wes.* **55**, 521–526.
Hoffmann E. C., Reyes H., Chu F.-F. et al. (1991). Cloning of a factor required for activity of the Ah (dioxin) receptor. *Science* **252**, 954–958.
Hoffman R. E., Stehr-Green P. A., Webb K. B. et al. (1986). Health effects of long-term exposure to 2,3,7,8-tetrachlorodibenzo-p-dioxin. *JAMA* **255**, 2031–2038.
Höpker W.-W., Burkhardt H.-U. (1984). Unsinn – und Sinn? – der Todesursachenstatistik. *Dtsch. med. Wschr.* **109**, 1269–1274.
Huff J. E., McConnell E. E., Haseman J. K. et al. (1988). Carcinogenesis studies: results of 398 experiments on 104 chemicals from US national toxicology program. Ann. N.Y. *Acad. Sci.* **534**, 1–30.

Hülster A. (1994). *Transfer von polychlorierten Dibenzo-p-Dioxinen und Dibenzofuranen (PCDD/PCDF) aus unterschiedlich stark belasteten Böden in Nahrungs- und Futterpflanzen.* Verlag U. E. Grauer, Stuttgart.

Johnson E. F. (1991). A partnership between the dioxin receptor and a basic helix-loop-helix protein. *Science* **252**, 924–925.

Kimmig J., and Schulz K. H. (1957). Berufliche Akne (sogenannte Chlorakne) durch chlorierte aromatische zyklische Äther. *Dermatologica* **115**, 540–546.

Kociba R. J., Keyes D. G., Beyer J. E. et al. (1978). Results of a two-year chronic toxicity and oncogenity study of 2,3,7,8-tetrachlorodibenzo-p-dioxin (TCDD) in rats. *Toxicol. Appl. Pharmacol.* **46**, 279–303.

Kociba R. J., Keyes D. G., Beyer J. E. et al. (1979). Long-term toxicologic studies of 2,3,7,8-tetraxchlorodibenzo-p-dioxin (TCDD) in laboratory animals. *Ann. N.Y. Acad. Sci.* **320**, 397–404.

Manz A., Berger J., Dwyer J. H. et al. (1991). Cancer mortality among workers in chemical plant contaminated with dioxin. *Lancet* **338**, 959–964.

May G. (1973). Chloracne from accidental production of tetrachlorodibenzodioxin. *Brit. J. Industr. Med.* **30**, 276–283.

Mocarelli P., Brambilla P., Gerthoux P. M. et al. (1996). Change in sex ratio with exposure to dioxin. *Lancet* **348**, 409.

Mocarelli P., Marocchi A., Brambilla P. M. et al. (1986). Clinical laboratory manifestations of exposure to dioxin in children. A six year study of the effects of an environmental disaster near Seveso, Italy. *JAMA* **256**, 2687–2695.

Müller H. E. (1991). Die realen und irrealen Gefahren der Dioxine. *Dtsch. med. Wschr.* **116**, 786–793.

Müller R. K. (1993). Umwelt und Gesundheit. *Versicherungsmedizin* **45**, 54–58.

Neubert D. (1985). Teratogenes Risiko durch Dioxine? Eine differenzierende Beurteilung. *Dtsch. Ärztebl.* **82**, B821–829.

Oliver R. M. (1975). Toxic effects of 2,3,7,8-tetrachlorodixbenzo-1,4-dioxin in laboratory worker. *Brit. J. Industr. Med.* **32**, 49–52.

Pluim H. J., Koppe J. G., Olie K. et al. (1992). Effects of dioxins on thyroid function in new-born babies. *Lancet* **339**, 1303.

Prochaska H. J., Santamaria A. B., and Talalay P. (1992). Rapid detection of inducers of enzymes that protect against carcinogens. Proc. Natl. Acad. Sci. USA 89,2394–2398.

Rall D. P. (1988). Laboratory animal toxicity and carcinogenesis testing. *Ann. N.Y. Acad. Sci.* **534**, 78–83.

Rier S. H., Martin D. C., Bowman R. E., Dmowski W. P., and Becker J. L. (1993). Endometriosis in rhesus monkeys (Macaca mulatta) following chronic exposure to 2,3,7,8-tetrachlorodibenzo-p-dioxin. *Fundam. Appl. Toxikol.* **21**, 433–441.

Roberts L. (1991). Dioxin risks revisited. *Science* **251**, 624–626.

Stone R. (1995). Panel slams EPA's dioxin analysis. *Science* **268**, 1124.

Triebig G. (1991). Is dioxin carcinogenic? *Lancet* **338**, 1592.

Vos J. G., Moore J. A., and Zinkl J. G. (1973). Effect of 2,3,7,8-tetra-chlorodibenzop-p-dioxin on the immune system of laboratory animals. *Environ. Health Perspect.* **5**, 149–162.

Wartenberg D, and Chess, C. (1992). Risky business. The inexact art of hazard assessment. *The Sciences*, March/April 17–21.

Zack J. A., and Gaffey W. R. (1983). A mortality study of workers employed at the Monsanto Company plant in Nitro, West Virginia. Environ. Sci. Res. 26, 575–591.

Zhang Y, Kensler TW, Cho C-G. et al. (1994). Anticarcinogenic activities of sulforaphane and structurally related synthetic norbornyl isothiocyanates. *Proc. Natl. Acad. Sci. USA* **91**, 3147–3150.

Zhang Y, Talalay P, Cho C-G, and Posner GH. (1992). A major inducer of anticarcinogenic protective enzymes from broccoli: isolation and elucidation of structure. *Proc. Natl. Acad. Sci. USA* **89**, 2399–2403.

Zober A., Messerer P., and Huber P. (1990). Thirty-four year mortality follow-up of BSAF-employees exposed to 2,3,7,8-TCDD after the 1953 accident. *Int. Arch. Occup. Environ. Health* **62**, 139–157.

III Science policy

10 Public policy and public health: coping with potential medical disaster

B. M. Craven and G. T. Stewart

Summary

Blame avoidance seems to have been the main aim of the politicians responsible for handling public health panics such as Legionnaire's disease, Swine 'flu, Chernobyl, BSE, and AIDS. Actions taken to avoid blame in response to a panic are unlikely to improve matters.

Introduction

Despite wars, floods, famines and ethnic cleansing, public health is improving in most parts of the world, as is shown by conventional indices such as mortality of infants, children and pregnant women, expectation of life, and freedom from major epidemics. This shifts the problem from coping with recurrent, lethal epidemics of former generations to dealing with less predictable hazards to health of the present day.

Day and Klein (1989) analysed the policy responses to the AIDS epidemic in the light of its feared social consequences and difficult moral dimension. They argued that the policy response required that the government be seen to be acting. Day and Klein suggested that the government's reliance on expert advice and values may be generalizable as a model to be applied to other cases.

In this paper, we examine five other events of this type to see if there is indeed any consistency of policy response to what Day and Klein identified as the 'unexpected event', i.e. one for which no contingency plan could be made. This rules out disasters such as earthquakes, oil tanker break-ups and airline crashes which are unpredictable but foreseeable. There are other disasters which were unforeseeable for which no contingency plan could be made. These include the sinking of the 'Herald of Free Enterprise', the deaths from crushing in the Heysel and Hillsborough stadia and, the explosion of the space shuttle, 'Challenger'. But none of these events qualify for this study because the damage was immediate and contained; the events were projectable, obviating the possibility of over- or under-reaction; finally, the event did not bring out contentious moral issues which would limit the extent to which policy responses can be clear and unambiguous.

We have selected six events (including a re-evaluation of AIDS policy outcome) which could, prima facie, satisfy most of these criteria. In chronological order, these are Legionnaire's disease, Swine 'flu, Chernobyl, contaminated blood and blood products, AIDS and BSE. Each of these illustrates a different kind of emergency.

In the case of BSE, no emergency was foreseen until, on scientific advice, it was believed that a link with Creutzfeldt-Jakob disease (CJD) in humans could not be ruled out. Many countries had banned British beef before the European Union (EU) world-wide ban. The dilemma then echoed that of the possibility of a repeat epidemic of Swine 'flu, presumed to be caused by the virus responsible for the deaths of 500 000 US citizens in 1918. The scope of BSE and its link, if any, with CJD was not projectable. The Chernobyl accident was unpredictable and unprojectable but not entirely unforeseeable. Legionnaire's disease was unpredictable, unforeseeable and initially unprojectable but was soon realized not to be a disaster. In the cases of Swine 'flu, Legionnaire's disease and Chernobyl there were no moral difficulties. The appearance of AIDS and contamination of blood and blood products by HIV was for a short time – between the identification of AIDS and the presumed causal agent HIV – unpredictable, unforeseeable and unprojectable. There were moral issues in the practical need to discriminate blood donors by life-style and sexual preference.

Of these candidates, Chernobyl is the only one which is non-medical in origin and is the poorest fit in not being unforeseeable. But none of the candidates fit exactly. Some of the qualifying characteristics are stronger in some and weaker in other cases. All this weakens the argument that these issues are generalizable in the ways suggested by Day and Klein. In a wider study of policy fiascos, Bovens and 't Hart (1996) also concluded that different disasters had too many differing qualities to recommend generalized policy response.

Weaver (1986), in an influential work, argued that voters are more sensitive to potential losses than they are to gains. In these circumstances, politicians will find that blame avoidance is less risky than claiming credit. Weaver identified four situations leading to blame-avoiding behaviour. The first is where there is a zero-sum conflict among the policy-makers' constituents when the benefits are concentrated and high, and the costs concentrated, high and imposed on a different constituency. The second situation is where there is a negative sum game such as in the administration of budget cutbacks. The third occurs when constituency opinion dominates one side of an issue; blame avoidance is the only strategy a politician can adopt if for instance, he or she is found to be a child abuser. The fourth situation is one where the policy interests of the policy-maker and clientele are opposed. Weaver cites the case of congressional pay increases.

It is part of our thesis that Weaver has neglected an important fifth category, namely potential medical and social catastrophe, where blame avoidance is a particularly appropriate response. Considerations of public health policy are deemed to rise above party politics because of the widely-held romantic

Table 10.1. Similarities and differences between Chernobyl, Swine 'flu, AIDS, BSE, contaminated blood and blood products, and Legionnaire's disease.

	Chernobyl	*Swine Influenza*	*AIDS*	*CJD*	*Contaminated blood and blood products*	*Legionnaires' disease*
Assumed causal agent	nuclear radiation	Swine influenza virus	HIV and other pathogens	Prion protein	Hepatitis, HIV	*Legionella* bacteria
Transmission	radiation leak from Chernobyl nuclear reactor	air/droplet borne	sex/drugs misuse	consumption of beef and beef products	transfusion	water cooled air conditioning systems
Interest groups to gain	anti nuclear pressure groups, non nuclear power generators	pharmaceutical companies	virologists, immunologists, health education	EC states ex UK, EC farmers	pharmaceutical companies	manufacturers of air conditioning systems
Interest groups to lose	nuclear industry	none	none	farming industry	none	none
Immediate taxpayer costs	small compensation to farmers	hundreds of millions of $	billions of all currencies	billions of £	several millions of £	several millions of £
Regulatory policy	none	none	health education	changing animal feed	screening blood, purifying blood products	increased regulatory requirements
Political response to crisis	reassurance	general immunization	yield to activists and professional advisers	slaughter part of UK herd	accept responsibility but not blame	reassurance
Political policy strategy	blame avoidance	blame avoidance	claim unearned credit	blame avoidance	blame avoidance	blame avoidance

notion that 'good health is priceless and no amount of money is too large to conserve life'. Legionnaire's disease, Swine 'flu, Chernobyl, HIV/AIDS, contaminated blood and blood products, and BSE have some further common characteristics (Table 10.1). The first is that the threat, at the time, was new. Second, there was no consensus among experts about either the likely magnitude of the threat or the identity of those likely to be adversely affected. Third, in the majority of cases there was a long period between exposure to the agent and its effect on the individual. Fourth, and perhaps most important, there was a public perception or belief that large sections of the community would suffer serious damage to health, and that there would be fatalities. The initial policy response in all cases invited a blame avoidance strategy. Credit claiming is possible in these circumstances only long after the issue has ceased to be a societal threat and when political value is low.

Blame-avoidance strategies are coupled with a belief that political action to prevent or limit a threat, even when the passage of time shows such strategies to be unnecessary, expensive and harmful, will be judged less harshly than political inactivity. Errors of omission are deemed more damaging than errors of commission. In the context of events examined in this paper, attempts to avoid blame were more likely to fail than to succeed and tended, therefore, to make circumstances worse. A consequence of increasing the emphasis on adopting blame avoidance strategies is greater regulatory control with a bias to over-regulation (Seldon 1988, Institute of Economic Affairs 1994, North 1993, North and Gorman 1990). Blame avoidance may be successful in reducing blame attributed to politicians but the reality is that it results in 'blame passing' with variable effects on public health.

The unexpected

Swine influenza

The most prevalent influenza strain in the world in early 1976 was Type A H3N2. It caused a relatively mild illness where outbreaks occurred. In January 1976 an eighteen-year-old private at the US Army training base Fort Dix collapsed and died shortly after attempting an overnight winter route march whilst suffering acute symptoms of influenza (Garrett 1995). The case was unusual in that influenza is a lethal threat to the elderly and those predisposed to bronchial ailments but rarely kills young healthy adults. There were fears immediately that the viral strain could be that which killed 500 000 Americans in 1918. It was noticed that people alive during the 1918 epidemic had antibodies to a Swine 'flu virus but that those born after this date did not. The virus was so called because it was thought capable of being passed from people to pigs (Shope 1936). Antibodies from other army recruits neutralized a sample of the Swine 'flu. From these two weak connections the conclusion was drawn that Private Lewis died from the virus (Type A, H1N1) which caused the 1918 epidemic.

Coincidentally at Rougemont, Switzerland, an international conference had discussed the appropriate public policy to be adopted in the event of a future 1918 type of epidemic (Selby 1976). It had been known that voluntary compliance with recommendations to inoculate was poor. In 1974 only 17.4 per cent of the elderly availed themselves of the vaccine. If a major epidemic was imminent, policy would require, at short notice, massive production of vaccine stocks and, in order to minimize the number of carriers, the compliance of about 80 per cent of the elderly population. Subsequently, in about 450 new influenza cases at Fort Dix, both virus types were identified with A/H3N2 being presumed, but never confirmed, twice as prevalent as A/H1N1. Politicians at this stage were advised by medical experts of the likelihood of hundreds of thousands of preventable deaths from a repeat of the 1918 epidemic unless they acted quickly (Silverstein 1981). But this epidemic did not occur. By mid-March 1976 influenza was in steep decline world-wide. The CDC stated that there was no epidemic in the USA (CDC 1976a) and still no Swine 'flu.

Nevertheless, on March 24 1976, after a meeting with Jonas Salk, Albert Sabin and CDC experts, President Ford announced on national television that he would ask Congress to appropriate $135m to finance vaccine production sufficient to inoculate everyone in the USA. The vaccines that were produced did not produce satisfactory immune responses in volunteers. Parke-Davis made 2 million vaccine doses against a wrong viral strain (CDC 1976b). The Pharmaceutical Manufacturers' Association then demanded exemption from liability concerning vaccine quality. There was also a minority of dissenting opinion which criticized the Government's forecasts of one million deaths, arguing inter alia that antibiotics, unavailable in 1918, would alleviate the main cause of deaths attributed to influenza which are secondary bacterial respiratory infections. Private Lewis might not have died if he had not taken part in the route march, and if he had not died it is unlikely that the words Swine 'flu would have been mentioned by the medical profession in 1976. In the event, the vast supplies of vaccine demanded by the forecasters were not forthcoming while, almost simultaneously, it was established that the only strain of Swine influenza anywhere in the world at that time was at Fort Dix (Dowdle and LaPatra 1983). Its source was never found.

Legionnaire's disease

The issue of Swine 'flu was overtaken by the emergence of Legionnaire's disease during the American Legion's convention of World War II veterans in Philadelphia in July 1976. Within a week there were 182 cases and 29 deaths with symptoms of fevers, muscle aches and pneumonia. Swine 'flu was the first suspect. This was medically disproved within four days but President Ford warned Congress that they would bear responsibility for millions of deaths in the event of a Swine 'flu epidemic, if they did not pass the bill granting immunity to vaccine manufacturers from quality liability. There was an atmosphere of fear and panic towards Legionnaire's disease which contrasted with the near-indifference of the public towards the fear of

Swine 'flu. The CDC was unable to isolate either a virus, microbe or chemical as the causal agent of Legionnaire's disease. As with Swine 'flu the infection remained confined to visitors and staff of the four convention hotels. Perhaps because the disease did not appear infectious or contagious the government reaction was, apart from frenzied efforts by the CDC to establish cause, passive. The cause of Legionnaire's disease remained unknown until January 1977 when the CDC claimed that it was caused by a bacterium named by them *Legionella*. The source of the infection was finally identified as one of the hotels' cooling towers which supplied the air conditioning influenza water source. Within a year the bacterium *Legionella* had explained a cluster of deaths in 1965 at a Washington psychiatric hospital and was subsequently associated with outbreaks in several other states (CDC 1977a, Broome et al. 1979, Marks et al. 1979) and three deaths at a hospital in Nottingham, England (CDC 1977b). The CDC was able to establish that the *Legionella* bacterium was not new and suggested that several thousand people in the USA had been dying from Legionnaire's disease every year for decades. Stricter regulations for air conditioning and water storage tanks was the policy response.

Blood and blood products

Blood is a commodity similar to many others but it has some unusual supply characteristics. For example, 'experience of countries in which a price mechanism (for blood) operates suggest strongly that persons who are happy to make themselves available as blood donors no longer come forward as blood sellers, but are replaced by the poorer and often less healthy members of society whose blood is also less healthy, who have every incentive to lie about their previous health record, and who are encouraged to sell their very substance on more occasions than is good for them or for their eventual recipients' (Cooper and Culyer 1968). Countries allowing blood donors to be paid, such as the United States and many Third World Countries, have had the worst experience with transfusion and haemophilia associated AIDS cases. 'Just those people who are most likely to be carrying multiple, concurrent infections are the people who have greatest incentive to donate their blood (for cash). The 'gift' of blood in this case places two people in intimate contact who otherwise would probably avoid one another at all costs' (Root-Bernstein 1993). The use of blood for transfusions has had an added significance because of the link with AIDS. As AIDS in haemophiliacs became a big public issue the cause of the immune-deficiency was attributed to (clotting) Factor VIII contaminated with HIV. [Transglutaminase, another immunosuppressive enzyme in Factor VIII, is also found, although in a different molecular form, in human semen (Ablin 1985). Blood transfusions in themselves are also immunosuppressive (Carr et al. 1984). There is also contradictory evidence of the possibility that changes observed in HIV seropositive haemophiliacs could be avoided by the use of high-purity blood products, i.e. that they may not be due to HIV. (Seremetis et al. 1990, de Biasi et al. 1991, Goedert 1994)]. The recycling of the plasma from the blood donated by homosexuals and drug users (before screening began in 1985) into the transfusable pool contained infectious agents such as cytomegalovirus (CMV), Epstein–Barr virus and hepatitis A and B virus which were unusually prevalent in this group of donors (Corey and Holmes 1980, Dietzman et al. 1977, Drew et al. 1981). The contamination of blood occurred when members of the homosexual community, who typically have a high level of antibodies to hepatitis B and sometimes of free virus in their blood plasma, were encouraged to contribute blood (Adams 1989), and when drug addicts did so for anonymous cash rewards. The homosexuals, who believed also that their blood was especially valuable to the manufacture of hepatitis vaccine, came forward in a strong public-spirited gesture which happened to coincide with self-interest. Components to make the hepatitis B vaccine were removed from the plasma and the remainder replaced in the general plasma pool. Some of this was used to manufacture Factor VIII – the blood clotting factor for type A haemophilia. In France the high rate of contamination was made worse by the use of blood from prisoners. It has been established in the French courts that there was a delay in the screening of blood following identification of HIV in 1984. Three health officials, including the former head of the state monopoly blood transfusion centre, have been given four-year custodial sentences for fraud and criminal negligence (Nau 1994).

Chernobyl

The first 'melt down' of a nuclear power station occurred in 1986 at Chernobyl in the Ukraine. This unique event could, at the time, have been considered a forerunner of similar disasters at other plants but political reaction was subdued. Decommissioning, at enormous cost, of all nuclear power plants to eliminate the possibility of a similar accident on domestic soil, and thereby eliminate any future deaths, was never seriously considered, despite fear in the minds of the public fuelled by reports in the media (Walker 1986). Conservative party politicians pointed out the shortcomings at Chernobyl of state planning and the neglect of safety issues; capitalist nuclear power stations were safe. The opposition Labour party in the UK questioned the wisdom of the UK nuclear programme.

All parties and interest groups accepted that the threat was real and irreversible. Environmentalists warned about the general dangers of nuclear power generation and disposal of nuclear waste. It emerged quickly that conventional thinking about wind dispersal of the radioactive emissions was incorrect. Contamination levels bore almost no relation to distance from the reactor. In this case the policy response was diverse. All parties accepted that the health risks were largely unquantifiable, would be difficult to detect at a statistically significant level and would not emerge or even be directly attributable to citizens in the UK for decades. Local agricultural areas of heavy contamination in Cumbria and Scotland were identified and animals and products slaughtered or banned. Threats to livelihoods were largely confined to dairy and sheep farmers. Losses and potential radiation damage here (Wynne et al. 1988) were sufficiently dispersed not to present any political embarrassment or to be directly attributable. At the time of the disaster in 1986, the Ukraine was part of the former USSR and the Cold War showed few signs of abating. Even without the friction of East/West relations, it is doubtful whether the authorities in Moscow had immediate, reliable, detailed information (Gubarev and Odinets 1988). The scope for internationally imposed formal regulation, therefore, was virtually non-existent. In addition to the almost inevitable delays in obtaining information, there was an incentive for the operators of the plant to discount fears to such an extent that their statements would border on deceit (Rubin 1987, Webster 1986). The response of the West was to supply financial and technical aid. By doing so western governments avoided future blame for any subsequent mishap at the reactor.

AIDS

There was a unified political response to the emergence of HIV and AIDS. The perceived harmful agent (HIV) is transmitted by sexual intercourse, especially by homosexual or bisexual men, by syringes shared by drug addicts and by blood or blood concentrates used for treatment, especially of haemophilia. Medical experts (Department of Health and the Welsh Office 1988, Cox et al. 1989) predicted the likelihood of a much more extensive spread in the general population by heterosexual transmission of HIV. All political parties were agreed that everything had to be done to minimize transmission but medical ethics and political correctness demanded that there should be no discrimination against homosexual men, the group at highest risk in developed countries, or those who injected drugs, renamed 'recreational' drug users. (There is doubt that government policies to contain the HIV/AIDS threat actually worked (Craven et al. 1993, Craven and Stewart 1995, 1996).) The AIDS epidemic was seen as an international problem and the unanimous international response was to finance, on a grand scale, health education programmes. It was also clear at an early stage that a policy strategy of yielding to activism about HIV/AIDS was propitious to an all party response. The crisis had not imperilled confidence in any domestic products nor was it a threat to jobs. The strategy response involved job-creating public expenditure especially in the area of AIDS prevention and indirectly through the capital building programme which modernized genito-urinary clinics. With the passage of time it has emerged that, in developed countries at least, AIDS has not and will not become the social catastrophe that was once feared. In 1996 AIDS is still confined to the original risk groups and the syndrome is now considered, from a medical perspective, more like any other long-term terminal illness (Danziger 1996). The politicians and their professional advisers were able to avoid blame with this policy and then claim spurious credit for averting the heterosexual epidemic which was not happening anyway.

BSE

On Wednesday 20 March 1996 the UK Health Minister announced to the House of Commons that the most likely explanation of the cause of 10 cases in the UK of an unfamiliar form of Creutzfeldt-Jacob Disease (CJD) was an exposure to Bovine Spongiform Encephalopathy (BSE, 'mad cow disease'). These cases were unusual in that they affected young people whereas CJD traditionally affects the elderly. Specified Bovine Offals known to contain BSE had been fed to many cattle herds. The first cases of 'mad cow' syndrome were noticed in 1985. The offal was banned in 1989. During this time practically the entire population of the UK would have been exposed to BSE. Of about 160 000 recorded cases of BSE world-wide, 99 per cent are in Britain. The World Health Organisation on 22 March 1996 ruled that British beef was safe and Dr Lindsay Martinez of the Emerging and Other Communicable Diseases division stated that 'the chance of any human being contaminated by this disease is absolutely remote.' There were two experts who were prominent dissidents. A microbiologist from Leeds University, Professor Richard Lacey, had repeatedly warned that BSE would pass to humans and that 'the worst case scenario is that half the country is vulnerable.' Another virologist Dr Narang argued that BSE is a slow virus taking decades to act. On 24 March John Pattison, Chairman of the Spongiform Encephalopathy Advisory Committee (SEAC), was reported as saying that he believed that 500 000 people were infected with CJD. The next day, March 25, the European Union's standing veterinary committee voted by 14–1 to impose an indefinite world-wide export ban on British beef and beef products. The Committee agreed that there was no scientific proof that BSE could be transmitted to humans but would not exclude the possibility and recommended a cull of older cattle. The epidemiological fact that there was no evidence of transmission to personnel at highest risk (abattoir workers, veterinarians, farm labourers) was ignored.

The consequences of this and of the announcement were profound. Food retailers such as McDonalds and Burger King announced that they would, with immediate effect, stop selling British beef in their products. Bass, the company which runs about 2 700 restaurants, suspended the serving of British beef products. Many schools and Local Authorities had already stopped serving British beef to pupils, customers and staff. Just as HIV became synonymous with AIDS so the distinction between BSE and CJD became blurred. The *Daily Telegraph* reported that a doctor attacked the Government's handling of the mad cow scare after his wife 'died from the human form of BSE' (26 March 1996). The British beef industry collapsed as veterinarians, abattoir owners and bureaucrats struggled to comply with EC demands for massive slaughtering of healthy cattle. Animal rights activists remained silent.

Perceptions and assessments of risks and uncertainties.

In all the cases discussed here the main informational problem is the absence of any reliable measure of the risk of danger from an unforeseeable event. There are two concepts to consider. The first is the probability estimation of risk and the second is the actual outcome in terms of persons affected. The popular tendency is towards individuals being protected against any risk, however remote and hypothetical. This is expressed by the public reaction on the Continent towards British beef. But in most day-to-day activities individuals always compromise. The probability of risk from common accidents in the home, factory, field, office, car or almost anywhere can be reduced by adopting prevention practices which are usually but not always costly. A point is eventually reached at which the cost of prevention together with the cost resulting from accidents is minimized; an optimal level of safety, or its converse the optimal number of accidents is aimed for or accepted. There are other risks, such as thunderbolts or earthquakes, for which the dangers are ignored by most people, presumably because the probability of risk is so small that any prevention cost is unacceptably high (Table 10.2).

Conceptual difficulties begin to arise when individuals accept a level of risk, which, in terms of probability of death, is many times greater in activities over which there is individual control, such as smoking, rock-climbing and hang-gliding, than in those where there is no control such as airline or train travel or

Table 10.2. Risk of an individual dying in any one year from various causes

Cause		
Smoking 10 cigarettes per day	1 in	200
All natural causes age 40	1 in	850
Any kind of violence or poisoning	1 in	3 300
Influenza	1 in	5 000
Accident on the road	1 in	8 000
Leukaemia	1 in	12 500
Playing soccer	1 in	25 000
Accident at home	1 in	26 000
Accident at work	1 in	43 500
Radiation working in radiation industry	1 in	57 000
Homicide	1 in	100 000
Hit by lightning	1 in	10 000 000
Release of radiation from nuclear power station	1 in	10 000 000

Source: BMA (1987)

nuclear radiation. This suggests that the acceptance of risk is connected with the perceived benefits from the risk and/or from the perceived probability estimation of the risk itself. For example, Peltzman (1975) studied accident rates in the USA attributed to regulations, such as the requirement to wear a seat belt, which were designed to reduce the risk of accidents and hence increase road safety. He found that drivers adjusted their behaviour by driving less carefully but keeping their risk levels constant, and concluded that 'autosafety regulation has not affected the death rate'. (For more information see Chapter 15.) Clearly, where there is known causation between activity and risk, as in the cases of smoking and respiratory disease or driving and road accidents, individuals make decisions about the perceived benefits associated with the perceived risk in deciding whether to undertake the activity. This assumes that the probability of the perceived risks can be assessed. If the probability can be assessed, as with the chance of an accident whilst driving, the risks may not be evenly distributed throughout the journey. The issue is still more complex. Thaler (1987) showed that even if precise probabilistic choices were offered in contrived experiments, economists and statisticians gave inconsistent answers. Similar arguments were examined by Kahneman eteal. (1982). An additional complication follows from the Yerkes–Dodson (Yerkes and Dodson, 1908) law which postulates a relationship between the quality of decision and the pressure or stress under which the decision is made. As pressure increases the quality of decision improves until a maximum is reached and the association is negative thereafter. In most, if not all, of the cases examined in this paper, the quality of expert and government decisions may have been compromised by political urgency and stress.

Yerkes and Dodson focus on low-probability risks. Examples include the near-meltdown at the Three Mile Island nuclear reactor, the spillages at Dounreay and Sellafield, an aircraft crash, a dam failure and an earthquake. These cases are unpredictable but not unforeseeable and contingencies are made. The cases examined in this paper are different in that not only is the event unpredictable but also unforeseeable. If there is no distribution of events and consequences, probabilistic risk measures cannot be made. AIDS, contaminated blood and blood products, CJD caused by BSE, Swine 'flu and Legionnaire's disease all belong to a special category which draws upon the unstable and unpredictable psychological and social responses to a threat which may range from the trivial to the catastrophic. They are all seen as potential mega-risks. (This terminology is not accurate. There is not a mega-risk, but a possibility of a mega-disaster (Lovins 1975, Royal Commission on Environmental Pollution 1976).) In the case studies used here the *ex-ante* perceptions of danger placed each in the potential catastrophe category. The *ex-post* evidence revealed that the societal risks were trivial. Except in AIDS (where all the original estimates were erroneous), there were no probability estimates associated with the original event. There is much uncertainty over how widespread the secondary damage will be.

Which way was the wind blowing at the time of the accident at Chernobyl? How many cattle will be affected by encephalitis and how many humans by CJD? Finally there is the temporal element. It takes time for radiation to disperse and for the adverse health consequences to show. It may be many years before CJD shows in humans and before AIDS escapes to the general population.

In such situations, policy-makers face extremely complex and inconsistent psychological issues that are further complicated by the fact that the perception of risk is often very different from the actual risk. Slovic et al. (1980) produced figures showing that students and members of a political group (the League of Women Voters - LOWV) considerably underestimated actual deaths from a variety of causes (see Table 10.3). This is not an unexpected or unusual finding. He also found a relatively small correlation between their own estimates of the likelihood of death resulting from particular causes and their own perception of risk. The authors concluded that 'we can reject the idea that laypeople wanted to equate risk with annual fatalities [but] were inaccurate in doing so' and in any case other considerations apply. In the case of meltdown at Chernobyl, estimates of deaths from a nuclear disaster were very high – more than 25 per cent expected more than 100 000 fatalities within a year – a figure shown to be grossly over-estimated. The International Atomic Energy Agency in its report stated there were 31 deaths.

Table 10.3 Ordering of per annum death estimates and perceived risk for various activities and technologies in the United States

	Actual deaths	*Rank*	*LOWV death estimates s*	*Students' death estimates s*	*Perceived risk ordering: LOWV*	*Perceived risk ordering: students*	*Perceived risk ordering: experts*
Smoking	150 000	1	6 900	2 400	4	3	2
Alcoholic beverages	100 000	2	12 000	2 600	6	7	3
Motor vehicles	50 000	3	28 000	10 500	2	5	1
Hand-guns	17 000	4	3 000	1 900	3	2	4
Electric power	14 000	5	660	500	18	19	9
Swimming	3 000	7	930	370	19	30	10
X-rays	2 300	9	90	40	22	17	7
Railroads	1 950	10	190	210	24	23	19
Commercial aviation	130	19	280	650	17	16	18
Nuclear power	100	20	20	27	1	1	20

LOWV: League of Women Voters.

Source: Slovic et al. (1979)

A similar situation occurs with HIV infection which can proceed to AIDS, but often, perhaps more often than not, fails to (Stewart 1995). And how do the risks of HIV compare with those of other sexually-transmitted infections, and other risks? AIDS was presented and accepted immediately as an unparalleled threat to public health in most First World countries. 'A 1987 Gallup Poll indicated that 68 per cent of American adults believed AIDS to be the nation's most serious health problem, five times more than named cancer and ten times more than named heart disease, even though cancer that year killed over forty times (about 475 000 deaths) as many Americans as died of AIDS, and heart disease killed seventy times (about 770 000 deaths) as many' (Fumento 1989). The passage of time has shown these perceptions and projections to be a gross overeaction.

The perception of high personal risk and potential disaster explains both the high risk profile of AIDS held by the public and the success of pressure groups and lobbies to acquire disproportionate funding (Craven et al. 1993, Craven and Stewart 1995, 1996, Whelan 1990). This same possibility was recognized by the Ford administration in the US in their fear of the Swine 'flu threat. Pharmaceutical companies were forbidden under the National Swine 'flu Immunization Programme of 1976 (Public Law 94–380) from making 'any reasonable profit' from the Swine 'flu vaccines.

The risk probabilities of heterosexual HIV transmission estimated by Hearst and Hulley (1988) were derived from partner studies (Padian et al. 1987a, 1987b, Peterman et al. 1988). Hearst and Hulley attempted to identify the risks of HIV infection between differentiated groups adopting different sexual practices. They found that the risk of a person becoming seropositive after a single sexual encounter with a partner from a low risk group (LRG) was 1 in 5 000 000. (Table 10.4) Two factors were important: first, the number of sexual encounters (as distinct from the number of partners) and, second, whether the partner was from a high risk group (HRG). Their findings have been disputed by Wittkowski (1989), who suggests that the ability to spot HRG partners and the compliance with advice will be imperfect. Further, he asserts that if 50 per cent of the blood samples cited by Hearst and Hulley came from lifetime monogamous individuals, then the incidence of HIV in the general population is doubled. When these factors are taken into account, the risk of HIV infection from a single unprotected sexual encounter with someone never tested for HIV is 1 in 20 000. Wittkowski also concludes that condom and suppository use may be much more effective than attempting to avoid high-risk partners.

In the case of AIDS, the perceived disaster was one of widespread heterosexual transmission of HIV seropositivity which would threaten all relationships at all levels. Again perception of risk is crucial. No medical evidence ever showed that the risk of HIV infection with an incubation period of ten years from sexual contact with low risk partners approaches the risk of death within a year from very common everyday risks such as smoking.

Unfortunately, the perception of potential disaster influencing the public's assessment of risk is not sufficient to explain the reaction to potential global

Table 10.4. Estimated risk of infection

	One sexual encounter		*500 sexual encounters*	
Partner never tested				
Not in high risk group				
using condoms	1 in	50 000 000	1 in	110 000
not using condoms	1 in	5 000 000	1 in	16 000
High risk groups (HRG)				
using condoms	1 in	100 000/10 000	1 in	210/21
not using condoms	1 in	10 000/1 000	1 in	32/3
Partner tested negative				
No history of HRB (a)				
using condoms	1 in	5 000 000 000	1 in	11 000 000
not using condoms	1 in	500 000 000	1 in	1 600 000
Continuing HRB				
using condoms	1 in	500 000	1 in	1 100
not using condoms	1 in	50 000	1 in	160
Partner tested positive				
using condoms	1 in	5 000	1 in	11
not using condoms	1 in	500	2 in	3

(a) HRB + High risk behaviour
Source: Hearst and Hulley (1988).

Swine 'flu epidemic. Here, projections of a million deaths among the elderly should have resulted in a public clamour for inoculation. But a Gallup poll indicated that whilst in September 1976, 93 per cent of the US adult population believed that the Fort Dix outbreak was caused by Swine 'flu, only 53 per cent intended to be inoculated. In contrast to this calm, almost complacent, approach there was huge over-reaction to the Legionnaire's outbreak. Garrett (1995) cites 'explosive outbreak,' 'mysterious and terrifying disease,' 'Legionnaire killer' being descriptions used by the press and made in television statements from politicians and the Philadelphia public. We conjecture that there is disproportionate fear among the public when a disease-causing agent is claimed to be 'new.' Influenza is popularly perceived as 'a very bad cold' to be treated with pain killers, hot drinks and a few days in bed. Radiation too is a well-known health risk. Experts consistently reassured that radiation from Chernobyl was always below health-threatening levels. These claims were checked by the Press. Experts claimed that the Chernobyl radiation would have been too dilute to threaten individuals directly in the UK and most other European countries.

Public policy and public health: an examination of blame avoidance strategies

Classification of unexpected events

This paper has addressed the issue of appropriate policy response to unforeseeable potential medical disasters. Although it has been argued (Day and Klein 1989) that policy could be generalized this is not the conclusion of this paper. The six events examined here all had, or have, the potential for medical disaster but none exactly fits the qualifying characteristics identified in the Day and Klein paper. For them, the event had to be unforeseeable so that no contingency plan could be made. A nuclear meltdown was widely considered a possibility after the accident at Three Mile Island so the inclusion of the case of Chernobyl is questionable. The event had to contain an element of moral evaluation which would limit the extent to which policy responses could be clear and unambiguous. Whilst this is clearly the case with HIV/AIDS there are no moral issues associated with Swine 'flu, Legionnaire's disease or Chernobyl. Political correctness and medical mis-judgement initially prevented blood donors being screened on the grounds of sexual preference or life-style. Moral issues concerning large-scale slaughter of cattle following the EU world-wide ban have not, ironically, concerned animal rights' groups. Whether lower beef prices and higher prices of substitutes adversely affect the poor has never been an issue. The only common characteristic of the events examined here is that of unprojectability. It seems unlikely that future crises can ever be classified according to pre-set criteria.

Unintended consequences of blame avoidance

If there is no history of party political involvement or threat to jobs and livelihoods and there is no international dimension then the appropriate strategy appears to be blame avoidance, involving all parties in talks, discussion and funding of policy. The strategy of blame avoidance was initially appropriate for funding research to counter the HIV/AIDS threat but it eventually resulted in excessively generous funding in comparison with other diseases. The power of rent-seeking interest groups and incrementalism prevented fundamental reassessment, a policy problem which has yet to be resolved. The Department of Health and others in receipt of tax funds subsequently claimed credit by asserting that earmarked expenditures prevented social catastrophe. The blame avoidance response to Swine 'flu also had quite unintended political consequences. In November 1976 the first of several hundred cases of Guillain-Barré syndrome were documented. The syndrome was usually rare and the main symptom was temporary paralysis lasting a few weeks. It was occasionally lethal. It was noticed and later confirmed that the syndrome appeared disproportionately often in those inoculated against Swine 'flu (Schonberger et al. 1979, Safranek et al. 1991). Besides permanent disabilities and deaths attributed to side effects of the vaccine the granting of immunity to the vaccine, manufacturers resulted in the government being blamed for setting 'precedents for government culpability in large-scale public health efforts' (Garrett 1995). In the 1980s the US Government was sued for polio vaccine allegedly causing polio in those inoculated with it. The government was therefore blamed for instigating an unnecessary and costly programme which damaged public health. Thus an attempt to avoid blame achieved the opposite of the result intended.

Avoiding blame by relying on expert advice

Day and Klein argued that in circumstances of catastrophe government would turn to expert information and advice. This has not proved to be a good general strategy. In the case of AIDS the passage of time showed that the forecasts and projections of experts were erroneous by orders of magnitude and resulted in huge unnecessary expenditures. The government stated that because experts considered there was no proven link between BSE and CJD, no culling was necessary. This was inconsistent since the only reason for having raised the alarm was the possibility of transmission of BSE from cattle to humans. It was this which prompted the further alarm and the imposition of the world-wide ban by the European Union on exports of all British beef, even from herds which had never had BSE, and of beef products. A short-lived suspension of co-operation between Britain and the rest of the EU followed.

Reliance on expert information proved disastrous in terms of the response of governments world-wide to the issue of contaminated blood and blood products. Risks from transfusions are increased when blood products (coagulation factors VIII and IX) are used. To prepare these concentrates, large pools of blood, sometimes from more than one country and from thousands of assorted donors, are required. There was no regulatory requirement for manufacturers, as in other industries producing biological products for public use, to advise users and recipients of this. There are parallels and differences with BSE-contaminated feedstuffs. The displayed list of contents on the packs of feedstuffs indicate cereal grains, oil seed products and by-products, in descending order of weight only. They did not reveal that, prior to 1988, protein rich cattle cake contained meat and bone meal made from the remains of sheep and cattle, some of which was contaminated with scrapie, a sheep disease similar to BSE. This was justified by the claim of UK Agricultural Supply Trade Association (the manufacturers' trade lobby), that full disclosure is rare because of the need to keep ingredients secret.

Initially, the process of concentration did not exclude viral contaminants, while the sources of supply and manufacture, mainly in the USA, ensured that there would be many drug addicts in the pools. Risks of infection with a variety of viruses existed therefore through the 1970s, long before the identification of HIV in 1984; recognition came after the event (Root-Bernstein 1993). The fact that not all recipients of infected material became seropositive to HIV, and that only a minority developed AIDS, is an aspect of the pathogenesis of the condition which still awaits debate.

The delivery of immunosuppressive material to persons already at risk was a danger which could and should have been foreseen by the manufacturers, licensing authorities and users. Experts could have been, should have been, and maybe were, aware of the likelihood of transmission but representations to government were not made (Craven and Stewart, 1996).

Blame avoidance and international conflict

If the domestic policy response to potential catastrophe becomes party political, the incumbent policy-makers are in a lose/lose game. Whatever 'blame avoiding' government action was taken in the case of BSE/CJD, jobs and livelihoods in the meat industry would be damaged either from the culling and/or the fall in consumer demand. This created potential for political gain from party differences over how to minimize the damage, something likely to occur when obvious losses fall on the community. But the domestic policy response to the BSE/CJD crisis showed that even if a unified political response had been possible in the UK, this need not hold internationally. The crisis demonstrated that foreign governments could restrict trade by playing the health card. The EU Standing Veterinary Committee, using romantic rationale, recommended that, so long as there remained the remotest risk of BSE causing CJD, all UK exports of beef and beef products were to remain banned world-wide until BSE had been eliminated. The EU policy was unique in insisting upon elimination rather than minimization or optimizing risk. This decision had little or nothing to do with public health and much to do with restriction of trade (Dnes 1996). Continental politicians could both claim credit and avoid blame.

Conclusion

In this chapter, we have examined the problems arising in recent years from the sudden occurrence of serious but unforeseen hazards to the public health, using as examples Legionnaire's disease, Swine 'flu, Chernobyl, AIDS and BSE. Although these occurrences are entirely different in origin and effects, they have common underlying and expressive features which determine a certain uniformity in the immediate response, and to some extent, in the aftermath. For the responsible authorities, there is an immediate problem and potentially a conflict if alarm, as in Chernobyl, exceeds information, or, at the other extreme, if unsupported, alarming information is given, as with

Swine 'flu. Action patterns differ accordingly. In the aftermath, while fuller explanation is being sought or becomes available, the strategy tends to be one of blame avoidance. Each of these occurrences had the potential for catastrophic consequences for public health. Informed projections were therefore required but, in each case, the outcome showed that experts could not be relied upon to give either objective or reasonably accurate projections and forecasts. Oblique pressures and shading from activists and vested interests supervene even when, as in each of these cases, the emergency is seen to be national and non-political in origin and impact. In two of the cases (Legionnaire's disease and Swine 'flu), where the feared danger to the public health did not materialize, the fault lay squarely with the expert advisers who got it all wrong. In the cases of AIDS and BSE, the crisis and aftermath were real and serious but the impact was social and economic with minimal danger to the health of the general population. With Chernobyl, the aftermath is one of uncertainty. In all of them, the expert advisers were more often wrong than right in their assessments. Executive action based on these assessments led inevitably to further errors which in turn favoured blame avoidance and 'muddling through' as policy responses. In the absence of a more effectual system for obtaining and checking the accuracy of advice given by experts in such matters, this might be inevitable. There would seem to be a need to repair defects in the processes of assessment, advice and executive action when unforeseen emergencies threaten the public health.

Acknowledgement

The authors would like to thank T. M. Hodgins and D. Wakelin. The usual disclaimer applies.

References

Ablin, R. (1985). Transglutaminase: Co-factor in Aetiology of AIDS?, *Lancet*, 6 April, 813–814.

Bootle, R. (1996). *The Death of Inflation*, Nicholas Brealey Publishing.

Bovens, M. and 't Hart, P. (1996). *Understanding Policy Fiascos*, Transaction Publishers, New Brunswick, New Jersey.

British Medical Association (BMA) Guide (1987). *Living With Risk*, J. Wiley and Sons.

Broome, C. V. et al. (1979). The Vermont Epidemic of Legionnaire's Disease, *Annals of Internal Medicine*, **90**, 573–577.

Carr, R., Edmond, E. and Peutherer, J. F. et al. (1984). Abnormalities of Circulating Lymphocyte Subsets in Haemophiliacs in an AIDS Free Population, *Lancet*, June 30, pp.1431–1434.

Centers for Disease Control (1976a). Current Trends: Influenza–United States, *Morbidity and Mortality Weekly*, **25**, 124.

Centers For Disease Control (1976b). Influenza Vaccine–Supplemental Statement,

Morbidity and Mortality Weekly, **25**, 221–227.
Centers for Disease Control (1977a). Follow up on Legionnaire's Disease – Ohio, *Morbidity and Mortality Weekly Report*, **26**, 308.
Centers for Disease Control (1977b). Legionnaire's Disease – England, *Morbidity and Mortality Weekly Report*, **26**, 391.
Cooper, M. H. and Culyer, A. J. (1968). *The Price of Blood*, Institute of Economic Affairs, Hobart Paper 41.
Corey, L. C. and Holmes, K. K. (1980). Sexual transmission of Hepatitis A in homosexual men: incidence and mechanism, *New England Journal of Medicine*, **302**, 435.
Cox, D. R. Anderson, R. M. and Hillier, H. C. (Eds.) (1989). Epidemiological and statistical aspects of the AIDS epidemic, *Philosophical Transcripts S. Soc. London (B)*, **325**, 37–187.
Craven, B. M. Stewart, G. T. and Taghavi, M. (1993). Amateurs confronting politicians: a case study of AIDS in England, *Journal of Public Policy*, **13** (4),305–325.
Craven, B. M. and Stewart, G. T. (1995). Corporate governance, financial reporting, regulation and the AIDS threat in Scotland, *Financial Accountability and Management*, **11**(3), 223–240.
Craven, B. M. and Stewart, G. T. (in press) AIDS, regulation and the law in the market for blood and blood products, *Medicine Science and the Law*.
Danziger, R. (1996). An epidemic like any other? Rights and responsibilities in HIV prevention, *British Medical Journal*, **312**, 26 April.
Day, P. and Klein, R. (1989). Interpreting the unexpected: the case of AIDS policy making in Britain, *Journal of Public Policy*, **9**(3), 337–353.
Department of Health and the Welsh Office (DoH) (1988). *Short term prediction of HIV infection and AIDS*, Report of a working group (Chairman: Sir D Cox) London: HMSO.
de Biasi, A. Rocino, E. Miraglia, L. et al. (1991). The impact of a very high purity factor VIII concentrate on the immune system of human immunodeficiency virus-infected haemophiliacs: a randomised prospective, two-year comparison with an intermediate purity concentrate, *Blood*, **78**(8), 1990–1921.
Dietzman, D. E., Harnisch, J. P. and Ray, C. G. et al. (1977). Hepatitis B surface antigen (HBsAG) and antibody to HGsAG: Prevalence in homosexual and heterosexual males, *Journal of the American Medical Association*, **238**, 2625.
Dnes, A. W. (1996). An economic analysis of the BSE scare, *Scottish Journal of Political Economy*, **43**(3), August, pp. 343–348.
Dowdle, W. and LaPatra, J. (1983). *Informed Consent: Influenza Facts and Myths*, Nelson-Hall, Chicago.
Drew, W. L., Mintz, L., Miner R.C., et al. (1981). Prevalence of cytomegalovirus infection in homosexual men, *Journal of Infectious Diseases*, **43**, 188–192.
Fumento, M. (1989). *The Myth of Heterosexual AIDS*, Basic Books, New York.
Garrett, L. (1995). *The Coming Plague*, Penguin.
Goedert, J. J. et al. (1994). Risks of immunodeficiency, AIDS, and death related to purity of factor VIII concentrate, *Lancet*, **344**, 791–792.
Gubarev, V. and Odinets, M. (1988). Chernobyl two years on, *International Pravda*, **2**(7), 26–27
Hearst, N. and Hulley, S. B. (1988). Preventing the heterosexual spread of AIDS, *Journal of the American Medical Association*, **259**, 16(22–29 April), p. 2429.
Institute of Economic Affairs, Special Issue, (1994) The Perils of Regulation, 14(4).

Kahneman, D., Slovic, P. and Tversky, A. (1982). *Judgement Under Uncertainty: Heuristics and Biases*, Cambridge University Press, London.
Lovins, A. B. (1975). *Nuclear Power: Technical Bases for Ethical Concern*, 2nd. edn, Friends of the Earth for Earth Resources Ltd.
Marks, J. S. et al. (1979). Nosocomial Legionnaire's Disease in Columbus, Ohio, *Annals of Internal Medicine*, **90**, 565–569.
Mills, T. C. (1992). *Predicting the Unpredictable*, Institute of Economic Affairs, Occasional Paper No. 87.
Nau, J-Y., (1994). New debate in French blood saga, *Lancet*, **344**, 398, August 6.
North, R. (1993). *Death by Regulation: The Butchery of the British Meat Industry*, Institute of Economic Affairs Health and Welfare Unit, Health Series, No. 12.
North, T. and Gorman, T. (1990). *Chickengate: An Independent Analysis of the Salmonella in Eggs Scare*, Institute of Economic Affairs Health and Welfare Unit, April.
Ormerod, P. (1994). *The Death of Economics*, Faber and Faber, London.
Padian, N. et al. (1987a). Male to female transmission of the human virus {HIV}: current results, infectivity rates, and the San Francisco population seroprevalence estimates, abstract THP. 3–48:171. Presented at the Third International Conference on AIDS., Washington , DC June 1–5
Padian, N. et al. (1987b). Male to female transmission of human immunodeficiency virus, *Journal of the American Medical Association*. **259**, 788–790.
Padian, N. et al. (1987c). Heterosexual transmission of acquired syndrome: international perspectives and national projections, *Review of Infectious Diseases*, **9**, 947–960.
Parker, D. and Stacey, R. (1994). *Chaos, Management and Economics*, Institute of Economic Affairs, Hobart Paper 125.
Peltzman, S. (1975). The effects of automobile safety regulation, *Journal of Political Economy*, **83**, 677–725.
Peterman, T. A., et al. (1988). Risk of HIV transmission from heterosexual adults with transfusion associated infections, *Journal of the American Medical Association*, **259**, pp. 55–58.
Root-Bernstein, R. S. (1993). *Rethinking AIDS*, New York: Free Press.
Royal Commission on Environmental Pollution (1976). *Nuclear Power and the Environment*, Cmnd. 6618, HMSO.
Rubin, D. (1987). How the news media reported on Three Mile Island and Chernobyl, *Journal of Communication*, pp. 42–57.
Safranek, T. J. et al. (1991). Reassessment of the association between Guillain-Barré syndrome and receipt of Swine influenza vaccine in 1976–1977: results of a two state study, *American Journal of Epidemiology*,**133**, 940–951.
Schonberger, l. B., et al. (1979). Guillain-Barré syndrome following vaccination in the national influenza immunization program, United States, 1976–1977, *American Journal of Epidemiology*, **110**, 105–123.
Selby, P. (1976). I*nfluenza: virus, vaccines and strategy*, Sandoz Institute for Health and Socio-Economic Studies, Academic Press, New York.
Seldon, A., (ed.) (1988). *Financial Regulation or Over-Regulation?* Institute of Economic Affairs.
Seremetis, S. L. M., Aledort, H. and Sacks (1990). Differential effects on CD4 counts of high or intermediate purity factor V111 concentrates in HIV+ haemophiliacs, *Blood and Coagulation Abstracts*, III, **76**, 46a, 1455.

Shope, R. E. (1936). Swine Influenza, *Harvey lectures*, **31**, 183 –213.
Silverstein, A. M. (1981). *Pure Politics and Impure Science, The Swine flu Affair,* John Hopkins UP, Chapter 3.
Slovic, P., Fischhoff, B. and Lichtenstein, S. (1980). Facts and fears: understanding perceived risk, In Schwing, R. C. and Albers, W. A. Jr. (eds), *Societal Risk Assessment: How Safe is Safe Enough?* Plenum Press.
Stewart, G. T. (1989). Uncertainties about HIV and AIDS, *Lancet*, **1**, 1325.
Stewart, G. T. (1992a). Epidemiology and transmission of AIDS, *The Society of Public Health Official Handbook and Members List*, pp. 19–24.
Stewart, G. T. (1992b). Changing case definition for AIDS, *Lancet*, **340**, 1414.
Stewart, G. T. (1993a). Errors in predictions of the incidence and distribution of AIDS, *Lancet*, **341**, 898.
Stewart, G. T. (1993b). AIDS Predictions, *Lancet*, **341**, 1287.
Stewart, G. T. (1994). Scientific Surveillance and the Control of AIDS: a call for open debate, *Health Care Analysis*, **2**, 279–286.
Stigler, G. J. (1971) The theory of economic regulation, *Bell Journal of Economics and Management Science*, **2**, 3–21.
Thaler, R. (1987). The psychology of choice and the assumptions of economics, In *Laboratory Experiments in Economics: six points of view* Roth A. E. (ed.), Cambridge University Press.
Walker, M. (1986). A stroll down gogol boulevard, *The Guardian*, 26 May.
Weaver, R. K. (1986). The politics of blame avoidance, *Journal of Public Policy*, **6**(4), 371–398.
Webster, D. (1986). How Ministers Misled Britain about Chernobyl, *New Scientist*, **9**, October, 43–46.
Whelan, R. (1991). The AIDS scandal, *Economic Affairs*, June.
Wittkowski, K. M. (1989). Preventing the heterosexual spread of AIDS: what is the best advice if compliance is taken into account? *AIDS* **3**, 143–145.
Wynne, B., Williams, P. and Williams, J. (1988). Cumbrian hill farmers' views of scientific advice, in *Chernobyl: The Government's Response*, UK House of Commons Select Committee on Agriculture, HMSO, pp 408–415, July, London, UK.
Yerkes, R. M., and Dodson J. D. (1908). The relationship of strength of stimulus to rapidity of habit formation. *J. Comp. Neurolog. Psychology*, **18**, 459-482.

11 How are decisions taken by government on environmental issues?

David Everest*

Summary

Politicians have at their disposal numerous national and international bodies whose primary purpose is to provide advice concerning the science relating to health and environmental issues. Despite the existence of these bodies, many decisions made by governments are not based on sound science. Two cases of particular interest are the gradual elimination of leaded gasoline and the control of sulphur emissions in Britain and Europe. In the first case, the scientific evidence suggested that lead in petrol was not responsible for the apparent decline of intelligence in some children. In the second case, the evidence did not suggest that sulphur dioxide from Britain was responsible for a significant proportion of the acidification of Scandinavian lakes, nor did it make environmental sense to retrofit power stations with flue gas desulphurization equipment.

Public concern over environmental issues is increasing world-wide, especially in the richer, developed countries, so that the importance of these issues is moving up both the UK and international political agendas.

The present discussion is concerned with scientific and economic analysis of environmental issues. In these instances it should be possible, at least in principle, to make a considered judgement of the costs and benefits of different measures required to reduce and correct observed environmental damage.

Factors influencing decisions on environmental issues

The importance afforded to particular environmental issues, the policies adopted by government, for their control or amelioration and the resources provided to meet these needs, are greatly influenced by public perception. In a democracy governments are sensitive to public opinion, which thus plays a major role in determining policy, including the distribution of scarce resources between different activities of government. Scientific analysis of environmental issues has played, and continues to play, an important part in policy formulation although, as will become apparent, policy is mainly determined by political rather than scientific considerations.

---

* Sadly, Dr Everest passed away in July 1998.

As discussed in the next section, uncertainty is an inherent part of the scientific process. Consequently, scientists tend to hedge their advice on particular issues with terms such as 'probably' or 'possibly'; certainty is not a concept that scientists typically invoke. This reduces the perceived utility of their advice to the politicians endeavouring to formulate policy to meet the demands of a concerned public who yearn for certainty on environmental issues. Senator Muskie, in the USA, spoke for many politicians when he called for 'one-armed' scientists: advisers who will not say 'On the one hand their evidence is so, but on the other hand...' (David 1975, Ashby, 1993). This has resulted in increasing pressure on scientists to provide 'certain' opinions when giving scientific advice orientated to policy requirements, rather than the hedged advice that is proper to scientists.

Along with scientists, economists are increasingly being called upon to give advice on environmental issues, for example the report by Pearce et al. (1989) commissioned by the Department of the Environment (DoE). On the whole, economists seem to be less ready than scientists to qualify their advice, a point which endears them to politicians. Undoubtledly, this is a factor behind the growing influence of economists and economic analysis in the development of environmental policy. An additional advantage for economists is that whilst scientific concepts are often poorly understood by the general public – and indeed many politicians – everybody understands the concept of money. Some economists have suggested that economic instruments should be used in addition to or even in place of conventional environmental regulations. Examples of the extensive literature on this topic are given by Kohn et al. (1996), Barker (1995) and Strassbourg and Heck (1996).

An important contribution of economics to environmental policy has been to point out, and attempt to measure, the cost to society of the hitherto uncosted use of air, water and land ('the global commons'), resources which are being exploited by humans. Much work has been published in the economic and environmental literature on the global commons, for example Pearce et al. (1989, 1991), Pearce and Maddison (1996). However, whilst this type of economic analysis may be correct in principle, there are real problems in its implementation.

Economic analysis of environmental issues is generally on reasonably firm ground when considering the costs of pollution control and how best to organize control systems to minimize economic costs. This reflects the fact that control measures usually involve established technologies with known economic costs. However, economists are usually on weaker ground when considering the cost of environmental damage, an essential part of the global commons approach. For most, possibly all, environmental problems, it is not possible to identify quantitative scientific links between exposure to pollutants and the degree of damage caused. This is a consequence of the complexity and incomplete knowledge of environmental phenomena. (Efforts by different authors to produce estimates of damage usually lead to wide variations in quoted damage costs. Therefore, the perceived importance of the environmental problem also varies between analysts who, in turn, determine the nature and scope of the control measures to be adopted.) To overcome this problem scientists interested and active in the development of policy have increasingly tried to provide consensus values of the degree of environmental damage and thus damage costs.

The role of science in determining environmental policy

In principle, the UK bases its environmental protection policy on considered scientific and economic analysis of the relevant cause-and-effect relationships. The Government position is illustrated by the following quotations from a 1990 UK Government White Paper.

> There needs to be a major and growing scientific effort to understand [the cause and effect relationships] fully and to identify the most effective and appropriate ways of intervening to protect the environment.

> We must act on facts, and on the most accurate interpretation of them, using the best scientific and economic information.

> However, when there are significant risks of damage to the environment, the Government will be prepared to take precautionary action to limit the use of potentially dangerous materials or the spread of potentially dangerous pollutants, even when scientific knowledge is not conclusive, if the balance of likely costs and benefits justifies it.

The principle of precautionary action noted above applies particularly when there are considered to be good grounds for judging either that action taken promptly at comparatively low cost may obviate the need for more costly damage control action later, or that irreversible effects may follow if action is delayed. Put in more environmentalistic terms: when the state of our planet is at stake, the risks can be so high and the costs of corrective action so great that prevention is better and cheaper than cure.

The principle of precautionary action is often quoted as a reason for bringing forward environmental controls and regulation. Clearly there is much scope for differences of opinion as to the point at which precautionary action should be taken. This judgement will depend on: 1) the strength of public concern about the particular environmental issue, and on whom the economic burdens of precautionary action will fall; 2) the differences of view regarding the extent of environmental damage and the strengths of the links between perceived damage and the claimed cause; 3) on moral considerations such as the need for equity between present and future generations. These are political rather than scientific considerations.

Hitherto in the UK a scientific approach to environmental issues, particularly those related to pollution, has been seen as a means for developing robust policies capable of delivering cost-effective environmental protection and achieving a good level of public acceptance. Indeed scientific analysis, supported by appropriate research programmes, provides a means, probably the only means, for achieving true understanding of the majority of environmental issues. However, the tentative, and, to the lay person, the apparently uncertain opinions often given by scientists on environmental questions represent a real problem for both the government policy maker and the general public. The former is because of the difficulty of incorporating scientific uncertainty into legally based environmental regulations, whilst the latter is led to doubt the utility and appropriateness of science in the development of environmental policy. They lose faith in the role of science in forming a firm base for this policy, because these uncertainties are contrary to the popular conception that science only delivers certainty.

However, it must be emphasized that the scientific and economic aspects of environmental problems are only two of the inputs to the formation of environmental policy. In practice, wider political considerations, often reinforced by apparently strong public perception of environmental risk and damage, usually have greater weight than scientific considerations in the formation of environmental policy. In this respect the treatment of environmental questions by government does not differ fundamentally from the generality of issues that governments must address. This is illustrated in the two specific environmental issues discussed later in this paper.

Basic principles

In considering the role of science in establishing environmental policy it is necessary first to understand the fundamental limitations that apply to the concept of 'certainty' in science. It must be recognized that the role and aim of science are to devise models of the real world (often with a strong mathematical base), and to analyse and test these models in the light of experience in that world. As scientific knowledge develops, the shortcomings of these models become apparent, leading to the development of improved models which better account for the behaviour of the real world. This is a continuing and probably never-ending task. It must be emphasized that the degree of scientific understanding depends on the quality of the underlying model and its ability to account for observed natural behaviour.

Scientific theories must be regarded only as models of the reality of nature, all of which are subject to continual modification in the light of experience and experiment. Similar and related changes in scientific perception occur frequently, for example in chemistry, geology and the life sciences. Sometimes, these changes in scientific perception are fundamental.

The reliability and effectiveness of any model depends on the accuracy and reliability of its experimentally derived information base. This is particularly strong in the field of physics–dynamics, astronomy and atomic scale phenomena. It was the scale and reliability of this information that underlay the advances in classical physics and quantum theory, for example. However, the information bases underlying other scientific fields are not so extensive or reliable as those in physics.

Uncertainties in the cause-and-effect relationships in environmental science persist while the time needed for basic research is being supplanted by the demands of the political process.

The scientific treatment of problems in the environment field, and in areas such as agriculture or medicine, also involves the development of models. These models constitute a key element in the development of relevant policy. As with the main line physics models noted above, the models developed in the environmental field are subject to continual modification and change of interpretation. The size and frequency of the changes in environmental models tend to be large, reflecting the relatively uncertain state of the environmental sciences. This gives scope for considerable divergence of view regarding the nature and interpretation of the output of any model among interested parties.

There is a further limitation to the application of the scientific method to the assessment of environmental problems. As discussed in Everest (1990), environmental matters, particularly those which are associated with what is termed 'risk', are now being recognized as being 'trans-scientific'. This means that although such problems are amenable to the scientific approach, they cannot be resolved fully by scientists and so must incorporate an element of public and political judgement. As discussed by Weinberg (1972), this relationship between science and judgement can occur in three rather different senses.

In the first case (low level insult) science is inadequate because to get answers would be too expensive. Weinberg uses as an example that the number of mice required for a statistically valid experiment to determine the x-ray dose as a possible public standard (150 million) that would double the spontaneous mutation rate would be 8 000 000 000 mice. Thus, he says, the question cannot be answered solely by direct scientific investigation. (See Chapter 6 for a fuller discussion of this issue.)

Second, in the area of the social sciences the subject matter is usually too subjective to follow scientific methods.

In the third case (choice in science), science is inadequate because the issues themselves involve moral and aesthetic judgements; they deal not with what is true but with what is valuable. Some authors question whether the latter moral and factual distinction is justified, because they consider that knowledge, including scientific knowledge, is inextricably entwined with ethics (e.g. Tipler 1994).

Generally, environmental risk issues involve a fusion of the natural and social scientific approaches, because a technically evaluated hazard cannot be fully evaluated unless it is judged in its social context (Douglas 1987). This again emphasizes the importance of political and public perception considerations in the formulation of environmental policy.

How is expert scientific advice provided to the UK Government on environmental issues?

A range of sources of expert advice is available to governments (Everest 1989, 1990). The first of these is the internal staff resources available in government departments and their associated research establishments that provide confidential advice to ministers.

A second source of information and advice is official agencies. In the UK, the Government Statistical Service provides information on topics such as air and water quality, radioactivity, solid wastes, etc. Reports are written by government agencies such as the Environment Agency and its various pollution inspectorates or the Health and Safety at Work inspectorates.

A third source of information and advice comes from publicly funded bodies such as the research councils, in particular the Natural Environment Research Council (NERC), from the universities, and from quasi-autonomous non-governmental organizations (QUANGOs) such as English Nature or the National Radiological Protection Board (NRPB). The information bases of these bodies reflect their own research activities and those placed with them by their customer departments. An example is the first report of the Scientific Group on Decommissioning Offshore Structures (NERC 1996) which the Energy Minister of the Department of Trade and Industry (DTI) commissioned from NERC.

A fourth source of information comes from scientific expert study groups set up by governments to review current knowledge on particular environmental topics. One example is the Steering Committee on Environmental Lead Monitoring, which followed changes in UK blood levels over the period in which the concentration of lead in petrol was reduced to 0.15 g/litre (DoE 1987, 1988, Quinn and Delves 1989).

A fifth source is that provided by formal specialist advisory committees. Advice provided by expert civil servants, by external research organizations or by the wider scientific community is of great utility, but cannot replace the authoritative policy-orientated advice that a well-established standing expert committee can provide. The latter are particularly useful in providing policy-makers with advice in areas of scientific doubt and uncertainty, although the need to supply policy-orientated advice can lead to partial abandonment by the committee of a strictly scientific approach.

UK expert advisory committees

In the UK, the Royal Commission on Environmental Pollution (RCEP) is the most important of the advisory committees. It was established by Royal Warrant in 1970 with terms of reference 'to advise on matters, both national and international, concerning the pollution of the environment; on the adequacy of research in this field; and the future possibilities of danger to the environment' (RCEP 1988). It has some autonomy, in particular concerning the selection of study topics and in the timing and content of its reports. It has wide access to information because it can demand written evidence and to cross-examine witnesses. As outlined in its first report (RCEP 1971) the strategy of the RCEP is to select problems that require more attention than they are getting and, after enquiry, to recommend action by government, local authorities, industry or the public.

Members of the RCEP are appointed by the Queen on the advice of the Prime Minister and are drawn from a variety of backgrounds, including academia, industry, public life (including the trade unions) and have a wide range of expertise in science, medicine, engineering, law, economics and business; traditionally at least one member of the Commission has been a parliamentarian.

The Radioactive Waste Management Advisory Committee (RWMAC) is a typical example of a Departmental Advisory Committee. The task of RWMAC is to advise ministers on the management and treatment of wastes from all stages of the nuclear fuel cycle, including the long-term management of spent fuels. It also endeavours to inform and to take account of public perception in this field. These matters are covered by the annual reports of the RWMAC (HMSO from 1980). The RWMAC considers that its main task is the provision of expert advice to Government. However, from its inception the RWMAC has attached importance to the need to inform the public about the facts, as the committee sees them, on the treatment, disposal and environmental impacts of radioactive waste. But, despite their efforts to inform the public, there is little evidence that the RWMAC has achieved much success in increasing public confidence in national radioactive waste policy.

International advisory bodies

A recent innovation is the international expert advisory body that provides advice to national governments and to international bodies such at the United Nations Environment Programme (UNEP). The most prominent of these is the Intergovernmental Panel on Climate Change (IPCC), which continues to play an influential part in the development of international policy to counter the perceived risk of global warming resulting from the build-up of greenhouse active gases in the atmosphere, in particular CO_2. It is noteworthy that in no other environmental field has a similar international advisory system evolved.

The IPCC was set up in 1988 under the joint sponsorship of the World Meteorological Organisation (WMO) and UNEP. It was agreed that the IPCC should identify uncertainties and gaps in present knowledge about climate change and its potential impacts. The WMO was concerned that UNEP might attempt to 'run' the global warming issue without a significant scientific input and they were assured that when the IPCC was set up WMO would be very much in control. The choice of WMO as the lead organization in the operation of the IPCC ensured a major scientific presence in the climate change debate, with its world-wide network of meteorological offices and senior staff involved in research in the atmospheric and related sciences. It provided the national and international links essential both for policy-making and for implementation. The WMO link was especially important for the IPCC because it also brought contact not only with the International Council of Scientific Unions (ICSU) and national research bodies, but also with national meteorological offices and hence government departments. A summary of the background institutional history of the IPCC, including its relationship with the UN General Assembly, is given by Boehmer–Christiansen (1993, 1996).

Discussion of specific environmental issues

Given below is a discussion of how policy was made on two particular environmental issues during the period that the author was an official in the UK Department of the Environment (1979–1986). These examples are chosen both on the basis of their own intrinsic interest and importance, and to illustrate some of the factors which underlie formulation of policy. The emphasis is on how these problems were seen and considered during this period, and discussion of subsequent developments is kept to the minimum required to support and illustrate the arguments. In both examples science played an important contributory role, particularly in understanding of the scope and boundaries of the problem, but the decisions made were essentially of a political rather than a scientific or conventional environmental nature.

Lead in petrol

Lead and its compounds are known toxins and are widely used in industry and in domestic applications. During the 1970s and early 1980s there was a considerable increase in public concern over the environmental health effects of long-term exposure to low levels of lead, that is, at levels below those usually considered toxic. Particular concerns were expressed over the neuropsychological effects in children (Rutter 1983, MRC 1985, 1988). There are a number of pathways by which lead reaches human beings (Davies et al. 1990) lead plumbed water supplies, old lead paint and the dust to which this gives rise, industrial emissions, and the use of lead additives in petrol.

The single-issue environmental group, the Campaign for Lead-free AiR (CLEAR) mounted a strong and partisan campaign to remove lead from petrol. CLEAR claimed that lead originating from petrol was a cause of brain damage to young children. This was a particularly strong rallying point for any single-issue environmental pressure group, so it seems appropriate to summarize the salient facts regarding the health effects of lead and the degree of exposure of the population.

The signs of clinically recognizable lead poisoning include anaemia and alimentary symptoms. If poisoning is unchecked, vomiting becomes persistent, muscle co-ordination is affected and there may be impairment of consciousness, leading to coma. Such lead poisoning, now rare, was usually associated with extreme occupational exposures, from dust caused by the stripping of old lead paint or from cosmetics with a high lead content.

Measurements of blood lead concentrations are normally taken as the means of estimating the extent of exposure, although it would be misleading to give a particular blood lead concentration above which symptoms of clinical lead poisoning are found in man. For both adults and children there appears to be considerable variation. Cases have been observed where blood lead concentrations exceed 150 g/dl, and claims have been made that symptoms of clinical lead poisoning have been encountered at blood lead concentrations as low as 60 g/dl (RCEP 1983).

The European Union directive regarding atmospheric lead (77/3/12/EEC) sets three reference levels and requires that periodic surveys are taken of the public's blood-lead levels. Surveys in 1979-80 and 1981 highlighted some localized breaches, which were remedied. Overall 3.1 per cent of the inner city population and 1.9 per cent of the outer city population were estimated to have blood lead concentrations in excess of 25 g/dl (i.e. within the second reference level), being higher in men, in heavy smokers and drinkers, in older people and those living in older dwellings (DoE 1981, 1982a, 1989). Blood lead concentrations for children in the first survey showed levels of around 16 g/dl between birth and 10–12 years (RCEP 1983).

Early action

In the period 1973–83 the UK Government, as a precautionary measure, endeavoured to keep lead emissions constant, whilst petrol consumption was rising, by means of a number of small cuts in the permitted level of lead in petrol (the 'salami-slicing' policy). This process culminated in 1981 in the decision to make a substantial reduction in the level of lead in petrol from 0.4 to 0.15 g/l, to take effect from the start of 1986 (RCEP 1983). The medical and scientific basis for this policy was reinforced by the report of the working party set up under the chairmanship of Professor Lawther by the former Department of Health and Social Security. Its objective was to review the overall effects on health of environmental lead from all sources – in particular its effects on the health and development of children (Lawther 1980, RCEP 1983).

The Lawther Report concluded that, for the majority of the population, food contributed most to the body burden of lead and that airborne lead (including that derived from petrol) was usually a minor contributor. The Report considered that the small proportion of the population whose blood lead fell into the upper part of the blood lead concentration derived their relatively raised blood lead not from lead in petrol but from a miscellany of sources, such as lead-containing paint. The report recommended progressive reduction in emissions of lead into the environment from all sources. This included reducing the lead content in petrol, tap water (in areas where this was a problem), paints, food, cosmetics and toys. However, the Working Party was unable to form a definite conclusion on the effects of small amounts of lead on the intelligence, behaviour and performance of children, for which it was criticized by CLEAR. It did recommend that when a child had a blood lead concentration of 35 g/dl steps should be taken to identify the source of lead and to reduce exposure.

Overall, the Lawther Working Party, from its interpretation of the available evidence, concluded that there was uncertainty about the health effects of lead even in the range 35–80 g/dl. However, in 1982 the UK Health Department recommended that if a blood lead level for a person, particularly a child, exceeded 25 g/dl as a result of environmental exposure, then steps should be taken to reduce exposure (DoE 1982b). This represents a considerable reduction as compared with the views expressed in the Lawther Report, and appeared to be primarily a political response to the considerable public criticism expressed over the Lawther Report, rather than a professional reappraisal of the issue.

It must be emphasized that these twin concerns (viz. the importance of lead in petrol as a pathway to man and the postulated effect of low levels of lead on children's intelligence) were strongly linked in the public's perception of the whole environmental lead issue. Public concern on these two aspects of the environmental lead issue, vociferously championed by powerful single-issue groups, undoubtedly had a major impact on Government policy on the lead issue.

The RCEP Report

This study was initiated as a result of the major and continuing concerns over the health effects of environmental lead contamination expressed both by the public and parts of the scientific community (RCEP 1983).

The RCEP Report (chapter VIII) acknowledged that humans ingest lead from many sources, with the contribution of different components depending heavily on environmental factors and personal habits. Food and drink form the major lead pathway for most people in the UK, although there is considerable uncertainty as to the relative contributions of several sources of lead to this pathway. Exceptional factors can increase the significance of particular pathways. For example, people living in areas with plumbosolvent water or lead lined water tanks may receive more than half their lead uptake from water. (Uptake signifies lead absorbed into the blood; intake is lead inhaled or ingested.) The importance of lead from petrol is difficult to evaluate since it reaches the body through several pathways, including indirectly via food which, in its primary production stage, is exposed to lead originating from petrol emissions. The Commission considered on the evidence available to them that, for a number of adults in the UK, this pathway contributes at least 20 per cent of their uptake; for the majority it contributes less. For young children lead in dust carried to the mouth by hands or objects placed in the mouth may constitute half their uptake. The lead content of dust was considered to arise from soil, paint, industrial emissions and emissions from petrol engines.

The Commission made 29 recommendations for action on the lead issue, including minimizing the hazards associated with old leaded paints, the problem of plumbosolvency, and the need for more research on pathways and the health effects of lead, particularly regarding young children. These were important but relatively uncontroversial recommendations. Politically, the most important recommendation was that the reduction of lead in petrol to 0.15 g/l, due to take place in 1986, should only be regarded as an intermediate stage in the total elimination of lead additives. The Government was invited to start negotiations within the European Community to secure the removal of the minimum level for lead in petrol that had been set by Directive 76/611/EEC. It was also invited to begin discussions with the UK oil and motor industries to agree a timetable for the introduction of unleaded petrol.

The Commission assumed that 'substantial continuing improvements in engine efficiency make it unlikely that any overall energy penalty will actually be experienced as a result of the switch from 97 RON petrol with 0.15 g/l of lead to 92 RON unleaded and the consequential change to lower compression ratios. A potential net energy saving (equivalent to about 2.5 per cent of crude oil requirements) will, however, have to be forgone' (RCEP 1983). In view of the subsequent increase in oil consumption of the transport sector this must be considered a not insignificant economic and environmental penalty for removing lead from petrol.

The RCEP also noted that the development of engines to run on unleaded petrol would remove an impediment to the control of other gaseous emissions by the fitting of exhaust catalysts, a requirement already being considered by the European Communities. Such catalysts are poisoned by lead, at least using current technology, and thus require the use of unleaded petrol.

The Government response

The lead-in-petrol issue divided UK Government Departments. The introduction of unleaded petrol was opposed, in particular, by the Department of Energy (DOE) because of the resulting increase in oil consumption. It was opposed also by the Department of Transport (DoT) because the use of unleaded petrol would increase fuel consumption and thus motoring costs. For the same reason the DoT was also strongly opposed to the introduction of exhaust catalysts for controlling emissions of nitrogen oxides, carbon monoxide and unburned hydrocarbons – for a discussion of the techno-political aspects of the vehicle emission issue see Boehmer–Christiansen (1990). The then Department of Industry was on balance opposed to the introduction of unleaded petrol. It was concerned that its introduction might have detrimental impacts on the UK car manufacturing industry, and would gravely affect the prospects of the manufacturers of lead additives with their strong export performance. However, this opposition was limited because UK industry, with support from the Department of Trade and Industry (DTI), had developed a strong position in the manufacture of exhaust catalysts.

The introduction of unleaded petrol was supported by the DoE who strongly argued its case in the Whitehall inter-Departmental debate that accompanies the development of government policies. However, this support was not based on a careful analysis of the scientific or health considerations linked to the importance of this particular pathway and the role of low levels of lead on children's behaviour and intelligence. Indeed there were reservations on both these counts by scientific and medical professionals in both the DoE and the Health Departments. The policy drive was based primarily on the political necessity for the DoE to be seen to take a positive role in promoting environmental protection, both at home and in the European Community. This political necessity reflected:

- the growing strength of popular environmental sentiment;
- the strength of the single issue pressure groups, in particular CLEAR, and their ability to exploit the emotive issue of the effect of lead on children's intelligence;
- the imminent general election of June 1983.

The RCEP lead report (RCEP 1983) was published in April and its recommendations were unusually rapidly accepted by the Government. This fact reflected the considerable political skill of DoE Ministers in manipulating the Whitehall system against the objecting Departments. As a result the Government went into the 1983 election with a populist 'environmental victory' to its credit. It was also a plus point for the Government that it was able to go to Europe with an environmental initiative – namely the introduction of lead-free petrol – so gaining European environmental credibility. As noted in the next section, the UK was perceived by its European partners as not having a progressive policy on environmental issues. However, the UK has often had a very good case in its opposition to many aspects of proposed European environmental legislation.

Postscript

An estimate of the possible effects of moving to lead-free petrol was obtained by the study undertaken of the effects on air lead and blood lead concentrations of reducing petrol lead from 0.4 to 0.15 g/l that occurred in 1986. This reduction in the lead content was greater than that obtained by going from 0.15 g/l to lead-free petrol. It resulted in a 50 per cent reduction in air lead concentrations, and was accompanied by a 16 per cent decrease in blood levels for children and 9–10 per cent for adults (Quinn and Delves 1989). However, these reductions must be considered in the context of a continuing long-term fall in blood lead concentrations of about 4–5 per cent a year, presumably from other pathways by which lead reaches adults and children. This downward trend is probably more significant for reducing lead body burden than the expensive once-for-all benefit achieved by removing lead from petrol.

Sulphur dioxide emissions and acid rain

Sulphur dioxide (SO_2) emissions from the combustion of high-sulphur oil and coal are a long-established environmental problem, particularly the high and health-damaging concentrations that can occur in the immediate vicinity of emission sources. One of the most effective means of controlling this problem is to discharge effluents through a high stack, so dispersing the SO_2 and thus keeping ground-level concentrations below health-damaging levels.

During the 1970s and 1980s concerns were growing over the long-range transport of SO_2 and other acid-forming emissions such as oxides of nitrogen (NO_x). Wet deposition (acid rain) of these emissions was said to be causing damage to acid-sensitive ecological systems, particular concerns being expressed by the Scandinavian countries where the soils are considered vulnerable to acidification. These countries considered that the bulk of their acidic depositions arose from emissions occurring outside Scandinavia. The UK is a major source of SO_2 emissions that could be transported to Scandinavia by the prevailing westerly air stream. The UK was thus named by the Scandinavians as an important contributor to their unwelcome acidic imports.

It is not the intention here to give a full discussion of the scientific and political aspects of the SO_2 acid rain issue; the interested reader is directed to Boehmer–Christiansen and Skea (1991) and the references they quote for a full account of this issue. Attention will concentrate on the activities of the UK in the first half of the 1980s which were related to the adoption of the then draft Large Plant Directive (eventually adopted in 1988 as 88/609/EEC – see DoE, 1990), and to negotiations to reduce SO_2 emissions which were held under the auspices of the United Nations Commission for Europe (UNICE). A politically influential development in this period was the '30 per cent Club'. This originated in 1984 as an informal group of ten countries, who pledged to achieve a 30 per cent reduction in their SO_2 emissions by 1993 compared with 1980 levels. The UK refused to join this club, instead offering to achieve the 30 per cent reduction by the year 2000.

There was much criticism, both nationally and internationally, of the UK's failure to join the 30 per cent Club; its offer to achieve this 30 per cent reduction by 2000 was considered to be weak and to reflect a 'lack of environmental will'. Particular criticism was directed at the failure of the UK to fit flue gas desulphurization technology (FGD) to remove SO_2 from the large coal-burning power stations that formed the backbone of the UK's national power supply in the mid-1980s. This lack of action was contrasted with the large-scale investment in FGD made by Germany (specifically West Germany) concerning their own coal-burning power plants. As with the UK, coal formed (and still forms) the mainstay of West German power supply.

UK Government and the 30% club

The view has been expressed (for example, Boehmer–Christiansen and Skea 1991) that UK opposition to emission controls, in particular the large-scale introduction of FGD, was justified on scientific grounds. This view reflected the then current UK scepticism regarding the strength of the links between UK SO_2 emissions and perceived environmental damage both at home and in Europe. This reluctance was criticized as being politically motivated, but the science was incomplete. Specifically, the 'critical load' of the receiving environment – the threshold level of exposure to SO_2 at which damage would occur – was not known.

The UK was classed as 'the dirty man of Europe' over its refusal to join the 30% Club, however, in order to join, a plan of emission reduction; would have had to be submitted. Against a background of coal-miners' strikes, a plan which favoured a switch from coal to natural gas or nuclear power (obviating the need for FGD) would probably have been poorly received. It is of interest that the UK virtually achieved the required 30% cut in SO_2 emissions by 1993 without major investment in FGD technology (DoE 1994).

From an environmental point of view it is probably better to reduce or even eliminate SO_2 emissions at source by using a clean fuel, rather than to attempt to remedy the impact of burning high-sulphur UK coal by means of retrofitting end-of-pipe FGD technology. The FGD process itself is not free from environmental criticism. It is a processing plant in itself, and involves the quarrying of large amounts of limestone, often from areas of high natural beauty, which is transported over long distances and finally becomes a very large waste-disposal problem. In theory the calcium sulphate formed in the FGD process could be used as a substitute for natural gypsum in making plaster board. However, the need for good quality gypsum would increase process costs, whilst the very large amounts of gypsum potentially available from extensive adoption of FGD technology would far exceed demand.

It certainly appears that the substitution of clean natural gas for coal in power generation is a more acceptable environmental solution to the emission problem than the coal burn–FGD approach. A good summary of the adverse environmental impacts resulting from the mining and combustion of coal is given in Cope (1993).

It is interesting how little environmental considerations appeared to enter the SO_2 acid rain debate, even in the views expressed by the environmental movement. The latter gave emphasis to the FGD option, apparently because they saw expansion of nuclear power as the perceived alternative to coal for power generation. There was also considerable sympathy in parts of the environmental movement for the miners cause during the period surrounding the miners strike, apparently based on a common opposition to what was seen as the actions of a remote and uncaring 'establishment'.

Scientific considerations did not determine policy nor, as some critics suggested, were they used as a means of delaying decisions. Genuine doubts existed in official UK circles regarding the utility, from an economic and environmental perspective, of the policy of general adoption of FGD for limiting SO_2 emissions. What science supplied was an understanding of the scope of the problem, some improvement in our still limited understanding of the ecological impacts of SO_2 and acid deposition generally, and an appreciation that there is no simple catch-all answer to this, or indeed most other, environmental issues.

References

Ashby, E. (1993). Foreword to book *Environmental Dilemmas: Ethics and Decisions*, ed. R. J. Berry, Chapman Hall.

Barker, T. (1995). 'Taxing pollution instead of employment: greenhouse gas abatement through fiscal policy in the UK', *Energy & Environment*, **6**, 1.

Black, Sir Douglas (1984). *Investigation of the Possible Increased Incidence of Cancer in West Cumbria*, HMSO.

Boehmer–Christiansen, S. (1990). *Energy & Environment*, **1**, 1.

Boehmer–Christiansen, S. and Skea, J. (1991). *Acid politics: Environmental and Energy Policy in Britain and Germany*, Belhaven Press, London.

Boehmer–Christiansen, S. (1993). *Energy & Environment*, **4**, 362.

Boehmer–Christiansen, S. (1996). *Energy & Environment*, **7**, 362.

Boyle, S. (1987).'Radioactive waste – how to break the log jam, *Energy World* **151** Oct.

Cope, D. (1993). The missing factor in the coal debate – environment. *Energy & Environment*, **4**, 307.

David, E. E. (1975). *Science*, **189**, 891.

Davies, D. J. et al. (1990). *Total Environ*, **90**, 13.

Department of Environment (1981). *European Community Screening Programme for Lead: UK Results for 1979–80*, Pollution Paper **10**.

Department of the Environment (1982a). *The Glasgow Duplicate Diet Study, Pollution Report No. 11*, HMSO.

Department of Environment (1982b).*Lead in the Environment* Department of the Environment and Welsh Office. Circular 22/82 HMSO.

Department of Environment (1987). Pollution Report No. 24 HMSO.

Department of Environment (1988). Pollution Report No. 26 HMSO.

Department of Environment (1989). *Digest of Environmental Protection and Water Statistics*, HMSO.

Department of Environment (1990). *The United Kingdom's Programme and National Plan for Reducing Emissions of Sulphur Dioxide (SO_2) and Oxides of Nitrogen (NO_x) From Existing Large Combustion Plants*, Scottish Development Department, Welsh Office and Department of the Environment for Northern Ireland HMSO.

Department of the Environment (1994). *Digest of Environmental Protection and Water Statistics*, HMSO.
Douglas, M. (1987). *Risk Assessment According to the Social Sciences*, Russell Sage Foundation, Chicago, USA
Environment Select Committee (1986). *Report on Radioactive Waste Management*, session 1984/85, paragraphs 268–71, HMSO.
Everest, D. A. (1989). *The Provision of Expert Advice to Government on Environmental Matters: The Role of Advisory Committees*, Research report no. 6, Environmental Risk Assessment Unit, University of East Anglia, Norwich, NR4 7TJ, UK
Everest, D. A. (1990).*Science and Public Affairs*, **4**, 17.
Everest, D. A. (1991). Environmental and natural resources. In *Science and Technology in the United Kingdom* (R. Nicholson, C. Cunningham, and P. Gummett, eds) Longman.
Everest D. A. (1993). Case study: the Government sector. In *Environmental Dilemmas: Ethics and Decisions*. Chapman & Hall.
Her Majesty's Government White Paper (1990). *This Common Inheritance*. HMSO London.
IPCC, WG 1 (1990). *Climate Change: The IPCC Scientific Assessment*. (J. T. Houghton, G. J. Jenkins, and J. J. Ephraums, eds) Cambridge University Press.
IPCC, WG 2, (1996). *Climate Change 1995. Impacts, Adaptations and Mitigation of Climate Change: Scientific –Technical Analysis*. (R. T. Watson, M. C. Zinyowera and R. H. Moss, eds.) Cambridge University Press.
IPCC, WG 3, (1996). *Climate Change 1995. Economic and Social dimensions of Climate Change*. (J. P. Bruce, P. H. Lee and E. F. Haites, eds) Cambridge University Press.
Lawther (1980). *Lead and health: the report of a DHSS working party on lead in the environment*, HMSO London.
Kohn, M. et al. (1996). Energy, environment and climate: economic instruments, *Energy & Environment*. **7**, 147.
Medical Research Council (1985). *The Neuropsychological Effects of lead in Children – A Review of Recent Research 1979–83*. London.
Medical Research Council (1988). *The Neuropsychological Effects of lead in Children – A Review of Recent Research 1983–1988*. London.
Mulder, J. et al. (1990). Effects of air pollutants on man-managed and natural eco-systems. In *Stockholm Environment Institute Air Pollution in Europe: Environmental Effects, Control Strategies and Policy Options*, Stockholm.
Natural Environment Research Council (NERC) (1996). First report of the Scientific Group on Decommissioning Offshore Structures.
Pearce, D. W. Markandya, A. and Barbier, E. (1989). *Blueprint for a Green Economy*. Earthscan, London.
Pearce, D. W., Barbier, E., Markandya A. et al. (1991). *Blueprint 2: Greening the World Economy*. Earthscan, London.
Pearce, D. and Maddison, D. (guest eds). (1996). *Valuing Air Pollution Damage, Energy Policy*, Special Issue.
Quinn, M. J. and Delves, H. T. (1989). *Human Toxicology*, **8**, 205.
Royal Commission on Environmental Pollution (RCEP) (1971). *First Report*. Cmnd. 4585, HMSO.
Royal Commission on Environmental Pollution (RCEP) (1983). *Lead in the Environment*. Cmnd 8852, HMSO.
Royal Commission on Environmental Pollution (RCEP) (1988). *A Short Guide to the Commission and its Activities*. HMSO.

Rutter, M. (1983). *Low Level Lead Exposure Sources, Effects and Implications.* In *Lead versus Health: Sources and Effects of Low Level lead Exposure.* (M. Rutter and R. Jones, eds). Wiley, Chichester.
Stratospheric Ozone Review Group. (Second Report, 1988). HMSO.
Strassburg, W. and Heck, V. (1996). Self-Commitments as an Effective Instrument for a World-Wide Climate Protection. *Energy & Environment*, **7**, 169.
Tipler, F. J. (1994). *The Physics of Immortality*. Chapter XIII, Macmillan, London
United Kingdom Photochemical Oxidant Review Group (Second Report, 1990).
Weinberg, A. (1972). *Minerva*, **10**, 209.

IV Commentaries

12 Should we trust science?

Terence Kealey

Summary

Professional scientists are privy to no intellectual or cultural secrets from which the rest of us are excluded; their only mystery lies in their factual knowledge. Moreover, scientists are human, so when they see the possibility of funding, they have an incentive to ensure that their research comes up with the 'correct' answers (that is, the answers which will ensure more money). Scientists use their specialist knowledge to advantage, sometimes engaging in distortion and obfuscation – often in unconscious ways – so that their results are pleasingly massaged into shape.

The supposed 'decline' in British science and 'brain drain' of scientists to America and elsewhere are good examples. The 'decline' is only relative: Britain now produces more scientific publications than it did a decade ago, but a smaller proportion of all scientific publications. The 'brain drain', far from being dramatic, is minuscule. Distortion of the facts of these cases led to unnecessary furore and was a waste of academic energy.

Similar distortions occur frequently in the search for funding of medical research. For example, charities raising money for research into cancer and vascular disease frequently tell us that these two diseases are the biggest killers in the Western world. Yet if both diseases were eradicated average life expectancies would only rise by about five years. Individual epidemiologists are also culpable in this respect.

Whatever happens, people will distort facts to suit their own interests, but these distortions are – paradoxically – worst when there is no competition for funding; when the scientific community is united in its belief in a particular doctrine, competing ideas do not receive the attention they deserve and funding becomes contingent on conformity to the received wisdom with the process of scientific discovery stifled. Perhaps the most extreme example of this occurred in Nazi Germany, where it was politically correct to believe in eugenics – the 'science' of eliminating 'undesirable' types, usually through murder and sterilization. Today, it is politically correct to believe that passive smoking causes lung cancer and that mankind's emissions of carbon dioxide will result in runaway global warming. The reason for the perpetuation of these dogmas is clear: scientists who conform to the received wisdom benefit from increased research funding while scientists who publish contradictory studies are punished with rejected grant applications.

In the spring of 1997 Mr Ian Taylor MP, the British government minister for science, launched National Science Week with the question: 'Why should we trust science'? The real question should be: 'Why should we trust scientists?' Of course we should trust science. 'Science' is the wonderful, abstract process by which observations are meticulously and objectively made, hypotheses are fairly tested, and experiments are honestly performed. But science in the real world is the process by which scientists earn their salaries.

Let us first destroy a myth. 'Science' in the abstract is not the preserve of scientists. Indeed 'science' or the scientific method of observation, induction and experimentation, is universally practised in the Western world. My wife, who has no formal scientific qualifications to her name, is a scientist. When she ventures forth to buy a new set of curtains she does so because she observes that the current lot has faded, she hypothesizes that, while green would go nicely with our wallpaper and sofas, she should experiment with a range of colours, and when she discovers that pink actually goes better, she buys curtains in that colour.

And when a door lintel falls on her head she does not rush to the witch doctor to propitiate evil spirits, she may even sue the builder. As Thomas Huxley remarked in his *Man's Place in Nature* (1863), 'Science is nothing but trained and organized common sense', and he could say this because common sense, universally in the West, has long been scientific in its adherence to the principles of objective observation, induction and experimentation.

Professional scientists, therefore, are privy to no intellectual or cultural secrets from which the rest of us are excluded; their only mystery lies in their factual knowledge. The scientist knows things that most other people do not but the same applies to all professional and trained persons. And the scientist is a human being. Public choice theory – the argument that individuals and professional groups will act in their own interests – applies as much to scientists as to other people. Indeed, it applies even more strongly to scientists than to most other groups because scientists have spun a semi-divine aura of rectitude and purity around themselves that protects them from the usual criticism that most professional groups encounter. Everyone 'knows' that politicians, lawyers and estate agents lie; everyone 'knows' that scientists worship truth. When, therefore, scientists' lies are exposed, the shock is almost palpable.

In 1992, for example, at the height of the controversy over the 'decline' of US science, the Investigation Subcommittee of the House (US Congress) Science Committee reprimanded the National Science Foundation (NSF) for publishing a report that was 'pseudoscience' and 'nonsense'. The NSF is a government-funded – to the tune of $2 billion a year – but (quasi-)autonomous, scientist-run agency for supporting research.

Earlier, in 1987, the NSF had claimed that a study by one of its policy analysts, Peter House, predicted that by the year 2005 the US would be short of 675,000 scientists. But the House Subcommittee found that the so-called shortfall amounted to no more than unwarranted extrapolations of inadequate statistics. Damningly, NSF statisticians had already exposed the study's faults privately, but internal NSF memoranda had tried to cover up these faults, referring to the need to 'protect the foundation from damage' and worrying about 'losing this discussion'. The chairman of the US Congress Subcommittee, Rep H. Wolfe (Democrat, Michigan) found that NSF officials had peddled misleading statistics to obtain federal funds. He commented, 'nobody expects NSF to play that game. Everyone around here assumes that NSF's numbers are good science' (Merris 1992).

One reason people believe in scientists is that they do not fully understand them. Ever since Rousseau dismissed scientists as deficient in human warmth or spirituality, scientists have been widely regarded as in some sense deficient (the current nouns might be 'nerds', 'geeks' or 'wonkers'). A 1962 survey of American college students' attitudes to scientists revealed that they are perceived as intelligent but socially withdrawn and possessed of a 'cold intellectualism'(Beardslee and Dowd 1962).

This is a false perception. Scientists are deeply passionate, not nearly cold enough in their intellectualism, but often lacking in self-knowledge. For example (try it yourself) most scientists will deny that they are motivated by any other passion than the search for truth. They see themselves as impervious to the pressures of self-interest. Adam Smith once commented that 'people of the same trade seldom meet together, even for merriment and diversions, but the conversation ends in conspiracy against the public', but scientists will almost uniformly deny that such strictures apply to them. Their denial is sincere, which makes it all the more dangerous.

Good science is difficult. It requires an absolute dedication to an idea, a capacity for vast effort, a strong sense of the possible, and a great sense of self. These are the qualities of the artist, not the desiccated calculating machine of the 1962 college survey. Science, moreover, demands money, and that forces scientists into politics.

Consider, for example, the public pronouncements of the charities that raise money for research into cancer and vascular disease (heart attacks and strokes). The charities constantly remind us that two thirds of us will die from one or other disease, but how often do we hear that, if cancer were cured tomorrow, average life expectancies would only rise by two years – a relatively small increase when viewed historically? If vascular disease were eradicated, average life expectancies would only rise by two and a half years.

Worse, if both diseases were eradicated tomorrow, it would only condemn many of us to death by lingering decrepitude and urinary incontinence. Of course we must research these diseases, if only because all knowledge is important, and because we wish to prevent tragic young deaths, but the partiality of the public pronouncements of these important charities illustrate that science is political, and therefore partial, in its public statements.

Individual scientists, too, are partial in their personal campaigns for research money, but that does not matter as long as scientists compete – molecular biologists against epidemiologists, principal investigators against unprincipled ones – the danger only arises when the scientific community is united.

A celebrated example of collective scientific error was the craze over eugenics of the first half of the twentieth century. This craze was extraordinary, with distinguished researchers, thinkers and politicians advocating special breeding programmes for the intelligent and the sterilization – or worse – of the less gifted. It is truly astounding, for example, to read H. G. Wells advocating the gassing of those who have inherited mental or physical disabilities – but Wells was only popularizing the scientific consensus.

I will not describe the eugenics issue here, as it has already been so well chronicled (Carey 1993), but remember this: by 1933, before Adolf Hitler had come to power and so made it advisable, more than half of all German academic biologists had already joined the National Socialist Party (Craig 1991). This was partly because the Nazis had so comprehensively adopted eugenics. Dr Mengele, for example, was an internationally respected eugenicist.

Eugenics provides a good example of public choice theory at work, because it glorified the role of scientists. It was the scientists who both identified the problem and provided the solution (do you see the similarity with climate change?) and who powered the political and public debate.

And the Nazis were the National Socialist party. *Socialism* is a philosophy that glorifies the state. (A very high proportion of German academics who were not Nazis were Communists during the 1920s and 1930s – the democratic scientist was a rare beast under Weimar.) Since scientists believe (wrongly, see Kealey 1996) that only the state will fund science, they will exploit their authority to push electorates into socialist or state corporate policies (see, for example, the letters written by the British scientific establishment to *The Times* on 23, 25 and 27 March 1992, just before the 1992 general election). The environmental movement, of course, is anti-free market and pro-regulation, which harmonizes with the scientists' perceived self-interest in a powerful state.

One stunning recent example of the capacity of the scientific community, collectively, to mislead, was provided by the scare over the so-called 'decline' of British (and US) science during the 1980s and 1990s. In 1981 Mrs Margaret Thatcher's government cut the budget of the University Grants Committee (UGC) by 8 per cent (the University Grants Committee then provided the universities with their infrastructural and teaching support). The universities responded with outrage, initiating a campaign of ruthless denunciation (the *Oxford Magazine*, for example, compared Mrs Thatcher with Hitler) which reached its apotheosis in Oxford University's denial, in 1985, of Mrs Thatcher's honorary degree (a denial only extended to one other individual, Mr Bhutto of Pakistan, a known murderer) but which grumbles on today through the activities of pressure-groups like Save British Science.

The 8 per cent cut of 1981 was, of course, justified because Mrs Thatcher's government was desperately trying to cut public expenditure. British universities then had higher staff:student ratios than any others in the world apart from those of Holland (which was also trying to reduce them), so it was reasonable to try to approach the ratios of the US, France or Germany, countries then much richer than Britain.

But the British academic and scientific communities behaved, as public choice theory predicts, selfishly. Not only did they ignore any wider issues than their own perquisites, they even abandoned academic and scientific rigour in launching their campaigns denouncing 'decline'.

Mr Ben Martin and his colleagues in the Science Policy Research Unit at Sussex University, published several papers in such prestigious journals as *Nature* and *New Scientist* under titles like 'The Writing on the Wall for British Science' and 'Charting the Decline in British Science' arguing that researchers in Britain had published 10 per cent fewer papers in 1982 than they had in 1973. In a series of reports, the Royal Society validated these findings, which Professor Denis Noble FRS (a most distinguished and honourable scientist) then widely publicized through the pressure group he had helped found, Save British Science. *Nature, New Scientist*, the *British Medical Journal* and other professional journals then ran leaders with titles such as 'Britain over the Hill', 'Sorry, Science has been cancelled' and 'Bye-bye Britain'. The popular media took up the cause (for references to these articles see Kealey 1996).

However, all that SPRU had actually shown was that Britain's science was in a relative decline of 0.9 per cent a year. What they failed to notice was that, world-wide, science publication was growing annually at 4 per cent , as it has done since 1650 (this, the most famous fact in science policy, was first popularized by the father of science policy, Derek de Sola Price who, in his Little Science, Big Science (1963, New York) remarked that the statement that '90 per cent of the world's scientists are alive today' had been true for every year since 1650). Since science internationally was growing at 4 per cent a year, and Britain's share of papers was falling by 0.9 per cent a year, Britain's science was actually growing in real terms.

In Japan, the Far East and other emerging countries, science is growing so fast that their share of papers in the scientific journals is expanding – as it should. Britain only constitutes 1 per cent of the world's population, yet publishes 8 per cent of the world's scientific papers. Britain's absolute annual growth in published scientific papers, but relative decline, will ultimately ensure a more even distribution of expertise. Moreover, Britain still publishes many more papers per capita, and wins more prizes, than any of the major science countries except the US, and Britain is still second only to the US in terms of the total number and quality of the papers it publishes.

Oddly, Martin also missed other indications of British scientific growth, such as the fact that British universities were employing 7,000 more researchers in 1982 than they had in 1973, or the fact that the private funding of academic science had doubled over the decade.

It was not difficult to point out these and the large number of other statistical mistakes that Ben Martin made, and between 1987 and 1997 I ran a solitary campaign of rebuttal (see Kealey 1991). It was only in 1997 that Sir Robert May, the Government Chief Scientist, finally confirmed my figures (Kealey in print).

Nobody wanted to believe my rebuttals. If I, single-handed, working in my spare time, could destroy the collective statistical wisdom of the

Royal Society, the Science Policy Research Unit and Save British Science, others could have done so too, but no one perceived it to be in their interests, so they did not. However, what is more worrying is that some publicly-funded academic bodies behaved poorly.

During the 1980s the Royal Society (for science), the British Academy (for the humanities) and the Economic and Social Research Council (for the economic and social sciences) co-ordinated a tripartite survey of the terrible 'brain drain' Britain was then supposed to be suffering. When the British Academy and Economic and Social Research Council found there wasn't any 'brain drain' in their fields, they suppressed their findings. Only the Royal Society published, but their 'drain' was tiny.

US science activists today are also confusing (deliberately, I fear) a relative decline in their share of papers with an absolute one. They should be exposed (for details see Kealey 1996).

The lesson we must all learn over the 'decline' fiasco, like the one we must learn over eugenics, is that scientists, when they collude, can no more be trusted than anyone else. They are no worse and no better than other people. If they want to 'see' decline, they will suspend all critical judgement over statistics that are absurdly misconceived. If they want to 'see' global warming they will.

I am not arguing that climate change and gaps in the ozone layer are fictions; I am arguing that the scientists engaged in demonstrating these phenomena have invested so much in them that we, as purchasers of science, can no more trust the science activists to be impartial than we can trust any other salesmen. The average shopkeeper is not a cheat but, nonetheless, the wise housewife never forgets the ancient injunction of *caveat emptor*.

References

Beardslee, D. C. and O'Dowd, D. D. (1962). The college student image of the scientist. In *The Sociology of Science*, B. Barker and W. Hirsch (eds), New York.
Carey, J. (1993). *The Intellectuals and the Masses*, London and New York.
Craig, G. W. (1991). *The Germans*, London.
Merris, J. (1992). *Nature* **356**, 553.
Kealey, T. (1996). *The Economic Laws of Scientific Research*, London and New York.
Kealey, T. (1991). *Scientrometrics* **20**, 369–394.
May, R. (1997). *Science 275, 793–796.*

13 The proper role of science in determining low-dose hazard, and appropriate policy uses of this information

Fritz Vahrenholt

Summary

According to an OECD report the German chemical industry leads the field in ecological matters. But the German public continues to demand more stringent environmental regulations. Numerous reasons may explain this phenomenon, but it doubtless has something to do with the lack of interest in technology displayed by political parties, associations, the public and the media. It also doubtless has something to do with the desire to have everything while taking no risks.

Where trust has broken down, and fewer and fewer people understand the increasingly complex interplay of technology, a sincere but irrational fear creeps in. This fear is one of the chief reasons for the German public's opposition to sophisticated technology. What we need is more research, more innovation, more technology, more readiness to change in order to face environmental problems. The irrational technophobia that is becoming increasingly widespread must at last be recognized as ecologically counterproductive and socially dangerous. We must learn to make progress; the modernization of our society matters once more. No progress towards solving the problems of the future is possible without taking risks.

'Spasem planetu' is Russian, and means 'let us protect the Earth'. In St Petersburg pedlars sell t-shirts printed with this slogan and the Greenpeace emblem.

When a Hamburg delegation came back from St Petersburg in early 1994 it was greeted by headlines in the German press declaring: 'From t-shirts to the sewage works – how dioxins get into drains'. Chemists at the University of Bayreuth had discovered an alarming clue that was the beginning of the trail. Dioxins from textiles found their way into sewage treatment plants via washing machines and the water from baths and showers. In Hamburg we have now identified this as one of the main sources of dioxins in sewage sludge that we had been seeking in vain for so long.

And it presented us with a new problem: will the public, that has become so acutely aware of environmental toxins, believe us? Won't they object, saying that they thought dioxin came from chemical manufacturers, from effluent pipes and chimneys and from incineration processes, especially those in public waste incinerators? They will say, 'We thought it was because of these industrial processes that our vegetables and our breast milk are contaminated. Now you tell us its in our clothes?' '*Seveso is Everywhere*: wasn't that the title of a book the present German Minister for the Environment wrote a few years ago?'

To some sections of the media only bad news is good news. But for the sciences and for environmental policy – and surely for industry, too – new information should always be 'good news' because it opens up new approaches to problems – to reduce emissions, for example. But one of the major problems lies in convincing the public, because discussion of risks is carried out erratically instead of on a scientific basis. All too often, stereotyped beliefs and prejudices prevent people from noticing change.

In the 1970s, the expression, 'Seveso is everywhere' may have been an apt description of the seemingly ubiquitous dioxin problem. From the production and use of herbicides to wood preservatives based on pentachlorophenol (PCP), as an emission from unfiltered waste incinerators and from motor-vehicle exhaust: dioxin pollution was so severe in the 1980s that the permitted intake of 1 pg per kg a day for babies through the mother's milk was greatly exceeded, and the limit for adults was exceeded as well.

A campaign to stop the sources of dioxin began in (West) Germany in the mid-1980s. In 1984, Boehringer was forced to close its dioxin-releasing pesticide production facility in Hamburg. In 1985, PCP production at Dynamit-Nobel in Rheinfelden was stopped, and a short time later the use of PCP as a wood preservative was banned in Germany. Thus, by 1986 the biggest sources of dioxin from German industry had been eradicated. Dioxin was (almost) banished from the exhaust of motor vehicles too. The introduction of unleaded petrol made it possible to do without the 'scavengers' dichloroethane and dibromoethane in the fuel; the purpose of these was to volatilize the lead from the engines, but they formed dioxin as a by-product. The 'scavengers' were banned by German law in 1992.

The last major step was taken in the early 1990s with the introduction of a new generation of waste incinerators that reduced the dioxin emissions of these plants (MVAs) by over 95 per cent. Dioxin pollution is now no longer detectable in the fences around incinerators. Indeed, every passenger who travels by ferry from Hamburg to Harwich or Newcastle causes more dioxin pollution than our new, modern MVAs. Since dioxin-contaminated products from the 1960s and 1970s, such as old wood and textiles, are still being thrown out of houses, a kilogram of dioxin enters Germany's household waste stream every year. But because modern waste incinerators actually destroy dioxin, total levels of dioxin in Germany have actually declined precipitously over the past five years. This is reflected in surveys of blood dioxin levels.

In German surveys, dioxin levels in blood fat show that dioxin intake by human beings has just about halved in the past five years. But our own survey in Hamburg, the most comprehensive in Germany, suggests that human dioxin contamination persists, mostly because of continued contamination of other parts of Germany and abroad. Hamburg is still a net importer of dioxin (not counting waste, a certain amount of which still has to be deposited in other German regions).

The subject of 'net imports from abroad' brings us back to the T-shirts. At most, 10 per cent of the dioxin content of domestic sewage sludge has been washed down from the roofs of houses or come from human excreta, in other words, indirectly from the food chain. But we wear up to 20 per cent on our own bodies before we wash it into the sewers – with the aid of strictly phosphate-free detergents, of course. This proportion comes out of the washing machines of private households, and gets into them from imported cotton clothing that has been treated with PCP. But PCP does not only come from South-East Asia; it is still produced in France as well. And the German government has not brought itself to prohibit imports.

Indeed, the whole situation reeks of sins of omission at the political level and in industry – in this case the chemical industry – as well as problems of communication. Dioxin is not a natural ingredient of cotton; whenever its source has been traced, it has been found to have been introduced artificially, first from pesticides and defoliants at the growing stage, then from bleaches and dyes containing chlorine for the semi-finished or finished product, and finally from fungicides intended as protection against mould during shipment. The paths that some articles of clothing take from the manufacturer to the shopkeeper are devious, especially those produced in the so-called low-wage countries, making control difficult. The same applies to the routes taken by some chemicals – this time in the opposite direction. Chemicals that are banned in Western Europe – for good reason – are still used in Eastern Europe and developing countries, and continue to turn up in Western Europe.

All this is unfortunate, not least because there is scarcely an industry that is so much in need of trust as the chemical industry. The average citizen does not understand the interplay of complex scientific, technical and ecological factors. He must trust chemists, the managers of chemical companies and government institutions not to deceive him. But to present the public with examples of aggressive marketing that show no heed for the long-term effects on health and the environment is to play into the hands of those who are hostile to the entire industry already.

And then there is always a host of charlatans waiting to join the public debate. Sometimes they come in the shape of semi-informed journalists, occasionally in academic robes. In Germany one professor announced a 'red alert' and proclaimed to the public, concerning an increased incidence of leukaemia on the North Sea coast, 'Carcinogenic dioxin has been measured in the surf off the island of Sylt ... The substances form a film on the surface of the water ... I am worried about the people of Sylt. And there is a possible hazard to long-stay visitors'.

So we have, on the one hand, manufacturers playing down the situation and, on the other, charlatanism with academic credibility – and both are served up simply as news to the public. Leukaemia, a terrible disease for which it is so enormously difficult to find the real causes either in individual cases or when it occurs in clusters, is always good for stirring up fears and raising suspicions to the level of facts. The long-term operation of nuclear power stations are blamed, although numerous investigations by independent institutions have failed to reveal any irregularities and it now seems rather unlikely that there is any connection. Nevertheless, in a local newspaper circulated in South Germany, far away from Hamburg, a reader has claimed in letters that other cases of the same disease have been caused by dioxin-polluted ash from a waste incinerator in Hamburg. After being mixed with filter dust – which was certainly not a good idea – this ash was used years ago to make ... a car park in the village concerned.

Of course, every scientist knows that to play down the risks from dioxin would be a most inappropriate thing to do. In the 1970s these compounds were indeed a hellishly dangerous example and symbol of ecologically and socially unacceptable chemical production. But is it right to characterize dioxin as the 'poison to beat all other poisons'? When we had to redevelop an estate built in Hamburg in the early 1950s and re-house some of the residents, it was the presence of dioxin in the soil – with concentrations in the nanogram range – that was the decisive factor. The findings caused a great hullabaloo in the city. But when it was discovered that the site on which the new houses were to be built was contaminated by DDT – with concentrations in the milligram range – there was hardly any reaction. The site had previously been used by a tree nursery that had sprayed it with this undoubtedly carcinogenic brew. DDT in the milligram range! It made the experts' hair stand on end, but the journalists and the public took no notice. After all, there was no dioxin.

What can be done to change this? How can we ensure that the safety of chemical plants in Germany has improved considerably. Emergency plans, systems for warning the neighbourhood, containment of highly toxic substances – all demands ridiculed by critics back in the 1970s – are now taken for granted. And it is also taken for granted that Hamburg's copper mill is the most modern and environmentally sound in the world, having a 99.998 per cent recovery rate, or that atmospheric emission from chemical and other industrial plants have been reduced by more than 70 per cent overall. According to an OECD report the German chemical industry leads the field in ecological matters. But if that is the case, why is this change of heart not acknowledged by the public? It is doubtful how far this is a specifically German phenomenon, but here (in Germany) it doubtless has something to do with the lack of interest in technology displayed by political parties, associations, the public and the media. It also must have something to do with the egocentric way of life that is becoming more and more common, and the desire to have everything while taking no risks. And it is also connected with the phenomenon described in the following words by *Der Speigel*, an influential political magazine:

'Incapable of solving the real environmental problems such as mass tourism and transport, the Germans hunt for the last molecule of a dangerous substance hidden behind the wallpaper'.

But it is also a fact that no generation in the history of mankind has had to come to terms with so much scientifically and technically induced change as our present one. And unscrupulous use of the achievements of civilization has shaken confidence in industry and the state. Where trust has broken down and fewer and fewer people understand the increasingly complex interplay of technology, sincere but irrational fear creeps in. This fear is one of the chief reasons for the German public's opposition to sophisticated technology. But what we need is more research, more innovation, more technology, more readiness to change in order to face the approaching global crises. The irrational technophobia that is becoming increasingly widespread must at last be recognized as ecologically counterproductive and socially dangerous.

We must learn to value progress. No progress towards solving the ominous problems of the future is possible without taking risks. And yet many people are willing to take very considerable risks in their leisure time and to impose them on others as well, such as dangerous driving and smoking in confined spaces.

That is yet another aspect of the risk question that is difficult to discuss rationally in any case. But surely we must feel surprised, at some stage, that parsley and basil would not stand a chance of approval as synthetic foods because of the traces of carcinogenic substances (e.g. psoralens) they contain. Or that in Germany, at least, the railway would be declared an infernal machine and never allowed a route for practical trials if it were invented today. When new technologies are introduced, the question of safety takes infinitely higher priority nowadays than it did in the past. And rightly so; but that is no help to us because the *demands* for safety are increasing out of all proportion.

It is becoming clear how much we have to do to draw the media out of their backward, unscientific attitudes every time lecturers stand in front of chemistry students at the University of Hamburg and try to bring the two halves of their brains together: the social prejudices in one half and their scientific knowledge in the other.

Honesty, openness and transparency – taken for granted in the sciences and, we hope, to an increasing extent in industry as well – are not enough when it comes to breaking down hardened prejudices. But we have no choice. To our knowledge the search for dioxin made in Hamburg and documented by the results of the Hamburg dioxin survey is without equal. And it is still true that the toxicity of these compounds justifies such comprehensive research.

But science must not restrict itself to observing what is obvious anyway and satisfying the curiosity of a mainstream that is motivated by self-interest.

It is only of use to society if it finds the courage and patience to distinguish between risks that are really high and those that are just highly discussed, between relevant and negligible input paths, and to make this so transparent that it is possible to form sensible rules of personal behaviour for our daily lives.

V Perception

14 Mass media and environmental risk: seven principles

Peter M. Sandman

Summary

Does media coverage sometimes exaggerate environmental risk? If so, what can be done via the media to put risks into context?

1. Surveys suggest that the amount of coverage accorded an environmental risk topic is not neccesarily related to the seriousness of the risk in health terms. Instead, it relies on traditional journalistic criteria like timeliness and human interest.
2. Within individual risk stories, most of the coverage isn't about the risk. It is about blame, fear, anger, and other non-technical issues about 'outrage,' rather than 'hazard.'
3. When technical information about risk is provided in news stories, it has little if any impact on the audience.
4. Alarming content about risk is more common than reassuring content or intermediate content except, perhaps, in crisis situations, when an impulse to prevent panic seems to moderate the coverage.
5. Exactly what information is alarming or reassuring is very much a matter of opinion. The media audience tends to be alarmed even by information the experts would consider reassuring.
6. Reporters lean most heavily on official sources. They use more predictably opinionated sources, industry and experts on the 'safe' side, only using activists and citizens on the 'risky' side when they need them.
7. Although the competition for journalistic attention is tougher for sources seeking to reassure than for those seeking to alarm, coverage depends even more on a different distinction: skilful sources versus inept ones.

1. *The amount of coverage accorded an environmental risk topic is not related to the seriousness of the risk in health terms. Instead, it relies on traditional journalistic criteria like timeliness and human interest.*

The observation that journalism focuses more on big controversies than on big health risks is neither novel nor likely to be challenged. There is a niche for public-service features about smoking, seat belts, and radon, but in the absence of a news peg these perennials are bound to get less media attention than a hot local toxic waste 'Superfund' fight. Journalists are in the news business, not the education business or the health protection business.

By way of example, we did a content analysis of network evening news coverage from January 1984 to February 1986 (Greenberg et al. 1989a, 1989b). Using the Vanderbilt University Television News Index and Abstracts rather than the coverage itself, we identified 564 environmental risk stories during the period studied, 1.7 per cent of the total air time in the evening newscasts. During this period, the networks ran only 57 stories about tobacco, and an astounding 482 stories about airplane safety and accidents. Based on the number of fatalities, there should be 26½ minutes of tobacco coverage for every second of airplane accident coverage. Instead, the ratio was 7:1 in the wrong direction. Acute environmental accidents like Bhopal received plentiful coverage (and deserved it); chronic environmental problems like asbestos contamination received much less, typically requiring an 'acute' news peg (new and timely information) on which to base the story. Geographical proximity was also a major factor. During the study period, Alabama, Louisiana, Mississippi, and West Virginia had about the same number of oil spills as California, Massachusetts, New York, and Texas. Yet almost three times as many spill stories were reported from the latter states (where the networks have bureaus and many viewers) than from the former (where they do not).

Seriousness (or 'consequence') is only one of a host of traditional journalistic criteria for newsworthiness. Most of the others – timeliness, proximity, prominence, human interest, drama, visual appeal, and the like – make a big controversy intrinsically newsworthy even if it is not a serious health threat. In some of my non-media writing, (Sandman 1987, 1991, 1993) I have used the terms 'hazard' and 'outrage' to refer, respectively, to the technical seriousness of a risk and its non-technical seriousness (a composite of such factors as control, fairness, familiarity, trust, dread, and responsiveness). In these terms, the mass media are in the outrage business. They don't create outrage, as my clients sometimes suppose. But they do amplify it.

2. *Within individual risk stories, most of the coverage isn't about the risk. It is about blame, fear, anger, and other non-technical issues about 'outrage,' rather than 'hazard.'*

In 1985, we asked the editors of New Jersey's 26 daily newspapers to send us their best environmental risk news stories or series from the previous year. The 248 stories that were submitted were content analysed for risk information (Sandman et al. 1987 pp. 6– 37). A full 68 per cent of the paragraphs had no risk information at all. Another 15 per cent dealt with whether the potentially risky substance was present or absent, and only 17 per cent of the paragraphs dealt with whether the substance was risky or not. A panel of one environmental reporter, one activist, one industry spokesperson, and one technical expert was convened to assess the stories more subjectively (ibid. pp. 38– 51) The panelists who disagreed about most things emphatically agreed that environmental risk information was scanty in these stories. Technical content was especially lacking. What risk information was provided came mostly in the form of opinions, not evidence.

In a parallel study, we asked reporters to specify which information they would need most urgently in covering an environmental risk emergency (*ibid.* pp. 61–98,) Most reporters said they would want only the most basic risk information in order to meet their deadline; technical details would be used, if at all, for a possible second-day story. The same minimal interest in technical information emerged in a role-playing exercise in which a panel of reporters interviewed a panel of experts after a simulated pesticide spill (SIPI 1985) What happened, how it happened, who's to blame, and what the authorities are doing about it all command more journalistic attention than, say, data on toxicity.

Many factors doubtless contribute to the scarcity of technical risk information in risk stories, among them the relative inaccessibility of technical sources and the 'technophobia' of many reporters, editors, and audiences. It is easier, more comfortable, and more productive to cover environmental politics than environmental risk.

3. *When technical information about risk is provided in news stories, it has little if any impact on the audience.*

Most of my corporate and government clients find these first two points very frustrating, believing fervently that the media should concentrate on data and should cover risk issues in proportion to the hazard, not the outrage. Some of them work hard to persuade journalists to pay more attention to hazard, and especially to hazard data. They even persuaded the Rutgers program to help most visibly with a technical primer for journalists on environmental risk (Sachsman et al. 1988, West et al. in press).

It turns out, however, that getting technical information into the media isn't only difficult; it is also close to useless. In a 1991 study, for example, we wrote news stories about a hypothetical perchloroethylene spill, systematically varying three dimensions of the coverage (Johnson et al. 1992, Sandman 1993, Salomone 1992): (1) The level of outrage (whether neighbours were angry or calm, whether the agency was helpful or contemptuous, etc.); (2) The seriousness of the spill (how much PERC was spilled, how many drinking water wells were nearby, etc.); and (3) The amount of technical information in the story. Experimental subjects were asked to read one story and answer questions about their reactions to the risk. The results: outrage had a substantial effect on risk perception; hazard (five orders of magnitude worth!) had a modest effect; technical information had no effect at all.

Technical information might be expected to reassure people that the experts are on top of the situation; or it might frighten them with all those polysyllabic words and scary possibilities; best of all, it might reassure them when the hazard was low and frighten them when it was high. Instead, it simply doesn't matter or, at least, we have yet to find a way to make it matter. In their focus on outrage rather than hazard, journalists are at one with their audience.

4. *Alarming content about risk is more common than reassuring content or intermediate content except, perhaps, in crisis situations, when an impulse to prevent panic seems to moderate the coverage.*

In the New Jersey content analysis, 10 per cent of the paragraphs asserted risk, while only 3 per cent denied risk and 4 per cent adopted an intermediate or mixed position (Sandman et al. 1987). Similarly, 10 per cent of the paragraphs said the risky substance was present; 2 per cent said it wasn't and 3 per cent were intermediate or mixed. Only 29 per cent of the articles contained even a single paragraph to the effect that the situation is not risky; by contrast, 57 per cent had at least one paragraph saying it is risky, and 45 per cent had at least one paragraph in the middle.

There are two points to note here: alarming content outweighs reassuring content, and opinionated or extreme content outweighs intermediate or mixed content. (There are limits to the second tendency. Advocates of the most extreme viewpoints like 'hazardous waste is a CIA plot' get crackpot coverage or none at all.) The tilt toward the alarmist side is not, I think, sensationalism or substantive bias; it is news judgment. Missing an issue is a much greater journalistic sin than overstating one (this is like the difference between Type I error and Type II error in research methodology). The possibility that X is dangerous thus makes the story worth covering. The claim that X is safe is newsworthy only because someone else claims it isn't. And so the dangerous side naturally gets more attention. As for the middle, how do you make an interesting story out of 'further research is needed'?

Both tendencies may be considerably smaller when a crisis does occur. In 1979, I worked with more than a dozen other staff members of the President's Commission on the Accident at Three Mile Island (TMI) on a content analysis of the first week of TMI coverage in the three networks, the two wire services, and five newspapers (Public's Right to Information Task Force 1979). The coverage turned out more reassuring than alarming (and arguably more reassuring than it ought to have been, given that behind the scenes some Nuclear Regulatory Commission experts spent the week expecting a catastrophic hydrogen bubble explosion) (Sandman and Paden 1979). Of media passages whose thrust was clearly either alarming or reassuring, 60 per cent were reassuring. If you stick to the technical issues, eliminating passages about inadequate flow of information and general expressions of fearfulness from local citizens, the preponderance of reassuring over alarming statements becomes 73 per cent to 27 per cent .

5. *Exactly what information is alarming or reassuring is very much a matter of opinion. The media audience tends to be alarmed even by information the experts would consider reassuring.*

Content analysis notwithstanding, we had a tough time convincing the Commissioners that TMI was not a case study in media sensationalism. As a group, they had appropriately varied attitudes toward nuclear power, but with the exception of one housewife they were all 'important people,' accustomed to dealing with, and feeling stung by, journalists. They found it easy to believe that the media had made such a mistake over TMI.

Those who favoured nuclear power were especially inclined to think so, of course. Selective perception works weirdly on deeply committed people. While most of us tend to suppose media content to be more in tune with our beliefs than it actually is, the people who care most are vulnerable to the opposite distortion: neutral coverage looks biased against them. As I have already suggested, industry is usually right in its perception that media coverage leans toward the alarming side of the balance but my industry clients think a mildly alarmist story is incredibly alarmist, and even a balanced story strikes them as off-base.

Environmental activists commit the same distortion, with less reason. A recent 'boomlet' in the debunking of environmentalist claims, led by Keith Schneider of *The New York Times*, has triggered endless teeth-gnashing about an 'anti-environmental backlash' among activists and environmental journalists (groups whose values and concerns are surprisingly similar). Of course credulously reassuring news stories are no more admirable than credulously alarming ones and they are more dangerous. But they are also scarcer.

There is another sense in which industry's complaints are more supportable than activists'. A neutral story is alarming. Consider the following two-paragraph, carefully balanced 'news story':

> Some experts think that your toothpaste has very likely been contaminated with a deadly poison.
> Other experts think that the risk that your toothpaste is contaminated is exceedingly low.

Assuming you believe this story, its net effect on your mood as you prepare to brush your teeth tomorrow morning will not be neutral! Similarly, if you remember news coverage of Three Mile Island as very frightening, you're right. For most people the coverage was frightening even though 'objectively' reassuring passages outnumbered alarming ones.

A case study analysis of newspaper coverage of dioxin contamination at an abandoned factory in Newark, NJ found that 'alarming' and 'reassuring' are not really characteristics of the coverage itself; they are characteristics of the interaction between the coverage and the audience (Salomone and Sandman 1991, Salomone et al. 1992, Salomone and Hance 1993). Experts, for example, considered test sample results showing low levels of contamination to be reassuring content; many citizens, however, find the same content exceedingly alarming, focusing more on the presence of the contaminant than on its concentration. Advice on how people can protect themselves from exposure, on the other hand, was experienced by citizens as reassuring, although an expert might justifiably claim that such advice takes a small risk more seriously than it deserves and is thus alarming. Even more importantly, laypeople are much more responsive to outrage than experts are. Information that the authorities knew about a problem for many years before they took action, for example, may strike an expert as irrelevant to the size of the risk, but it outrages and therefore alarms many non-experts. Perhaps most dramatic was the finding that explicit statements by official sources minimizing the risk 'the levels are low,' 'it hasn't spread,' 'don't worry' were considered offensive, incredible, and therefore alarming by citizen readers. Such statements would of course be coded as reassuring by any formal content analysis.

Yet another study asked students to respond to hypothetical news stories about a chlordane spill (Salomone 1992). Once again, the amount of technical data in the stories had no effect on resulting risk perceptions. The tone of the stories – predominantly alarming, balanced, or predominantly reassuring – mattered more. Alarming stories yielded alarmed readers. Reassuring stories yielded reassured readers, but only if these readers were asked to assume that they lived near the site of the spill and faced practical, immediate decisions such as whether to evacuate. Subjects who were asked to assess pesticide risks in a more generalized way were alarmed by both the alarming and the reassuring story; the intermediate, balanced story produced the most positive responses. Apparently one-sidedly reassuring risk information is likely to strike readers as incredible and therefore produce a boomerang effect unless they face a decision about what to do, in which case their response may be much less sceptical.

6. *Reporters lean most heavily on official sources. They use more predictably opinionated sources industry and experts on the 'safe' side, only using activists and citizens on the 'risky' side when they need them.*

Government is the number one source of environmental risk news. This was especially clear in the New Jersey content analysis (Sandman et al. 1987). When unattributed paragraphs are eliminated, government officials accounted for 57 per cent of all paragraphs in the New Jersey study. Industry spokespeople, by contrast, accounted for 15 per cent of the attributed paragraphs; citizens accounted for 7 per cent , advocacy groups for 6 per cent , and experts for 6 per cent . On network television, government officials still led, but by much less (Greenberg et al. 1989a, b).They were 29 per cent of the on-air sources; industry spokespeople were 13 per cent , citizens were 25 per cent , advocacy groups were 7 per cent , and experts were 14 per cent . When the networks used only one source for a story, that source was a government official 72 per cent of the time. Two-source stories most typically paired government and industry, citizens and industry, or citizens and government. Activists and experts turned up most often in stories with three or more sources.

You can see the journalistic scavenger hunt at work here. For a minor story, reporters may need only one source; if a competing source has something to say, he or she can create a follow-up story another day. But for a more significant story, reporters typically start with a government official, who sets up the story. If the government says 'dangerous,' they look for an industry source or possibly an expert to say 'safe.' If the government says 'safe,' they look for a citizen or possibly an activist to say 'dangerous.'

This scavenger hunt takes place whether the 'truth' is alarming, reassuring, or somewhere in the middle. In the epistemology of routine journalism, there is no truth (or at least no way to determine truth); there are only conflicting claims, to be covered as fairly as possible, thus tossing the hot potato of truth into the lap of the audience. Investigative reporters may undertake to expose a larger truth. A general assignment reporter on a breaking story just wants to get somebody to say how bad the situation is and somebody else to say it isn't so bad and move on to another story.

Reporters know with some precision which content to expect from which sources. In the New Jersey study, for example, experts and individual citizens were most likely to address the riskiness issue; industry and government tended to talk about other things (Sandman et al. 1987). Not surprisingly, activist groups were the most likely to assert risk; they did so 33 times as often as they denied it. Industry sources, at the other extreme, denied risk 5 times as often as they asserted it. Experts, surprisingly, were the most 'certain' sources; they either asserted or denied risk 5 times as often as they came down in the middle. (One wonders what expert qualifiers never made it into the stories.) Advocacy groups and state government officials were among the least certain, asserting or denying risk only a little over twice as often as they expressed a middle view.

A similar pattern emerged when we asked source 'types' to comment on coverage of four case studies: dioxin in Times Beach, Missouri; methyl isocyanate in Institute, West Virginia; dioxin in Newark, New Jersey; and radon in Clinton, New Jersey. Panels of journalists, activists, industry spokespeople, government officials, and technical experts assessed the coverage, story by story (Salomone et al. 1990). Overall, activists gave the stories higher quality ratings than the other panelists, and scientists gave them lower quality ratings. All five types of panelists judged a story to be higher quality if they thought it was more accurate. All except the government sources thought it was higher quality if they thought it had more risk information and all were critical of the scarcity of technical information. But they differed on the relationship between quality and the alarm-reassurance dimension. The industry, government, and expert panelists all gave higher quality ratings to the stories they considered less alarming; journalists, on the other hand, rated alarming stories as higher quality than calming ones. Surprisingly, the activist panelists were unaffected by this dimension in their assessment of quality. On the whole, the findings suggested that there is at least as deep a desire among traditional industry, government, and expert news sources to support the status quo than there is among journalists and activists to undermine it.

7. *Although the competition for journalistic attention is tougher for sources seeking to reassure than for those seeking to alarm, coverage depends even more on a different distinction: skilful sources versus inept ones.*

For a variety of reasons, most journalists are naturally more allied with their alarming sources than with their reassuring ones. This is not mostly because reporters are anti-establishment activists in disguise. It is more because reporters are interested in their careers, and a scary story is intrinsically more interesting, more important – 'better' by journalistic standards – than a calming one.

There are also several kinds of culture conflict at work: 'word' people versus 'number' people; lovers of breadth who spend each day scratching the surface of a new topic versus lovers of depth who devote their lives to dotting technical i's and crossing technical t's; conflict-seekers versus conflict avoiders, etc.

Factors like these undoubtedly make journalists a little less hospitable to industry and technical sources than to activist and citizen sources. But only a little. After all, the scavenger hunt is what matters, and reassuring sources are on the list too. (Journalists don't particularly like government officials, either, but they still give them the lion's share of the coverage.) The main effect of the 'natural antagonism' between journalists and their reassuring sources is on the source side of the dialogue. Industry spokespeople and technical experts stereotype journalists far more negatively than journalists stereotype them. They anticipate much worse treatment than they get; they imagine mistreatment when it didn't happen, and they provoke mistreatment by acting defensively or by being demanding. Ultimately, this may be the biggest reason why the reassuring side of the risk debate gets inadequate coverage, a reason even bigger than the journalist's natural affinity for bad news. The sources of alarming information tend to be co-operative and canny, while the sources of reassuring information are, for the most part, lousy sources (Sandman et al. 1992). The latter sources can, and should, learn to do better.

From the perspective of those who want more data and more reassuring and intermediate content, media coverage of risk seems to have improved a little in the past year or two. Coverage of EMF and global warming, for example, has been more balanced and more data-focused than veterans of acid rain and dioxin would have expected. Perhaps technical sources are getting more skilled, or journalists less cowed by numbers. This is, in any case, a modest trend at best and there is no evidence that it has made much difference to the audience. In general, four biases still prevail, both in the coverage itself and in the response of readers and viewers:

- alarm over reassurance;
- extremes over the middle;
- opinions over data;
- outrage over hazard.

There isn't much that a source can do to adapt to the first bias. Each source's position on the risk in question is what it is, on the 'risky' side or the 'not risky' side, regardless of journalistic preferences. The other three biases, however, can be productively deferred to.

Avoid intermediate or mixed positions. Stake out a stance that is clearly pro or con. If you must peddle the middle, work hard to make it interesting.

Focus more on opinions than on data, more on anecdotes than on tables and charts, more on concrete nouns and active verbs than on jargon and abstractions. When you have a piece of data worth showcasing which is applicable much less often than you think use every strategy available to simplify it, personalize it, and put it into a human context.

Above all, focus on outrage. The most striking statements an environmental activist can make to the media are statements aimed at increasing, focusing, and mobilizing outrage. These are the statements that are most likely to get in, and most likely to affect the audience. Conversely, the most striking statements an industry spokesperson can make to the media are statements aimed at reducing outrage: acknowledgments of problems, apologies for misbehaviours, offers to share control, explanations of what the source is doing and what the audience can do to mitigate the risk, demonstrations of accountability in lieu of trust, etc. (Sandman et al. 1993). Sources who are convinced a risk is huge usually know how to manipulate the outrage. Sources who are convinced it is trivial, on the other hand, usually make the mistake of believing that the key task is to explain the data. They, and if they are right about the risk, the rest of us, too are paying heavily for this mistake.

Acknowledgement

An earlier version of this chapter originally appeared in *Risk: Health, Safety and Environment*, **5** Summer 1994, 251–260.

References

Greenberg, M. R., Sachsman, D. B., Sandman, P. M., and Salomone, K. L. (1989a). Network evening news coverage of environmental risk, *Risk Analysis*, **9**(1),119– 126.

Greenberg, M. R., Sachsman, D. B., Sandman, P. M., and Salomone, K. L. (1989b). Risk, drama and geography in coverage of environmental risk by network TV, *Journalism Quarterly*, Summer, 267– 276.

Greenberg, M. R., Sandman, P. M., Sachsman, D. B., and Salomone, K. L. (1988c). Network television news coverage of environmental risks, *Environment*, **31**(2), 16– 20, 40– 44.

Johnson, B. B., Sandman, P. M., and Miller, P. (1992). Testing the role of technical information in public risk perception, *RISK Issues in Health and Safety*, **3**, 341– 364.

President's (The) Commission on the Accident at Three Mile Island. (1979). *Report of the Public's Right to Information Task Force*, Staff Report to Washington, DC: US Government Printing Office No.052– 003-00734-7, October.

Sachsman, D. B., Greenberg, M. R., and Sandman, P. M. (eds) (1988). *Environmental Reporter's Handbook* (Newark, NJ: Hazardous Substance Management Research Center,

New Jersey Institute of Technology).
Salomone, K. L. and Sandman, P. M. (1991). *Newspaper coverage of the diamond shamrock dioxin controversy: how much content is alarming, reassuring, or intermediate?* Center for Environmental Communication, Rutgers University, New Brunswick, NJ.
Salomone, K. L., Hance, B. J. and Sandman, P. M. (1992). Toward an understanding of what constitutes reassuring information during controversies over low-risk hazards. Poster presentation at the annual meeting of the Society for Risk Analysis, San Diego, CA, December
Salomone, K. L., Greenberg, M. R., Sandman, P. M., and Sachsman, D. B. (1990). A question of quality: how journalists and news sources evaluate coverage of environmental risk, *Journal of Communication*, **40**(4),117– 130.
Salomone, K. L. (1992). *News Content and Public Perceptions of Environmental Risk: Does Technical Information Matter After All?* New Brunswick, NJ: Center for Environmental Communication, Rutgers University.
Salomone, K. L. and Hance, B. J. (1993). communicating reassuring information during environmental controversies: The Diamond Shamrock Case. Panel presentation at the annual meeting of the International Communication Association, Washington, DC, May.
Sandman, P. M. (1971). Environmental advertising and social responsibility. In David M. Rubin and David P. Sachs, eds, *Mass Media and the Environment* New York: Praeger.
Sandman, P. M. (1972). Who should police environmental advertising? *Columbia Journalism Review*, **January/February**, 41– 47.
Sandman, P. M. (1973). Madison Avenue vs. The Environmentalists, Journal of Environmental Education, **Fall**, 45– 50.
Sandman, P. M. and Paden, M. (1979). At Three Mile Island, *Columbia Journalism Review*, **July/August**, 43– 58
Sandman, P. M., Sachsman, D. B., Greenberg, M. R., and Gochfeld, M. (1987). *Environmental Risk and the Press.* New Brunswick, NJ: Transaction Books.
Sandman, P. M. (1987). Risk communication: facing public outrage, *EPA Journal*, **November**, pp. 21– 22.
Sandman, P. M. (1991). *Risk = Hazard + Outrage: A Formula for Effective Risk Communication* (videotape) Fairfax, VA: American Industrial Hygiene Association.
Sandman, P. M., Sachsman, D. B., and Greenberg, M. R. (1992). *The Environmental News Source: Providing Environmental Risk Information to the Media*, New Brunswick, NJ: Center for Environmental Communication.
Sandman, P. M. (1993). *Responding to Community Outrage: Strategies for Effective Risk Communication*, Fairfax, VA: American Industrial Hygiene Association.
Sandman, P. M., Miller, P. M., Johnson, B. B. and Weinstein, N. D. (1993). Agency communication, community outrage, and perception of risk: three simulation experiments, *Risk Analysis*, **13**(6), pp. 585– 598.
Scientists' Institute for Public Information (1985). environmental emergencies: are journalists prepared? *SIPIscope*, **13**(4), 1– 11.
West, B., Sandman, P. M., and Greenberg, M. R. *Reporter's Environmental Handbook* (New Brunswick, NJ: Rutgers University Press, in press).

15 Cars, cholera, cows, and contaminated land: virtual risk and the management of uncertainty

John Adams

'They should shoot the scientists, not cull the calves. Nobody seems to know what is going on.' Dairy farmer quoted in the Times (2.8.96)

Summary

Three categories of risk are identified:

- directly perceptible risks: e.g. traffic to and from landfill sites;
- risks perceptible with the help of science: e.g. cholera and toxins in landfill sites;
- virtual risks – scientists don't know/cannot agree: e.g. BSE/CJD and suspected carcinogens.

Most of the literature on risk management deals with the first two categories. Quantified risk assessments require that the probabilities associated with particular events be known or be capable of plausible estimation. When scientists cannot agree on the odds, or the underlying causal mechanisms, of illness, injury or environmental harm, people are liberated to argue from belief and conviction.

Many environmental hazards are 'virtual risks'. Individual responses to the risk posed by these hazards vary according to personality type. Attempts by regulators to control environmental hazards are likely to be complicated by the actions of two groups: egalitarians will demand the elimination of the hazard, regardless of the cost; individualists, on the other hand, view such regulations as an unnecessary intrusion into their lives, so they will protest against all attempts to regulate such hazards.

Even if regulations are imposed, the evidence suggest that people will adapt to the new regulatory framework and adjust their exposure to risk accordingly. When people are forced to wear seat-belts, for example, they tend to drive less carefully, so that the number of people injured remains relatively constant. Similarly, if people are prevented from smoking cigarettes in restaurants, they may respond by going out less, so that they can indulge their habit at home (surrounded, perhaps, by non-smoking guests all voluntarily being fumigated), or, as in France, simply by disobeying the law.

Cars and the risk thermostat

Some risks we manage by reacting to directly perceived danger. Riding a bicycle, crossing a road and driving a car are all activities that involve responses to directly perceived dangers. Figure 15.1, the risk thermostat, is a simplified description of the way such risks are managed. The model postulates that:

- everyone has a propensity to take risks;
- this propensity varies from one individual to another;
- this propensity is influenced by the potential rewards of risk taking;
- perceptions of risk are influenced by experience of accident losses – one's own and others';
- individual risk-taking decisions represent a balancing act in which perceptions of risk are weighed against the propensity to take risk;
- accident losses are, by definition, a consequence of taking risks; the more risks an individual takes, the greater, on average, will be both the rewards and losses he or she incurs.

The arrows connecting the boxes in the diagram are drawn as wiggly lines to emphasize the modesty of the claims made for the model. It is an impressionistic, conceptual model, not an operational one. The contents of the boxes are not capable of being objectively measured. The arrows indicate directions of influence and their sequence. The model demonstrates why, in situations in which people act upon perceptions of risk, it is impossible to

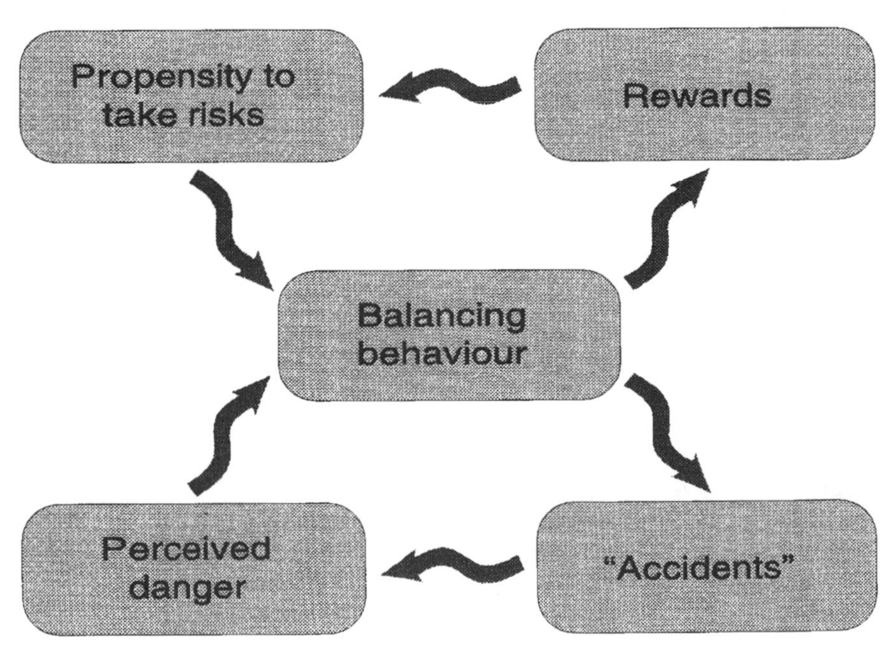

Figure 15.1. The risk 'thermostat'

devise objective measures of safety or danger; a dangerous road, for example, can have a good accident record if parents forbid their children to cross it, old people are afraid to cross it, and fit adults cross it quickly and carefully.

Figure 15.1 can be considered as a description of the behaviour of the driver of a single car going around a bend in the road. His speed will be influenced by the rewards of risk. These might range from getting to the church on time to impressing his friends with his skill or courage. His speed will also be influenced by his perception of the danger. His fears might range from death, through the cost of repairs and loss of licence, to embarrassment. His speed will also depend on his judgement about the road conditions – is there ice or oil on the road? how sharp is the bend and how high the camber? – and the capability of his car – how good are the brakes, suspension, steering and tyres?

Over-estimating the capability of the car, or the speed at which the bend can be safely negotiated, can lead to an accident. Underestimating these things will reduce the rewards gained. The consequences, in either direction, can range from the trivial to the catastrophic.

The balancing act described by this illustration is analogous to the behaviour of a thermostatically-controlled system. The setting of the thermostat varies from one individual to another, from one group to another, from one culture to another. Some like it hot – a Hell's Angel or a Grand Prix racing driver for example – others like it cool – a 'Mr Milquetoast' or a little old lady named Prudence. But no one wants absolute zero.

Risk: an interactive phenomenon

Figure 15.2 introduces a second car to the road to make the point that risk is usually an interactive phenomenon. One person's balancing behaviour has consequences for others. On the road other motorists can impinge on your 'rewards' by getting in your way and slowing you down, or help you by giving way. One is also concerned to avoid hitting them, or being hit by them. Driving in traffic involves monitoring the behaviour of other motorists, speculating about their intentions, and estimating the potential consequences of a misjudgement. If you see a car approaching at speed and wandering from one side of the road to another, you are likely to take evasive action, unless perhaps you place a very high value on your dignity and rights as a road user and fear a loss of esteem if you are seen to be giving way. During this interaction enormous amounts of information are processed. Moment by moment each motorist acts upon information received, thereby creating a new situation to which the other responds.

The spread of diseases, and attempts to limit them, also provide examples of this phenomenon. The behaviour of carriers of diseases, and those potentially infected, is influenced by perceived rewards and dangers. Campaigns to curb the spread of AIDS appear to have heightened awareness of the dangers of unprotected sex, but also, according to the critics of safe-sex

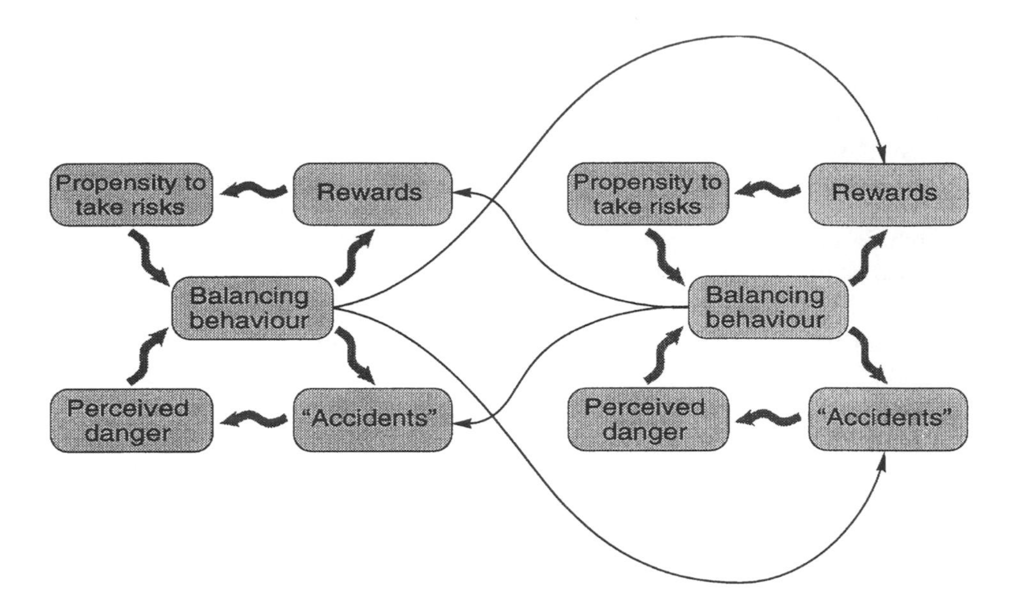

Figure 15.2 The risk 'thermostat': two drivers interacting

campaigns, to have promoted the activity by which the disease is spread by advertising its rewards/pleasures. BSE also provides numerous examples. The compensation scheme for BSE infected cattle, in which the initial level of compensation was set at half the value of healthy animals, is thought to have encouraged the concealment of sick animals. Government attempts at reassurance – including John Gummer's force-feeding a hamburger to his daughter – did not often achieve the intended effect; the public was reacting selectively to competing risk messages, and beef producers, retailers, civil servants and politicians were responding in turn.

Figure 15.3 introduces another complication. On the road, and in life generally, risky interaction frequently takes place on terms of gross inequality. The potential damage that a heavy lorry can inflict on a cyclist or pedestrian is great; the physical damage that they might inflict on the lorry is small. The lorry driver in this illustration can represent the controllers of large risks of all sorts. Those who make the decisions that determine the safety of consumer goods, working conditions or large construction projects are, like the lorry driver, usually personally well-insulated from the consequences of their decisions. The consumers, workers or users of their constructions, like the cyclist, are in a position to suffer great harm, but not to inflict it.

Problems of measurement

Risk comes in many forms. In addition to economic risks – such as those dealt with by the insurance business – there are physical risks and social risks, and innumerable subdivisions of these categories: political risks, sexual

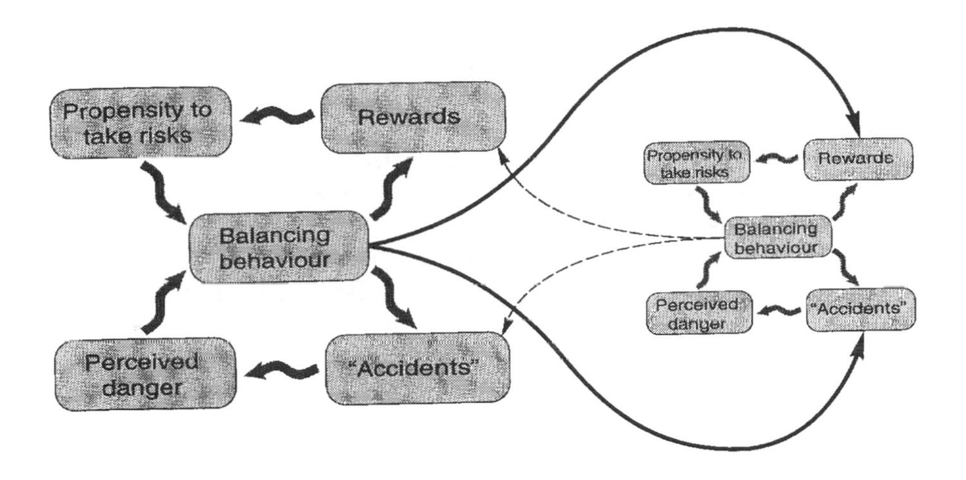

Figure 15.3 The risk 'thermostat': lorry driver and cyclist interacting

risks, medical risks, career risks, artistic risks, military risks, motoring risks, legal risks ... The list is as long as the number of adjectives that might be applied to behaviour in the face of uncertainty. These risks can be combined or traded. In some occupations people are tempted by danger money. Some people, such as sky-diving librarians, may have very safe occupations and dangerous hobbies. Some young male motorists would appear to prefer to risk their lives rather than their peer-group reputations for courage.

Although the propensity to take risks is widely assumed to vary with circumstances and individuals, there is no way of testing this assumption by direct measurement. There is not even agreement about what units of measurement might be used. Usually the assumption is tested indirectly by reference to accident outcomes; on the basis of their accident records young men are judged to be risk seeking, and middle-aged women to be risk averse. But this test inevitably gets muddled up with tests of assumptions that accidents are caused by errors in risk perception, which also cannot be measured directly. If Damon Hill crashes at 180 mph in his racing car it is impossible to determine 'objectively' whether it was because he made a mistake or because he was taking a risk and was unlucky.

Both the rewards of risk and accident losses defy reduction to a common denominator. The rewards come in a variety of forms: money, power, glory, love, affection, (self-) respect, revenge, curiosity requited, or simply the sensation (pleasurable for some) accompanying a rush of adrenaline. Nor can accident losses be measured with a single metric. Road accidents, the best documented of all the realms of risk, can result in anything from a bent bumper to death, and there is no formula that can answer the question – how many bent bumpers equals one life? The search for a numerical measure to acount for the harm or loss associated with a particular adverse event encounters the problem that people vary enormously in the importance that they attach to similar events. Slipping and falling on the ice is a game for children, and an event with potentially fatal consequences for the elderly.

Figure 15.4 is a distorted version of Figure 15.1 with some of the boxes displaced along an axis labelled 'Subjectivity–Objectivity'. The box which is displaced furthest in the direction of objectivity is 'balancing behaviour'. It is possible to measure behaviour directly. It is well documented that parents have withdrawn their children from the streets in response to their perception that the streets have become more dangerous. It is possible in principle to measure the decline in the amount of time that children spend in the streets exposed to traffic, but even here the interpretation of the evidence is contentious. Do children now spend less time on the street because they spend more time watching television, or do they spend more time watching television because they are not allowed to play in the streets? *All* of the elements of the risk compensation theory, and those of any contenders of which I am aware, fall a long way short of the objective end of the spectrum. Behaviour can be measured but its causes can only be inferred.

And risks can be displaced. If motorcycling were to be banned in Britain it would save about 500 lives a year. Or would it? If it could be assumed that all the banned motorcyclists would sit at home drinking tea one could simply subtract motorcycle accident fatalities from the total annual road accident death toll. But at least some frustrated motorcyclists would buy old bangers and try to drive them in a way that pumped as much adrenaline as their motorcycling, and in a way likely to produce more kinetic energy to be dispersed if they crashed. The alternative risk-taking activities that they

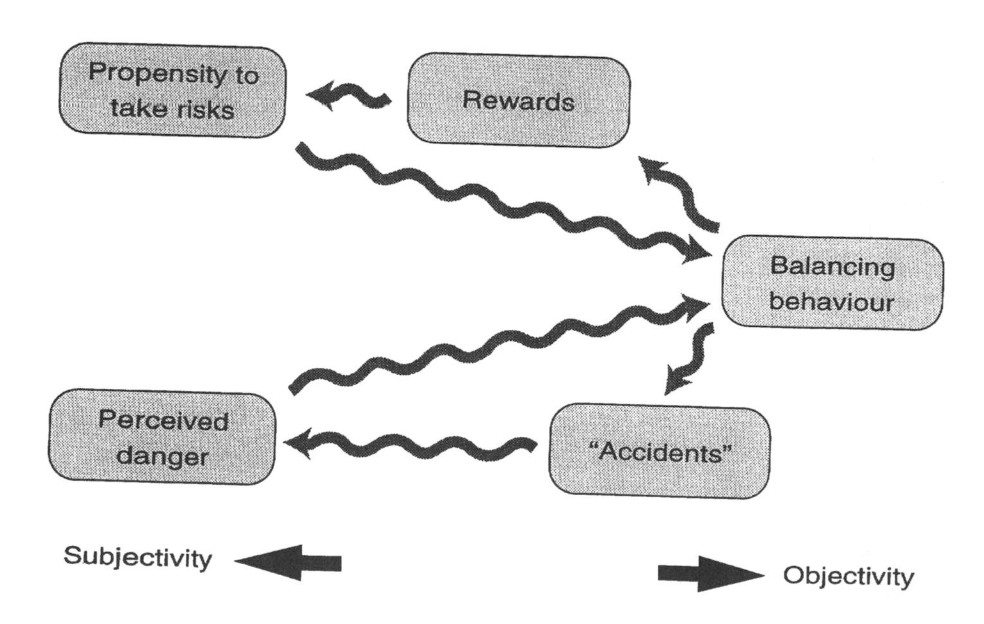

Figure 15.4 The risk 'thermostat' - stretched

might get up to could range from sky-diving to glue sniffing, and there would be no set of statistics that could prove that the country had been made safer, or more dangerous, by the ban.

Reliable knowledge: risks perceived through science

Figure 15.5, the dance of the risk thermostats, is an attempt to suggest a few of the complications that flow from the simple relationships depicted in Figures 15.1 to 15.4. The world contains over 5.5 billion risk thermostats. Some are large; most are tiny. Governments and big businesses make decisions that affect millions if not billions of people. Individuals for the most part adapt as best they can to the consequences of these decisions. The damage that they individually can inflict in return, through the ballot box or market, is insignificant, although in aggregate they can become forces to reckoned with. The slump in the market for beef in response to fears of BSE has not only caused losses to the beef industry, but set off a Europe-wide political chain reaction.

A small part of the dance can be observed directly in crowded local high streets on any Saturday morning, as cyclists, pedestrians, cars, lorries and buses all contend for the same road space. But not all the dangers confronting the participants in this dance are visible to the naked eye. Some cyclists can be observed wearing masks to filter the air. Lead, oxides of nitrogen, carbon monoxide, volatile organic compounds, and ultra-fine particulates are all invisible substances that some of the better-informed shoppers might be

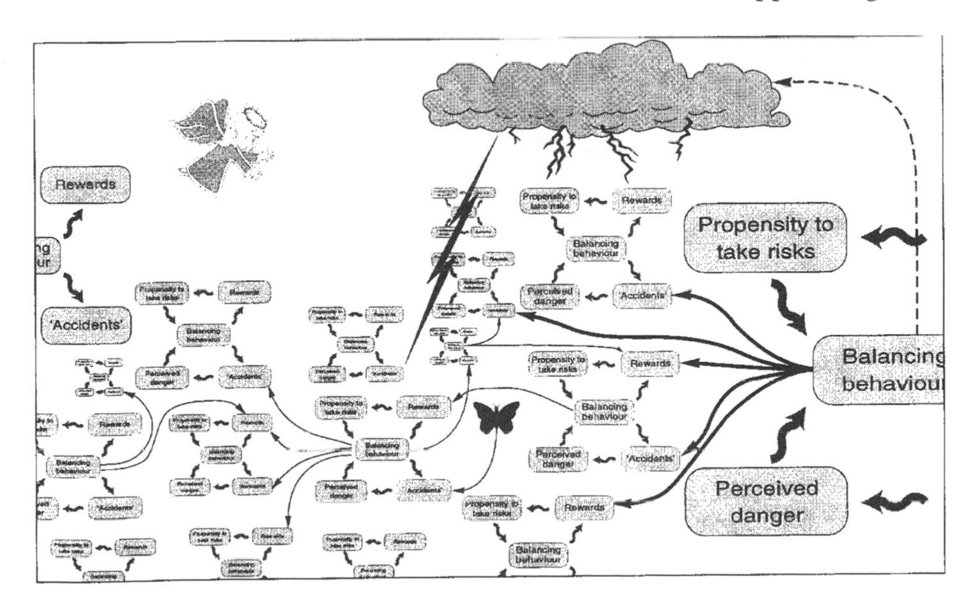

Figure 15.5 The dance of the risk thermostats

worrying about. Few of these worried people will be toxicologists capable of judging these dangers accurately. Their concerns will usually be based on scientific pronouncements filtered through the media, and perhaps augmented by the campaigns of environmentalists.

The dance of the thermostats, with its multiple connections, might also be viewed as a description of the way in which an infectious disease is passed amongst risk takers. Science has an impressive record in reducing certain dangers, perhaps most impressive in the field of infectious diseases. Cholera is an example of a disease where the science of epidemiology yielded useful insights more than 30 years before the responsible bacterium had been found by laboratory scientists. Dr John Snow mapped the latest cases of cholera in an outbreak in London in 1849. The focus of the cluster was the Broad Street well in Soho. As a dramatic gesture, the pump handle was removed although the outbreak had already started to subside.

Together they altered the perception of the risk in a way that led to behaviours – improved sanitation and vaccination – which have virtually eradicated the disease in the developed world. This is an example of an area of risk where science has produced 'reliable knowledge'.

This knowledge has reduced the risk of the illness in 'clean' countries to the point that vaccination is considered unnecessary, but the risk thermostat still plays a part. Many tourists, protected by vaccination now venture into parts of the world where they would previously have feared to go – thereby exposing themselves to other diseases and dangers. There are numerous other examples of science defeating risks only for people to reassert their determination to take risks. The Davy Lamp which operated at a temperature below the ignition point of methane was heralded as a safety device that reduced the danger of explosions in mines. But it permitted the expansion of mining into methane rich atmospheres and was followed by an increase in mining productivity, and in explosions and fatalities. Improvements in brake technology, when fitted to cars, usually result in drivers going faster, or braking later, the potential safety benefit getting consumed as a performance benefit.

Individuals versus institutions. In the dance of the risk thermostats one can distinguish between individual risk takers and institutional risk takers. The distinction can be important. Every *individual* performs the mental balancing act described in Figure 15.1 in his or her own head. *Institutions* – government departments or large commercial enterprises – usually assign the job of risk management to particular people or departments. The risk-decision process which in individuals confronting directly perceivable risks is usually conducted informally and intuitively, in institutions becomes explicit and formal. Figure 15.6 is taken from *A Guide to Risk Assessment and Risk Management for Environmental Protection* published by the Department of the Environment (HMSO 1995). It sets out the sequences of steps recommended in a formal risk assessment. Figure 15.7 describes a similar set of procedures used by a large pharmaceutical company to manage risk. The risk literature is replete with similar algorithms. When compared with Figure 15.1 above

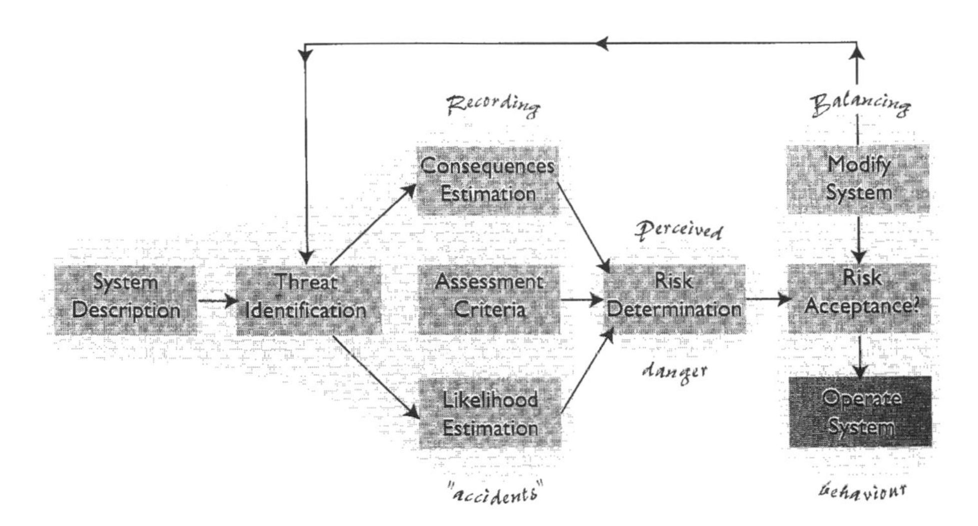

Figure 15.6 The risk assessment process

they can be shown to be more elaborate versions of *the bottom loop* of the risk thermostat model.

What is termed 'risk management' in institutional settings, with a few exceptions such as venture capital enterprises, turns out on inspection to be exclusively concerned with *risk reduction*. Institutional risk management models characteristically have no top loop; the 'rewards' loop is the responsibility of some other department – often marketing. This view was reinforced during a seminar I presented to the risk managers of a large private-sector concern, when one of the participants said, rather morosely, 'around here we're known as the sales prevention department.' As the following pronouncements from Shell Oil indicate, the objective of most institutional risk managers is the elimination of *all* accidents.

> The safety challenge we all face can be very easily defined – to eliminate all accidents which cause death, injury, damage to the environment or property. Of course this is easy to state, but very difficult to achieve. Nevertheless, that does not mean that it should not be our aim, or that it is an impossible target to aim for. (Richard Charlton (1991), director of exploration and production, Shell Oil.)

> The aim of avoiding all accidents is far from being a public relations puff. It is the only responsible policy. Turning 'gambling man' into 'zero-risk man' (that is one who manages and controls risks) is just one of the challenges that has to be overcome along the way. (Koos Visser (1991), Head of Health, Safety and Environment, Shell Oil.)

The single-minded pursuit of risk reduction by institutional managers inevitably leaves the pursuers disappointed and frustrated. Safety interventions that do not lower the settings of the risk thermostats of the

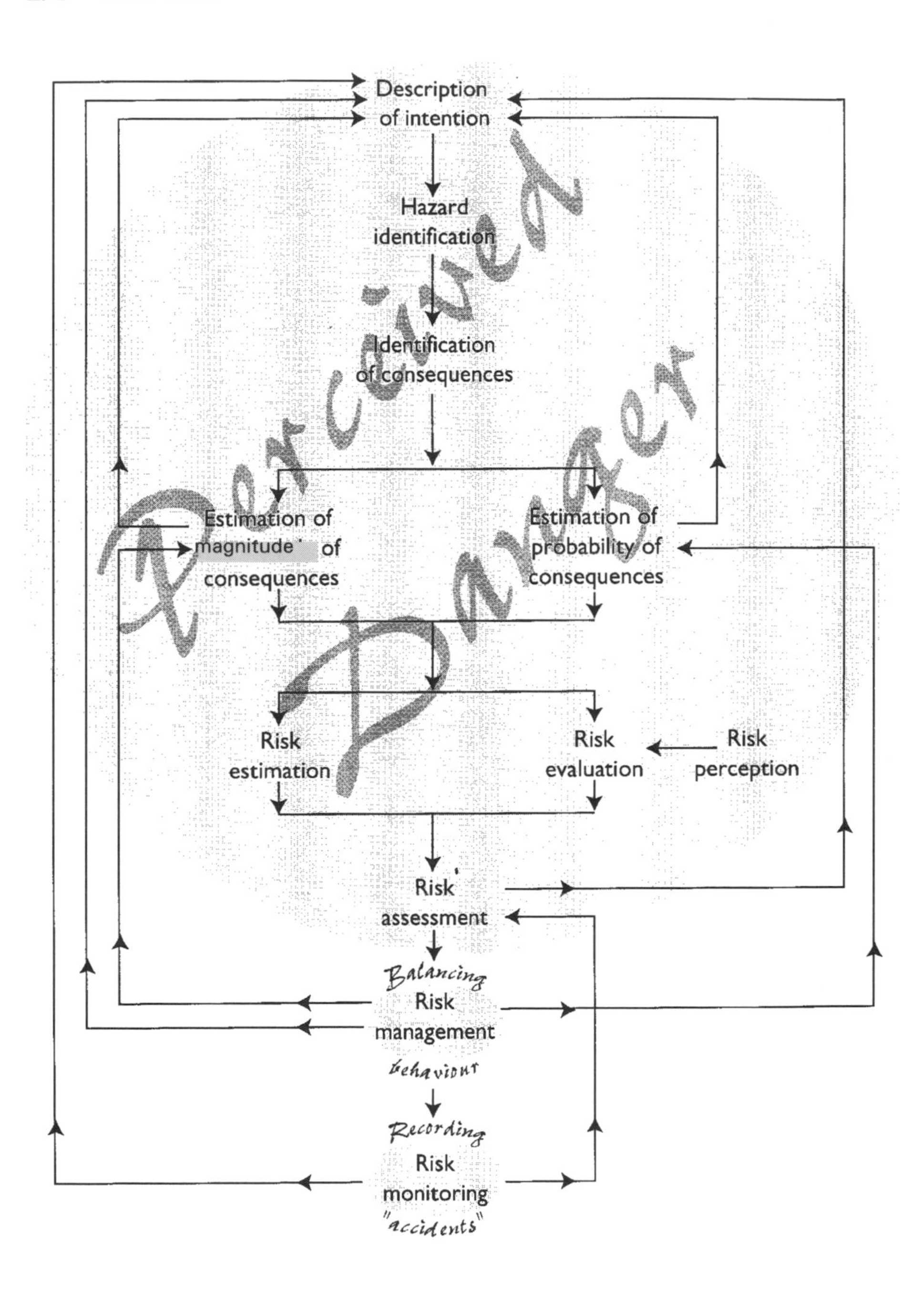

Figure 15.7 Risk assessment: a business model

individuals at whom the interventions are aimed are routinely offset by behavioural responses that reassert the levels of risk with which people were originally content with. This problem is compounded by the division of labour usually found in institutional risk management; different people or departments are commonly placed in charge of the top and bottom loops – with no one obviously in charge of the overall balancing act. What then happens when the problem is further compounded by a lack of reliable knowledge or agreement about the rewards and accident costs to be balanced? (Appendix 15.3 discusses the problems encountered by cost-benefit analysis - a method that does acknowledge the rewards of risk taking.)

Virtual risk – beyond reliable knowledge

Virtual reality is a product *of* the imagination which works *upon* the imagination. It is capable of simulating something real – as in the case of a flight simulator used to train pilots – or something entirely imaginary – as in the case of the Space Invaders of computer games. Virtual risks may, or may not, be real.

When scientists do not know or cannot agree about the 'reality' of risks, people are liberated to argue from belief and conviction. A typology of 'myths of nature', describes various preconceptions about nature that inform risk-taking decisions in such circumstances. The essence of each of the four myths is illustrated by the behaviour of a ball in a landscape; each myth is associated with a distinctive risk-management style (see Figure 15.8).

- *Nature benign*: nature, according to this myth, is predictable, bountiful, robust, stable, and forgiving of any insults humankind might inflict upon it; however violently it might be shaken, the ball comes safely to rest in the bottom of the basin. Nature is the benign arena of human activity, not something that needs to be managed. The management style associated with this myth is therefore relaxed, exploitative, laissez-faire.
- *Nature ephemeral*: here, nature is fragile, precarious and unforgiving. It is in danger of being provoked by human greed or carelessness into catastrophic collapse. The objective of management is the protection of nature from Man. People, the myth insists, must tread lightly on the earth. The guiding management rule is the precautionary principle.
- *Nature perverse/tolerant*: this is a combination of modified versions of the first two myths. Within limits nature can be relied upon to behave predictably. It is forgiving of modest shocks to the system, but care must be taken not to knock the ball over the rim. Regulation is required to prevent major excesses, while leaving the system to look after itself in minor matters. This is the ecologist's equivalent of a mixed-economy model. The manager's style is interventionist.

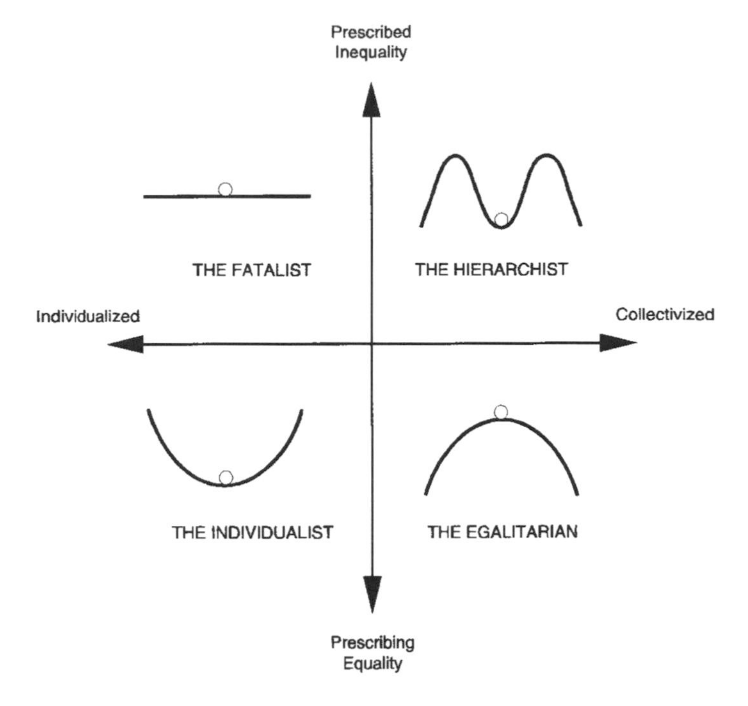

Figure 15.8 Four rationalities: a typology of bias

- *Nature capricious*: nature is unpredictable. The appropriate management strategy is again laissez-faire, in the sense that there is no point in management. Where adherents to the myth of nature benign trust nature to be kind and generous the believer in nature capricious is agnostic; the future may turn out well or badly, but in any event, it is beyond his control. The non-manager's motto is *que sera sera.*

Plural rationalities

These distinctive management styles have been associated by anthropologists with distinctive 'rationalities' (see Figure 15.8).

- Individualists are enterprising 'self-made' people, relatively free from control by others, and who strive to exert control over their environment and the people in it. Their success is often measured by their wealth and the number of followers they can command. The self-made Victorian mill owner or present-day venture capitalist would make good representatives of this category. They oppose regulation and favour free markets. Nature, according to this perspective, is to be *commanded* for human benefit.

- Egalitarians have strong group loyalties but little respect for externally imposed rules, other than those imposed by nature. Group decisions are arrived at democratically and leaders rule by force of personality and persuasion. Members of religious sects, communards, and environmental pressure groups all belong to this category. Nature is to be *obeyed.*
- Hierarchists inhabit a world with strong group boundaries and binding prescriptions. Social relationships in this world are hierarchical with everyone knowing his or her place. Members of caste-bound Hindu society, soldiers of all ranks and civil servants are exemplars of this category. Nature is to be *managed.*
- Fatalists have minimal control over their own lives. They belong to no groups responsible for the decisions that rule their lives. They are non-unionized employees, outcasts, refugees, untouchables. They are resigned to their fate and see no point in attempting to change it. The best you can do is *duck if you see something about to hit you.*

Coping with risk and uncertainty: the dose–response curve

Wherever the evidence in a dispute is inconclusive, the scientific vacuum is filled by the assertion of contradictory certitudes. There are numerous risk debates, such as that about BSE/CJD, in which for the foreseeable future scientific certainty is likely to be a rare commodity; issues of health, safety and the environment – matters of life and death – will continue to be decided on the basis of scientific knowledge that is not conclusive.

Just how remote is the prospect of scientific resolution, and how large is the scientific vacuum, can be illustrated with the help of some numbers taken from a report by the US National Research Council (1992) which notes that about 5 million different chemical substances are known to exist, and that their safety is theoretically under federal government regulatory jurisdiction. Of these, it points out, fewer than 30 have been definitely linked to cancer in humans and about 7 000 have been tested for carcinogenicity on animals.

These last two numbers greatly exaggerate the extent of existing knowledge. Given the ethical objections to direct testing on humans, most tests for carcinogenicity are done on animals. The NRC report observes 'there are no doubt occasions in which observations in animals may be of highly uncertain relevance to humans'; it also notes that the transfer of the results of these tests to humans requires the use of scaling factors which 'can vary by a factor of 35 depending on the method used', and observes that 'although some methods for conversion are used more frequently than others, a scientific basis for choosing one over the other is not established'. A further difficulty with most such experiments is that they use high doses in order to produce results that are clear and statistically significant for the animal populations tested. But for most toxins the dose levels at issue in environmental controversies are much lower.

Extrapolating from the high dose levels at which effects are unambiguous for animals to the much lower exposures experienced by the general human population in order to calculate estimates of risk for people in the real world requires a mathematical model. Figure 15.9 illustrates the enormous variety of conclusions that might be drawn from the same experimental data, depending on the assumptions used in extrapolating to lower doses. It shows that the estimates produced by the five different models converge in the upper right hand corner of the graph. Here, the five models agree that high dose levels produce high response levels. The supra-linear model *assumes* that the level of response will remain high as dose levels are reduced. The threshold model *assumes* that when dose levels fall below the threshold there will be no response. Below the dose levels used in the experiment one can but assume.

Beyond the problems of identifying the causes of morbidity and mortality and specifying dose–response relationships, there are four other sources of uncertainty of even greater significance. First, variability in susceptibility within exposed human populations, combined with the variability in their levels of exposure, makes predictions of the health effects of the release of new substances at low dose levels a matter of guesswork. Secondly, the long latency periods of most carcinogens and many other toxins – cigarettes and radiation are two well-known examples – make their identification and control prior to the exposure of the public impossible in most cases. Thirdly, the synergistic effects of substances acting in combination can make innocent substances dangerous, and the magnitude of the number of combinations that can be created from 5 million substances defies all known computers. And fourthly, the gremlins exposed by chaos theory (represented in Figure 15.5 by the Beijing Butterfly) will always confound the seekers of certainty in complex systems sensitive to initial conditions.

The BSE/CJD controversy contains all of these sources of uncertainty. The very existence of prions remains to be proved 'Nobody has proven that these prions really exist'. 'The prion hypothesis is the "cold fusion" of infectious disease - it's a very radical idea, and just like cold fusion it has some very appealing aspects. But because it's so radical it deserves a very high level of scepticism and scrutiny before it's adopted'. Robert Rohwer, quoted in *Science*, 12.7.96.

Whether the 'new strain' of CJD is a mutant version of CJD, or a new strain of BSE, remains in dispute. Scientific effort up until now has concentrated on demonstrating the possibility of BSE jumping the species divide from cows to humans. Estimation of the dose levels at which it would become a significant threat, and identification of the causes of variation in susceptibility, are projects to be worked on later. The long and variable latency period for spongiform diseases makes reconstruction of exposures to presumed causes and exploration of hypotheses about synergistic effects extremely difficult.

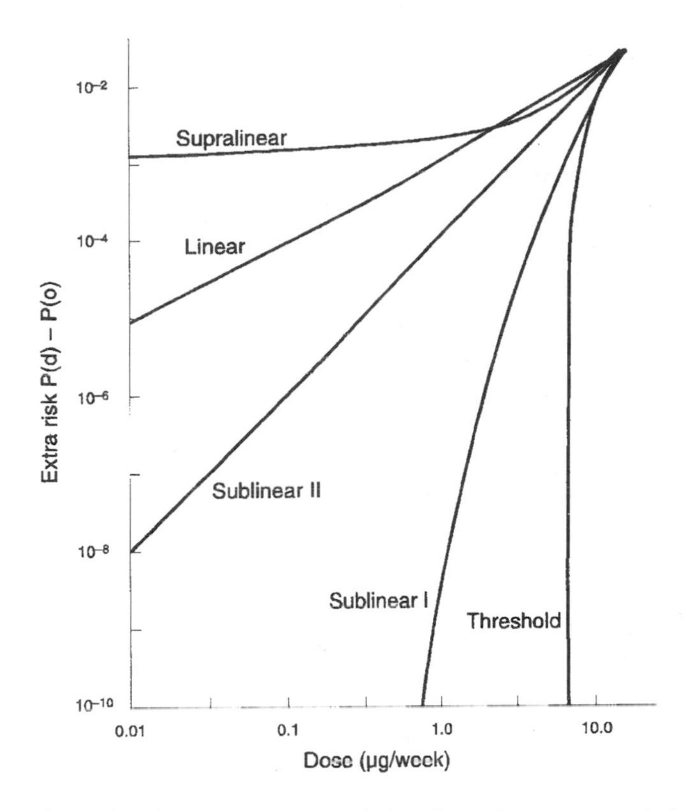

Figure 15.9 Alternative dose-response extrapolations from the same empirical evidence. Source: National Research Council (1992), p.263.

Figure 15.10 shows the risk thermostat fitted with cultural filters. The mythological figures of Cultural Theory are caricatures, but they have numerous real life approximations in debates about risk. Long-running controversies about large scale risks are long running because they are scientifically unresolved, and unresolvable within the time scale imposed by necessary decisions. This information void is filled by people rushing in from the four corners of Cultural Theory's typology to assert their contradictory certitudes. The clamorous debate is characterized not by irrationality but by plural rationalities.

The contending rationalities not only perceive risk and reward differently, they also differ according to how the balancing act ought to be performed. Hierarchists are committed to the idea that the management of risk is the job of 'authority' – appropriately assisted by expert advisers. They cloak their deliberations in secrecy because the ignorant lay public cannot be relied upon to interpret the evidence correctly or use it responsibly. The individualist scorns authority as 'the Nanny State' and argues that decisions about whether to wear seat belts or eat beef should be left to individuals. Egalitarians focus on the importance of trust: risk management is a consensual activity; consensus

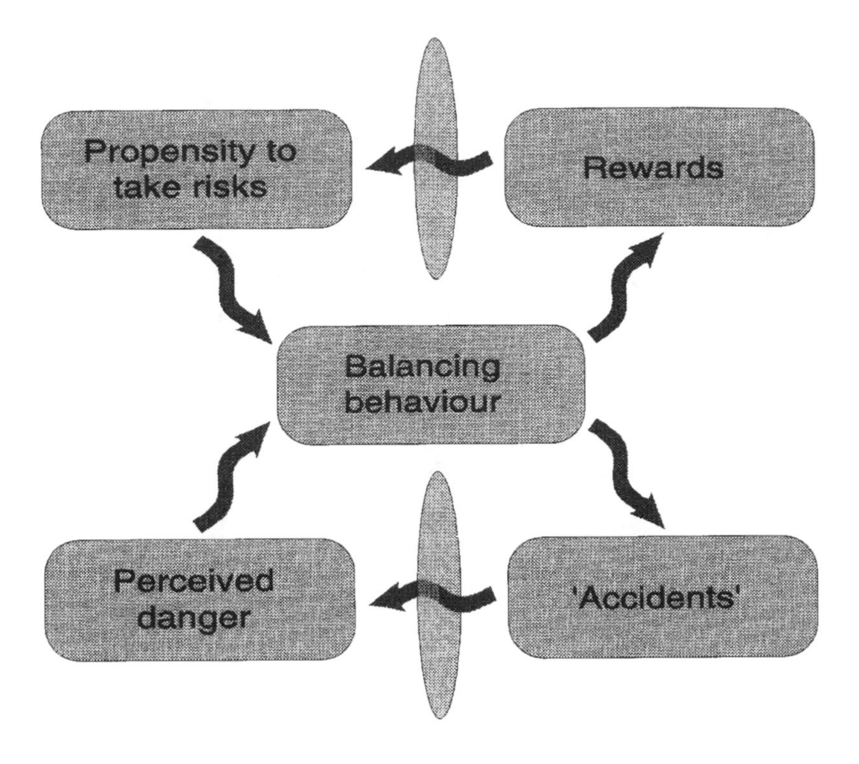

Figure 15.10 The risk thermostat fitted with cultural filters

building requires openness and transparency in considering the evidence.

These different styles of balancing act respond differently to uncertainty. Ignorance is a challenge to the very idea of authority and expertise. The response of hierarchists is to conceal their doubts and present a confident public face. Confession of ignorance or uncertainty does not come easily to authority. In the face of uncertainty about an issue such as BSE, they seek to reassure. The propensity of authority to cope with ignorance by denying its existence is described by Ravetz. Individualists are assiduous collectors of information – even paying for it – but are also much more comfortable with uncertainty. Their optimism makes them gamblers – they expect to win more than they lose. Markets in their view are institutions with a record of coping with uncertainty successfully. If the experts cannot agree about BSE, there is no basis upon which central authority can justifiably act; the risk should be spread by letting individuals decide for themselves. The egalitarian instinct in the face of uncertainty is to assume that authority is covering up something dreadful, and that untrammelled markets will create something dreadful. They favour democratizing the balancing act by opening up the expert committees to lay participation and holding public inquiries to get at the truth – which, when known, will justify the draconian intervention that they favour in the markets. The fatalist just carries on drinking beer, reading the 'Sun' and buying lottery tickets. Figure 15.11 presents a representative selection of comments about BSE categorized by the cultural theory typology of bias. The falling out between political and scientific authority manifest in the upper right hand corner is characteristic of the disarray into which hierarchy falls when its mask of authoritative knowledge is torn off.

BSE/CJD: should we follow a risk-averse environmental policy?

Who are 'we'? 'Risk-averse' and 'risk-seeking' are usually labels that people apply to others whose risk thermostats are fitted with different cultural filters. Those who argue for a more risk-averse policy are, in effect, saying that there is a discrepancy between the dangers that they perceive and the risks that they are prepared to take. The activities of environmental groups (egalitarians) lobbying for the precautionary principle can be seen as a collective behavioural response to this discrepancy. The environmentalist case rests on the conviction that growth processes – economic and demographic – are pressing against global limits. Perhaps the best exemplars of this conviction are Meadows et al. who argue in *Beyond the Limits* (1992) that:

> The human world is beyond its limits. The future, to be viable at all, must be one of drawing back, easing down, healing. ... The more we compiled the numbers, the more they gave us that message, loud and clear.

In the BSE debate the complementary message that is received and re-transmitted loud and clear by egalitarians is that BSE is a punishment for unnatural methods of agriculture. Modern intensive, high-energy production methods, veal crates, battery chickens, genetic manipulation, food preservation methods, pesticides and feeding meat to herbivores are all, according to this perspective, aspects of the same hubristic syndrome. The remedy? Nature is to be obeyed; we must (re)turn to more humane and extensive, organic, natural methods of production.

This message is countered by an individualist back-lash that views the environmental lobby itself as an environmental threat. Julian Simon, for example, insists that there is a positive correlation between indices of material wealth and an improving environment. With Herman Kahn he has argued:

> We are confident that the nature of the physical world permits continued improvement in humankind's economic lot ... indefinitely. ... there are always newly arising local problems, shortages, and pollutions ... But the nature of the world's physical conditions and the resilience in a well-functioning economic and social system enable us to overcome such problems, and the solutions usually leave us better off than if the problem had never arisen; that is the great lesson to be learned from human history.

This 'rationality', when confronted with the evidence of BSE/CJD, sees no evidence of serious harm. It points to the enormous benefits of intensive

Fatalist	**Hierarchist**
• "They should shoot the scientists, not cull the calves. Nobody seems to know what is going on." Dairy Farmer quoted in *The Times* (2.8.96)    • **"Charles won't pay for Diana's briefs"** Main headline in *The Sun* on 21.3.96, the day every other paper led with the BSE story.	• "We require public policy to be in the hands of elected politicians. Passing responsibility to scientists can only undermine confidence in politics and science." John Durant, *The Times Higher* 5.4.1996 • "As much as possible, scientific advice to consumers should be delivered by scientists, not politicians." *The Economist*, 21 March 1996 • "I believe that British beef is safe. I think it is good for you." (Agriculture Minister Douglas Hogg 6.12.95) "I believe that lamb throughout Europe is wholly safe." (Douglas Hogg, 23.7.96) • "I felt the need to reassure parents." Derbyshire Education chief quoted in *The Sun*, 21,3.96 • "I have not got a scientific opinion worth listening to. My job is simply to make certain that the evidence is drawn to the attention of the public and the Government does what we are told is necessary." Health Secretary Stephen Dorrel, *Daily Telegraph,* 22.3.96 • "We felt it was a no-goer. MAFF already thought our proposals were pretty radical." Richard Southwood explaining why he had not recommended a ban on cattle offal in human food in 1988, quoted by B Wynne, *Times Higher* 12.4.96
Individualist	**Egalitarian**
• "The precautionary principle is favoured by environmental extremists and health fanatics. They feed off the lack of scientific evidence and use it to promote fear of the unknown." T. Corcoran, *The Toronto Globe and Mail* • "I want to know, from those more knowledgeable than I, where a steak stands alongside an oyster, a North Sea mackerel, a bolied egg and running for the bus. Is it a chance in a million of catching CJD or a chance in ten million? I am grown up. I can take it on the chin." Simon Jenkins, *The Times,* quoted by J. Durant in *Times Higher,* 5.4.96 • " 'Possible' should not be changed to 'probable' as has happened in the past." S.H.U. Bowies, FRS, *The Times* 12.8.96 • "It is clear to all of us who believe in the invisible hand of the market place that interference by the calamity-promoting pushers of the precautionary principle is not only hurtful but unnecessary. Cost-conscious non-governmental institutions are to be trusted with the protection of the public interest." P. Sandor, *Toronto Globe and Mail* 27.3.1996 • "I shall continue to eat beef. Yum, *yum*." Boris Johnson, *Weekly Telegraph*, no 245.	• Feeding dead sheep to cattle, or dead cattle to sheep, is "unatural" and "perverted". "The present methods of the agricultural industry are fundamentally unsustainable." "Risk is not actually about probabilities at all. It's all about the trustworthiness of the institutions which are telling us what the risk is." (Michael Jacobs, The Guardian, 24.7.96) • "The Government ... choose to take advice from a small group of hand-picked experts, particularly from those who think there is no problem." Lucy Hodges, *Times Higher* (5.4.96) • "It is the full story of the beginnings of an apocalyptic phenomenon: a deadly disease that has already devasted the national cattle herd ... could in time prove to be the most insidious and lethal contagion since the Black Death." "The British Government has at all stages concealed facts and corrupted evidence on mad cow disease." "Great epidemics are warning signs, symptoms of disease in society itself." G. Cannon in the foreword to *Mad Cow Disease* by Richard Lacey • "My view is that if, and I stress if, it turns out that BSE can be transmitted to man and cause a CJD-like illness, then it would be far better to have been wise and taken precautions than to have not." Richard Lacey ibid.

Figure 15.11 BSE/CJD: a typology of bias

agricultural production: the freedom from toil and drudgery provided by modern machinery, improved nutrition and material standards of living enjoyed by both farmers and consumers, the vast choice now available to food shoppers. Their version of the precautionary principle sees all these benefits being placed in jeopardy by an over-reaction to tenuous scientific evidence about the cause of a very rare illness.

One side says that if you cannot prove it is safe you must treat it as dangerous. The other side says that such an approach would quickly bankrupt any imaginative government, and argues that if you cannot prove it is dangerous you should treat it as safe. Governments, the hierarchists, are caught in the middle. Committed to the idea that problems such as BSE ought to be manageable, and embarrassed by their manifest failure to do so convincingly, they seek to reassure the public that eating British beef is probably safe, and commission more research that they hope will confirm it.

And so the arguments continue. The precautionary principle can be shaped to support almost any cause. Environmentalists use it to argue for minimal interference with nature. Edward Teller makes use of it to argue for the development of more powerful H-bombs and delivery systems to enable the world to fend off asteroids – even if the odds of them being needed are only one in a million. All such arguments are about *virtual risks* – about the future, which does not exist except in people's imaginations.

The Sydney Smith dilemma

Julian Simon, an economist from the University of Maryland, after a robust display of optimism, observes that 'nothing has reduced the doomsayers' credibility with the press or their command over the funding resources of the federal government' (Meadows et al. 1992). Health and environment debates have a durable and predictable character. The specific issues may change but the same caricatures from the cultural theory typology reappear in each new debate. The BSE/CJD controversy is but the most recent instalment in a much larger, long-running debate. On all sides convictions appear to be as strongly held as ever, and as resistant as ever to counter evidence. Debates about Brent Spar (Appendix 15.1), the NHS (Appendix 15.2) and contaminated land (Appendix 15.3) all draw upon similar contending rationalities.

It has probably always been thus. Over 150 years ago the Reverend Sydney Smith was being taken on a conducted tour of an Edinburgh slum. Down a narrow alley between two tall tenements they came upon two women shrieking abuse at each other across the alley. Smith stopped, looked up, and listened. He then shook his head and walked on, lamenting 'they'll never agree; they're arguing from different premises'.

There is an enormous gulf between what scientists know or are likely to discover, and what needs to be known before decisions about environmental risks can be based on conclusive evidence. Scientists also have cultural filters – usually called 'paradigms'. The discovery of the Antarctic ozone hole was delayed by such a filter. US satellites failed to pick it up because they had been programmed to reject such ozone losses as anomalies far beyond the error range of previous predictive models. For scientists and lay-people alike our cultural filters are parts of our identities and essential to our sense of social solidarity. The persistence of contradictory rationalities built upon partial knowledge suggests that we are doomed, for the foreseeable future, to continue to argue from different premises. But the argument is likely to be more civilized, and our cultural filters less crudely selective, to the extent that we are sensitive to these differences and understand their causes and effects.

PS I eat beef and buy a lottery ticket every week.

Appendix 15.1

Fatalist

'As long as I've a pint and a quiet life, it is nothing to do with me.' (F. Arbuthnot, *Guardian*, characterizing and complaining about the apathetic response to Brent Spar of the British public, 19.6.96)

'Only 59 per cent of those questioned about Brent Spar were aware of the incident.' (Shell spokesman, P.J., 5.2.96). 41 per cent is a conservative estimate of the proportion of fatalists in the UK.

Hierachist

'Government permission has been given for the disposal of the redundant structure in a designated deepwater Atlantic site.' (Shell, *Daily Mirror*, 1.5.95)

'The British Government says that it has observed all the rules and that it must take a "balanced and proportionate approach".' (*Guardian*, 6.5.95)

'Most experts ... believe that the environmental implications of [on shore disposal] would be far greater than at sea.' (*Guardian*, 22.6.95)

'The authority of government remains an asset no nation can dispense with.' (Hugo Young, *Guardian*, 22.6.95)

'Militant pressure groups ... rush to judgement, exaggerating their case, and expressing themselves in simplistic terms designed for easy headlines. They undermine both balanced decision-making and parliamentary democracy.' (Michael Dobbs, *Times*, 13.9.95)

'Wimps.' (John Major's description of Shell for backing down on plans to sink Brent Spar – all national newspapers 22.6.95)

Individualist

'Greenpeace is an irresponsible single-issue lobby, and its business is to make a hysterical case against any pollution of the sea ... Shell has behaved with cowardice.' (William Rees-Mogg, *Times*, 22.6.95)

'Out of this there may well be all kinds of environmental technologies for dismantling and dealing with and detoxifying this whole thing. ... If they don't exist, we can invent them. Under stress innovation will rule supreme.' (R. Aspinwall, *Guardian*, 22.6.95)

'All industrial activities have environmental downsides, and the only issue is how to minimize the ecological costs of what we do. ... It is the old game of creating a Them and Us world, in which capitalists are villains and the public are victims.' (Richard North, *Evening Standard*, 21.6.95)

'Rival contractors vie for possible dismantling contract.' (*Financial Times*, 22.6.95)

'The US oil industry has dumped thousands of oil rigs into the Gulf of Mexico with the approval of environmentalists. That's because … man-made structures like sunken ships or oil rigs create habitats for marine life quite similar to that of a coral reef.' (*Wall Street Journal*, 6/9/95)

Egalitarian

'Our argument was always that Shell was wrong in principle to seek to dump the installation at sea.' (postscript to apology by Peter Melchett to Shell for overestimate of the amount of oil in Brent Spar, *Guardian*, 6.9.95)

'it would set a precedent …' (M. Corcoran, *Scotsman*, 3.5.95)

'The Government are allowing oil companies to get away with murder … they have raked in millions from North Sea oil and now they won't pay the bill to clean up.' (Greenpeace, *Daily Mirror*, 1.5.95)

'Thousands of jobs are at stake' (Jimmy Airlie of AEEU, *Times*, 22/6/95)

'The Government negotiates secretly with the … corporation and both keep under tight wraps the data on which the rest of us might base a reasoned opinion.' (S. Holt, *The Ethical Consumer*, 6.2.96)

'In the depths strange species lurk and though we may never see them, we feel in our hearts that they should be left alone. Why must they share the great dark, deep with bits and bobs from a dismembered oil platform?' (Suzanne Moore, *Guardian*, 22.6.95)

'It is wrong to dump old cars in the village pond, and it is wrong … to treat the sea as a dump.' (Greenpeace, *Dundee Courier*, 6.2.96)

Appendix 15.2

Fatalist

- Smokes, eats junk food, drinks and takes drugs, resorts to faith healers
- Low expectations
- '**Encouragement to eat healthy foods has small response**.' (Headline, *Western Morning News*, 10.7.96)

Hierachist

- 'We work closely with the trusts to monitor their performance and to encourage them to improve services. We welcome the high standards revealed in the performance tables and look forward to building on this in the future.' (Ron Spencer, Chief Executive, Cornwall and Isles of Scilly Health Authority, *Cornish Guardian*, 4.7.96)
- Orthodox, high tech, allopathic medicine
- Pursues immortality within budgetary constraints. (40 per cent of Americans die as a result of intensive care being switched off)
- Emphasis on cure rather than prevention
- Pills for mental problems
- Invasive procedures – surgery, chemical and radiation therapies, X-rays, organ transplants
- Large scale – big hospitals, large bureaucracy
- Authoritarian management headed by unelected quangos
- Many accountants, risk managers, cost–benefit analysts
- Hold many conferences
- 'HIV and Aids are self-inflicted diseases contracted through immoral behaviour.' Dr Adrian Rogers, cons. Parliamentary candidate for Exeter explaining opposition to Disability Discrimination Bill.

Individualist

- *Freedom* – to smoke, drink, drive without seat belts
- Cryogenics
- Private health insurance, and health care
- 'Trivial' cosmetic surgery
- Pursues immortality to limits set by personal wealth
- Buys organs for transplanting
- Eats steak
- Sees health care as a business opportunity

- Hedonistic life style – lives life to the limit, and then prepared to spend a lot on repairing the damage
- 'Overwhelming evidence shows that the economic growth and technological advance arising from market competition have in the past two centuries been accompanied by dramatic improvements in health – large increases in longevity and decrease in sickness. ... Not a day goes by without charges that products of technology harm the human body and the physical environment. The very earth itself is said to be in serious danger ... it could be claimed that ours is the environmental age, the time in which technology ceased to be a liberating force and became, instead, a mechanism for self-enslavement, as if the things we created were destroying us.

  The claims of harm from technology, I believe, are false, mostly false or unproven.' (Aaron Wildavsky (1988) *Searching for Safety*)

Egalitarian

(a) Conventional

- 'What is going to make these GPs realize that the service which they are providing is totally inadequate.' (W. McIntosh, director Patient First, *Cornish Guardian*, 4.7.96)
- 'The health service internal market is costly and wasteful. It is undermining fundamental NHS principles of equity and treatment according to need. Managers are under intense pressure to fiddle the figures to show that it works.' (J. Drown, former finance director, Radcliffe Infirmary NHS Trust, *Guardian*, 24.4.96)
- Concern for fairness of provision
- Suspicious of secretive, unelected Quangos
- Prevention stressed before cure – emphasis on healthy living
- Small scale – local clinics
- Accessible by public transport

(b) Alternative

- Low tech and 'natural'
- Herbalism, kinesiology, acupuncture, homeopathy
- Meditation, analysis and therapy for mental problems
- Small scale – sitting room surgeries
- Vegetarian, organic foods, vigilant for pesticides, preservatives, genetically modified foods
- Natural span = 3 score years and 10
- Natural death
- Dignity in death

Appendix 15.3

Why cost–benefit analysis cannot settle arguments about the risks associated with contaminated landfill: a critique of Risk–benefit Analysis of Existing Substances: Guidance Produced by a UK Government/Industry Working Group, DoE/Chemical Industries Association (1995).

In institutional settings (in both government and industry) *risk management* is usually equated with *risk reduction*, and formalized risk management procedures commonly lack both a 'top loop' and explicit instructions for balancing the costs and benefits associated with risk taking. The *Guidance* produced by the DoE/industry working group is an important exception. It proposes a method, *cost–benefit analysis*, for performing the balancing act. It is a method with a record of inflaming environmental arguments rather than resolving them (Adams 1995, Ch. 6).

The *Guidance* presents the essence of the method, as currently applied by the US EPA, to the regulation of dangerous substances. The *costs* associated with regulating a dangerous substance are the costs to industry, to government and to society of compliance with regulation. The *benefits* are defined as the reductions in risk to health and the environment flowing from regulation; they are calculated by a seven-stage process.

1. Identify dose–response estimates or other data used to quantify the incidence effects.
2. Estimate baseline exposure.
3. Calculate baseline number of cases of each health effect.
4. Estimate post-regulatory exposure for each regulatory option.
5. Calculate residual number of cases of each health effect associated with each regulatory option.
6. Calculate cases avoided of each health effect for each regulatory option.
7. Monetize the benefits: calculated benefits of each option in dollars.

Stage 1 encounters the problems discussed earlier; dose–response estimates for the cocktail of ingredients in many toxic waste sites are basically anybody's guess. In the extensively quoted US EPA procedures it states, for example, that 'for cancer assessments, dose–response is usually modelled as a linear no-threshold relationship.' This, as noted in the main paper, is a hotly contested assumption.

The problems with stage 1 are frequently compounded in stage 2 by large uncertainties about exposure. Cocktail effects, leaching into groundwater, the routes by which leachates might reach people, and the sensitivity of the routes of airborne pollutants to turbulence and climatic variables, are but some of the variables upon which science has a precarious grasp. These uncertainties in stages 1 and 2 result in very large error bands associated with numbers in stage 3. Stage 4 requires assumptions about the efficacy of proposed regulatory interventions, requiring a further widening of the error bands for the numbers in stages 5 and 6.

But all these difficulties are trivial in comparison with the problems encountered in stage 7. How does one attach monetary values to the health effects of pollutants? The report cites the Department of Transport's estimate that a life is worth £750 000, but appears oblivious to the literature contending that such a number is a nonsense. Beyond fatalities, there is a huge range of other environmental and health effects whose monetization has generated acrimonious dispute.

The core of the problem of valuing environmental and health effects lies in the definition of costs and benefits. A cryptic reference in the report to 'the marked divergence in many studies between WTP and WTA' indicates an awareness of the existence of the problem, but the rest of the report displays a lack of awareness of its significance. WTP refers to a person's Willingness to Pay for a benefit, and WTA refers to a person's Willingness to Accept compensation for a loss. There are many problems associated with estimating WTP values for projects that affect health and the environment, but these pale into insignificance alongside the problem presented by WTA. The only way yet discovered for estimating the amount of money that a person would accept as compensation for a loss is to ask him. No one else, certainly not the economist, is permitted to answer for him.

Studies that have attempted to measure WTA values associated with threats to health and the environment routinely encounter people who say that no amount of money could possibly compensate them for certain losses – such as loss of life, cherished landscapes and endangered species. Such answers must be either discarded as 'irrational' or entered into the spreadsheet as infinities – and it takes only one infinity to wreak the whole exercise. As a consequence, cost–benefit analysts commonly resort to asking the wrong question. They ask people how much they would be willing to pay to prevent a loss. This has the advantage of eliminating large numbers, because the answers are constrained by ability to pay, but it also biases the results strongly against those who can afford to pay little or nothing to protect their health and environment.

The definition of 'costs' and 'benefits' determines the choice of measure adopted. Table A15.3.1, shows the way in which the legal/moral context of a problem can transform a cost into a benefit. It represents the possible bargains that might be struck during a train journey by two travellers – a non-smoker, and a smoker – sharing a compartment, depending on the rules of the railway company.

Under the permissive rule, which allows smoking, fresh air will be viewed by the non-smoker as a benefit – a departure from the status quo for which he expects to have to pay. The amount that he might pay will depend on the strength of his distaste for smoky air, and what he can afford. The amount that the smoker might accept to forego his rights could depend on the strength of his addiction or his income – or his compassion, the exercise of which

Table A15.3.1 Who pays whom?

	Smoker	*Non-Smoker*
Permissive rule	**Willingness to accept** compensation for foregoing the right to smoke	**Willingness to pay** for the benefits of a smoke-free journey
Restrictive rule	**Willingness to pay** for the right to smoke	**Willingness to accept** compensation for foregoing the right to fresh air

would produce 'payment' in the form of moral satisfaction.

Under the restrictive rule, which forbids smoking without the agreement of fellow passengers, the smoker's willingness to pay will be influenced by his income and the strength of his addiction, and the non-smoker's willingness to accept will be influenced by his degree of aversion to smoky air and how badly he needs the money. While it is difficult to imagine a civilized smoker requiring an extortionate sum of money to forego his rights, it is possible to imagine a desperately ill asthmatic refusing a very large sum of money to maintain his air supply in a breathable state. In any event, only in exceptional circumstances are a person's WTA and WTP likely to be the same.

With respect to real world environmental problems one can find analogous situations. It does sometimes make sense to ask how much people might be prepared to pay to prevent certain environmental losses. The threat to Venice by the rising waters of the Mediterranean is an example. There may also be cases in which everyone can agree that nature can be improved upon, and they are happy to pay for the benefit. But most current environmental controversies focus on human impacts on the environment. They might be characterized as disputes between 'developers' (representing the beneficiaries of proposed projects) and 'environmentalists' (representing the losers), and the choice of which measure to use to value the prospective losses stemming from the project is, in effect, a choice of rule. If, in the above illustration, the smokers represent polluting industry, and the non-smokers the defenders of the environment, then to ask the environmentalists how much they are willing to pay to prevent damage to the environment is to assume a permissive rule. It is tantamount to basing the cost-benefit analysis on a presumption in favour of 'development'. It is to assert that people have no right to clean air and water, to peace and quiet, to their architectural heritage, to cherished landscapes, or to habitats for endangered species; these are all transformed into privileges for which people are expected to pay out of limited budgets. It treats all those adversely affected as a group of non-smokers travelling in a smoking compartment.

The different perspectives of the participants in debates in which CBA is

Table A15.3.2 Cost–benefit analysis: argument from different premises

Fatalist
The fatalist has no opinion on any of these issues. He feels equally hard done by or neglected by the 'system' managed by the hierarchist, and by the individualistic market in which he is an unsuccessful competitor. He also has little faith in the ability of the improving projects of the egalitarian to make a difference.

Hierachist (the manager of the railway in Table A15.3.1)

CBA –	The hierarchist believes in CBA as a 'rational' management tool, and in the assumptions about social consensus upon which it rests.
Dose–response –	in arguments about the evidence, favours more research to reveal the 'truth'.
Burden of proof –	vacillates in the face of unclear evidence about danger of 'passive smoking'; more research needed.
Externalities –	favours regulation to contain harmful externalities.
Precautionary – *principle*	puts concern for dignity of management above the welfare of the passengers.
Discount rate –	the rate to be applied to future environmental costs should be estimated by economists who, with a bit more research, will uncover the 'true' rate.
WTP vs WTA –	by what rules should the railway be run? Dithers about this 'difficult problem', but in practice favours WTP because WTA simply unworkable.

Individualist (the smoker in Table A15.3.1)

CBA –	the individualist is pragmatic; if it produces the 'right' answer he is happy.
Dose–response –	in arguments about dose–response evidence, favours thresholds below which does can be assumed to be harmless.
Burden of proof –	lies with those who believe 'passive smoking' to be harmful.
Externalities –	invokes market solutions, such as tradeable permits, to deal with harmful externalities.
Precautionary – *principle*	invokes the principle to fend off any constraints on the free market that threaten economic growth.
Discount rate –	favours a market rate, which reduces to insignificance environmental costs more than a few decades into the future.
WTP vs WTA –	favours a presumption in favour of 'development' and hence the use of WTP to value environmental costs.

Egalitarian (the non-smoker in Table A15.3.1)

CBA –	the egalitarian is suspicious; CBA rarely produces the 'right' answer.
Dose response –	in arguments about the evidence, favours no-safe-dose model.
Burden of proof –	lies with those who believe 'passive smoking' to be harmless.
Externalities –	appeals to civic virtue to curb the behaviour that produces harmful externalities, and draconian sanctions for wicked offenders.
Precautionary – *principle*	invokes the precautionary principle to prohibit activities that might harm the environment.
Discount rate –	not appropriate to reduce future environmental damage to money but, if it is done a low or zero discount should be used.
WTP vs WTA –	if the environment is to be monetized the WTA measure should be used to value all adverse impacts attributable to human intervention.

used can be characterized using the same typology as that deployed in Figure 15.11 and Appendices 15.1 and 15.2. Table A15.3.2 sets out the very different premises that the participants would be likely to bring to the table in a negotiation about contaminated land. The *Guidance* concludes on a cautionary, but also confusing note.

> It may be tempting for risk benefit practitioners to believe that they can provide the answer to regulatory control. This temptation should be resisted. Risk–benefit is as an art as much as a science. ... There will often be room for debate about the measurement of particular parameters. In such circumstances one of the most important functions of risk–benefit analysis may be to raise the level of debate in international negotiations.

This *Guidance* reflects an ambivalence and ambiguity frequently found in presentations of CBA whether it is applied to road schemes, contaminated land or global warming. The steps by which the analysis is to be conducted are set out in meticulous detail. Certain difficulties are alluded to, although their magnitude is usually understated. Users of the method are advised to use it cautiously. And then it is suggested that it use will raise the level of debate in negotiations.

The use of the method in negotiations about global warming created an uproar that some feared would derail the scientific work of the Intergovernmental Panel on Climate Change. Economists' insistence on valuing an OECD life at £1.5 million and an African, Chinese or Indian life at £150 000 unsurprisingly offended those deemed cheap. 'COBA', the Department of Transport's version of CBA used to evaluate road schemes has not persuaded objectors to the Department's road building plans. For objectors it has become a term of abuse used to describe a Government con trick. Table A15.3.2 provides reasons for supposing it is unlikely ever to settle arguments about human impacts on the environment. Progress in negotiations involving participants arguing from different premises requires a willingness to explore, and compromise on, fundamental differences. CBA enshrines key hierarchist premises; it cannot serve as a framework for negotiation in such circumstances.

References

Adams, J. (1995). *Risk*, UCL Press

Adams, J. Interview on Big Science, BBC2 television 22.8.95. Washington DC: National Academy Press.

Benedick, R. E. (1991). *Ozone Diplomacy*, Harvard U.P.

Meadows, D. H,. Meadows, D. L. and Randers, J. (1992). *Beyond the Limits: global collapse or a sustainable future*. London, Earthscan.

National Research Council (1992) *Risk Assessment in the Federal Government: managing the process*. Washington DC: National Academy Press.

Ravetz, J. (1993). In The sin of science: ignorance of ignorance, *Knowledge*, **15**(2),157-165.
Thompson, M., Ellis, R. and Wildavsky, A. (1990). *Cultural Theory*, Boulder Colorado, Westview.

Glossary

AAF	acetyl amino fluorine
ABP	4-aminobiphenyl
ADP	adenosine diphosphate
AECL	Atomic Energy of Canada Limited
AHH	arlyhydrocarbon hydroxylase
AHR	aromatic hydrocarbon receptor
AIDS	acquired immune deficiency syndrome
ANLL	acute non-lymphoblastic leukaemia
ATP	adenosine triphosphate
BaP	benzo(a)pyrene
BEIR	American National Research Council Committee on the Biological Effects of Ionizing Radiation
BgVV	German Federal Institute for Health Protection of Consumers and Veterinary Medicine
BSE	bovine spongiform encephalopathy
CAT	computer-aided tomography
CBA	cost-benefit analysis
CCT	cotinine/creatinine ratio
CDC	US Centers for Disease Control
CHD	coronary heart disease
CIC	comparative index of carcinogenicity
CJD	Creutzfeldt-Jakob disease
CO	carbon monoxide
CO_2	carbon dioxide
CRS	US Congressional Research Service
DDT	dichloro-diphenyl-trichloroethane
DOE	UK Department of the Environment
DNA	deoxyribonucleic acid
EMF	electromagnetic fields
ETS	environmental tobacco smoke
EU	European Union
FDA	US Food and Drug Administration
FGD	flue gas desulphurization
FRDS	US Federal Reporting Data System
f/l	fibres per litre
GST	glutathione-S-transferase
HIV	human immunodeficiency virus
HSE	UK Health and Safety Executive
IARC	WHO International Agency for Research on Cancer
ICRP	International Commission on Radiological Protection
l	litre

LD_{50}	median lethal dose
kg	kilogram (10^3 grams)
MCLG	Maximum Contamination Level Goal
mg	milligram (10^{-3} grams)
ug	microgram (10^{-6} grams)
MS	mainstream smoke
mSv	milliSievert
NAT	N′-nitrosoanatabine
NCI	National Cancer Institute (USA)
NDEA	n-nitroso diethylamine
NDMA	n-nitoso dimethyamine
ng	nanogram (10^{-9} grams)
NNK	4-(nitrosomethylamino)-1-(3-pyridyl)-1-butanone
NNN	N′-nitrosonornicotine
NERC	UK Natural Environment Research Council
NIOSH	National Institute for Occupational Safety and Health
NO	nitrogen oxide
NOEL	No Observable Effect Level
NRC	US National Research Council
OECD	Organisation for Economic Co-operation and Development
OR	odds ratio
OSHA	US Occupational Safety and Health Administration
PAH	polyaromatic hydrocarbons
pg	picogram (10^{-12} grams)
ppb	parts per billion (10^9)
ppm	parts per million (10^6)
ppt	parts per trillion (10^{12})
PCDD	polychlorinated dioxins
PVC	polyvinyl chloride
QRA	quantitative risk assessment
RCEP	UK Royal Commission on Environmental Pollution
RR	relative risk
SDA	Seventh Day Adventist
SES	socio-economic status
SO2	sulphur dioxide
SS	sidestream smoke
TCDD	tetrachlorodibenzo-p-dioxin
TCDF	2,3,7,8-tetrachlorodibenzofuran
TD50	median toxic dose
TEQ	toxic equivalents
UNEP	United Nations Environment Programme
UNSCEAR	United Nations Scientific Committee on the Effects of Atomic Radiation
US EPA	US Environmental Protection Agency
VSD	virtually safe dose

WHO	World Health Organisation
WTA	willingness to accept
WTP	willingness to pay

About ESEF

Academic Members of ESEF August 1998

Prof. Tom Addiscott UK
Prof. Bruce Ames USA
Dr Sallie Baliunas USA
Dr Alan Bailey UK
Dr Robert C. Balling USA
Prof. A. G. M. Barrett UK
Dr Jack Barrett UK
Dr Sonja Boehmer-Christiansen UK
Prof. Norman D. Brown UK
Prof. Dr K. H. Büchel Germany
Dr John Butler UK
Mr Piers Corbyn UK
Prof. Dr. A. Cornelissen The Netherlands
Dr Barrie Craven UK
Mr Peter Dietze Germany
Dr A. J. Dobbs UK
Dr John Dowding UK
Dr Patricia Fara UK
Dr Oeystein Faestoe UK
Dr Frank Fitzgerald UK
Prof Dr Hartmut Frank Germany
Dr James Franklin Belgium
Dr Alastair Gebbie UK
Dr T. R. Gerholm Sweden
Prof. Dr Gerhard Gerlich Germany
Prof. D. T. Gjessing, Norway
Dr Manoucher Golipour UK
Dr Adrian Gordon Australia
Dr Vincent R. Gray New Zealand
Dr Gordon Gribble USA
Prof Dr Hans-Eberhard Heyke Germany
Dr Vidar Hisdal Norway
Dr Jean-Louis L'Hirondel France
Dr Sherwood Idso USA
Dr Antoaneta Iotova Bulgaria
Prof. Dr Zbigniew Jaworowski Poland
Dr Tim Jones UK
Prof. Dr Wibjörn Karlén Sweden
Dr Terrence Kealey UK
Prof. Dr Kirill Ya.Kondratyev Russia

Prof. Dr F. Korte Germany
Mr Johan Kuylenstierna Sweden
Dr Theodor Landscheidt Germany
Dr Alan Mann UK
Dr John McMullan UK
Prof. Dr Helmut Metzner Germany
Dr Patrick Michaels USA
Sir William Mitchell UK
Dr Paolo Mocarelli Italy
Dr Asmunn Moene Norway
Dr Brooke T. Mossman USA
Prof Dr Hans-Emil Müller Germany
Prof Dr Dr Paul Müller Germany
Dr Joan Munby UK
Mr Liam Nagle UK
Dr Genrik A. Nikolsky Russia
Dr Robert Nilsson Sweden
Prof. Dr Harry Priem The Netherlands
Dr Christoffer Rappe Sweden
Dr Ray Richards UK
Dr Michel Salomon France
Dr Tom V. Segalstad Norway
Dr S. Fred Singer USA
Dr Willie Soon USA
Dr G. N. Stewart UK
Dr Gordon Stewart UK
Dr Maria Tasheva Bulgaria
Dr Wolfgang Thüne Germany
Dr Alan Tillotson UK
Dr Brian Tucker Australia
Prof. Dr med. Karl Überla Germany
Prof. Dr H. P. van Heel The Netherlands
Dr Robin Vaughan UK
Prof. Nico Vlaar The Netherlands
Dr Horst Wachsmuth Switzerland
Dr Michael P.R. Waligórski Poland
Dr Gunnar Walinder Sweden
Dr Gerd-Rainer Weber Germany
Prof Donald Weetman UK
Dr Charlotte Wiin-Christensen Denmark
Dr Aksel Wiin-Nielsen Denmark
Dr James Wilson USA

Business Members

Dr Alfred Bader UK
Mr John Boler UK
Mr Charles Bottoms UK
Dr Francisco Capella Gómez-Acebo Spain
Mr Richard Courtney UK
Mr Michael Gough USA
Dr Claes Hall UK
Mr Richard Hallett UK
Mr Peter Henry UK
Mr Holger Heuseler Germany
Mr Graham Horne
Dr Warwick Hughes Australia
Dr Kelvin Kemm South Africa
Mr Peter Plumley UK
Dr John Rae UK
Dr Michael Rogers UK
Mr Peter Toynbee Australia
Dr Wynne Davies UK

ESEF Committee

Prof. Dr Frits Böttcher The Netherlands
Dr John Emsley UK
Mr Roger Bate UK

Administrator

Lorraine Mooney UK

Mission statement

The European Science and Environment Forum is an independent, non-profit-making alliance of scientists whose aim is to ensure that scientific debates are properly aired, and that decisions which are taken, and action that is proposed, are founded on sound scientific principles.

ESEF will be particularly concerned to address issues where it appears that the public and their representatives, and those in the media, are being given misleading or one-sided advice. In such instances ESEF will seek to provide a platform for scientists whose views are not being heard, but who have a contribution to make.

Members are accepted from all walks of life and all branches of science. There is no membership fee. Members will be expected to offer their services in contributing to ESEF publications on issues where their expertise is germane.

Purpose of ESEF

The European Science and Environment Forum is a Charitable Company Limited by Guarantee (No.1060751). It was established in 1994 to inform the public about scientific debates. Our chosen method for achieving this objective is to provide a forum for scientific opinions that are usually not heard in public policy debates.

Our primary role is to provide an independent voice to the media, the general public and the educators, and by doing so, we aim to provide balance on scientific issues. A secondary role is to contribute to the scientific debate itself. Many of our authors will simplify papers that they originally wrote for the peer reviewed scientific literature. ESEF's tertiary role is to advise scientists how to present their findings to the media, and how the media will perceive, and may use, the information. We hope that this will provide dialogue and understanding between these two important institutions.

How was ESEF formed?

ESEF was formed in 1994 by Roger Bate (Director of the Environment Unit at the Institute of Economic Affairs, London), Dr John Emsley (Science Writer in Residence at Imperial College London University) and Professor Frits Böttcher (Director of the Global Institute for the Study of Natural Resources in The Hague). The issue of climate change was the initiation for the meeting. All three thought that the debate had been unduly one-sided and they wanted to provide a forum for scientists to publish their arguments for public consumption. The media, and via them the public, tended to only hear the so-called consensus view presented by government and intergovernmental science panels.

Of course climate change is not the only issue where member scientists consider that the media debate is not balanced and that there are many environmental and public health issues which are not fully discussed in the public arena either.

ESEF decided on a mission to provide the media and the public with accessible first-hand research of leading scientists in their fields, as an alternative to reports received from specialist journals, government departments or single-issue pressure groups. As Einstein is reputed to have said: 'Make science as simple as possible, but no simpler'. Our aim is to provide science simplified as far as possible. Our members are from fields as diverse as nuclear physics, biochemistry, glaciology, toxicology and philosophy of science. We intend on liaising between the media and our expert members to provide an independent voice on subjects germane to various public policy debates.

To maintain its independence and impartiality, ESEF accepts funding only from charities, and the income it receives is from the sale of its publications. Such publications will automatically be sent to members. Copies will be sent to selected opinion formers within the media and within government.

Principles recommended for use in public policymaking

Much of the debate on science in the spheres of public health and environment centres around hazard assessment and the way that such assessment is used in public policy. We recommend adherence to the following principles for the use of science in public policy:

- Objectivity comes from open debate based on critical examination of evidence.
- Scientists will not always be able to provide unequivocal advice to policymakers; the boundaries of knowledge must therefore be clearly stated.
- Scientists, regardless of affiliation, should be able to provide policymakers with information to help in the definition of public policy priorities
- Scientific evidence should be judged on the consistency of the methodology and the accuracy of the data used, independently of the funding and explicit motivation of the scientist.

Authors' addresses

Dr John Adams
Geography Department
University of London
Gower Street
London
UK

Prof. Bruce N. Ames
401 Barker Hall
Division of Biochemistry and Molecular Biology
University of California
Berkeley CA92720
USA

Roger Bate
ESEF
4 Church Lane
Barton
Cambridge
CB3 7BE
UK

Dr Barrie M. Craven
Newcastle Business School
University of Northumbria
Newcastle upon Tyne
NE1 8ST
UK

Dr Marie-Louise Efthymiou
Centre for Toxicovigilance
Fernand Widal Hospital
Paris
France

Dr Alvan R. Feinstein
Sterling Professor of Medicine and Epidemiology
Yale University School of Medicine
New Haven Connecticut
USA

Dr Etienne Fournier
Professor Emeritus of Clinical Toxicology
Fernand Widal Hospital
200, rue du Fbg Saint-Denis
75475 Paris Cedex 10 – France

Dr Lois Swirsky Gold
Director of the Carcinogenic Potency Project
National Institute of Environmental Health Sciences Center
University of California, Berkeley, CA 94720
and Senior Scientist
Berkeley National Laboratory
Berkeley, CA 94720
USA

Dr Zbigniew Jaworowski
Central Laboratory for Radiological Protection
ul. Konwaliowa 7
03-194 Warsaw
Poland

Dr Terence Kealey
Department of Clinical Biochemistry
Addenbrooke's Hospital
Cambridge
CB2 2QR
UK

Dr Hans E. Müller
University of Bonn, Medical Faculty
Public Health Laboratory Braunschweig, (retired)
Germany

Dr Joan Munby
School of Health Sciences
University of Sunderland
Pasteur Building
Chester Road
SR1 3SD
UK

Professor Robert Nilsson
KEMI
National Chemicals Inspectorate
Box 1384
S-171 27 SOLNA
Sweden
Professor of Molecular Toxicology and Risk Assessment
Department of Genetic and Cellular Toxicology
Wallenberg Laboratory, Stockholm University
S-10691 Stockholm
Sweden

Dr Göran Pershagen
Institute of Environmental Medicine
Karolinska Institute
Stockholm
Sweden

Dr Peter M. Sandman
54 Gray Cliff Road
Newton Centre
MA 02150
USA

Dr Gordon T. Stewart
Glenavon
Clifton Down
Bristol BS8 3HT
UK

Dr Fritz Vahrenholt
The Senator
Free and Hanseatic
City of Hamburg
State Ministry of the Environment
Germany

Professor D F Weetman
3 Allenheads
Fatfield
Washington
Tyne and Wear
NE38 8PD
UK

Dr James D. Wilson
Senior Fellow
Resources for the Future
Center for Risk Management
1616P Street NW
Washington DC 20036
USA

Index